Neural Networks in QSAR and Drug Design

Principles of QSAR and Drug Design

NEURAL NETWORKS IN QSAR AND DRUG DESIGN

Edited by

James Devillers

CTIS, Lyon, France

ACADEMIC PRESS

Harcourt Brace & Company, Publishers

London San Diego New York

Boston Sydney Tokyo Toronto

This book is printed on acid-free paper

United States Edition published by
ACADEMIC PRESS INC.
San Diego, CA 92101

ACADEMIC PRESS LIMITED
24–28 Oval Road
London NW1 7DX

A catalogue record of this book is available from the British Library

ISBN 0–12–213815–5

Typeset by Florencetype Ltd, Stoodleigh, Devon
Printed and bound in Great Britain by Hartnolls Ltd, Bodmin, Cornwall

Contents

A colour plate section appears between pages 212–213

Contributors

S. Anzali, E. Merck, Department of Medicinal Chemistry/Drug Design, 64271 Darmstadt, Germany.
G. Barnickel, E. Merck, Department of Medicinal Chemistry/Drug Design, 64271 Darmstadt, Germany.
R.S. Boethling, Office of Prevention, Pesticides and Toxic Substances, 7406, US Environmental Protection Agency, 401 M St, SW, Washington, DC 20460, USA.
L. Buydens, Catholic University of Nijmegen, Laboratory for Analytical Chemistry, Toernooiveld 1, 6525 ED Nijmegen, The Netherlands.
M. Chastrette, Laboratoire de Chimie Organique Physique et Synthétique, URA 463, Université Claude Bernard-Lyon I, 69622 Villeurbanne CEDEX, France.
K.C. Chou, Upjohn Laboratories, Kalamazoo, MI 49007–4940, USA.
J. Devillers, CTIS, 21 rue de la Bannière, 69003 Lyon, France.
D. Domine, CTIS, 21 rue de la Bannière, 69003 Lyon, France.
C. El Aidi, Laboratoire de Chimie Organique Physique et Synthétique, URA 463, Université Claude Bernard-Lyon I, 69622 Villeurbanne CEDEX, France.
D.W. Elrod, Upjohn Laboratories, Kalamazoo, MI 49007–4940, USA.
T. Gallagher, School of Chemistry, University of Bristol, Cantock's Close, Bristol BS8 1TS, UK.
J. Gasteiger, Computer-Chemie-Centrum, University Erlangen-Nürnberg, 91052 Erlangen, Germany.
C. Guillon, CTIS, 21 rue de la Bannière, 69003 Lyon, France.
W. Karcher, CEC, JRC, Ispra Establishment, I-21020 Ispra, Varese, Italy.
M. Krug, E. Merck, Department of Medicinal Chemistry/Drug Design, 64271 Darmstadt, Germany.
D.J. Livingstone, ChemQuest, Cheyney House, 19–21 Cheyney Street, Steeple Morden, Herts SG8 0LP, UK; Centre for Molecular Design, School of Biological Sciences, University of Portsmouth, Portsmouth, PO1 2QF, UK.
G.M. Maggiora, Upjohn Laboratories, Kalamazoo, MI 49007–4940, USA.
D.T. Manallack, Chiroscience Limited, Cambridge Science Park, Milton Road, Cambridge CB1 4WE, UK.
J. Sadowski, Computer-Chemie-Centrum, University Erlangen-Nürnberg, 91052 Erlangen, Germany.

M. Wagener, Computer-Chemie-Centrum, University Erlangen-Nürnberg, 91052 Erlangen, Germany.
D. Wienke, Catholic University of Nijmegen, Laboratory for Analytical Chemistry, Toernooiveld 1, 6525 ED Nijmegen, The Netherlands.
C.T. Zhang, Upjohn Laboratories, Kalamazoo, MI 49007–4940, USA; Department of Physics, Tianjin University, Tianjin, China.

Preface

What are exactly artificial neural networks? What can they do? How do they work? How can we use them in practice? These and many similar questions are being asked by professionals involved in molecular modeling. Finding answers is a time-consuming and difficult task. Indeed, the literature on artificial neural networks is extensive. Numerous excellent books exist but they are often expressed in technical jargon, emphasizing the theoretical aspects of the artificial neural networks instead of providing didactic examples of applications.

Neural Networks in QSAR and Drug Design is the first book on the practical use of artificial neural networks in molecular modeling. It provides a systematic entry path for the professional who has not specialized in mathematical analysis but wants to use these powerful tools for solving numerous complex modeling problems in his/her specific domain of research.

The different chapters of the book allow readers to discover the principal neural network paradigms through different examples of applications in medicinal chemistry, pharmacology, perfumery, and environmental chemistry. An attempt has been made to stress the advantages and limitations of artificial neural networks in QSAR and drug design.

To ensure the scientific quality and clearness of the book, all the contributions have been presented and discussed in the frame of the *Second International Workshop on Neural Networks and Genetic Algorithms Applied to QSAR and Drug Design* held in Lyon, France (June 12–14, 1995). In addition, they have been reviewed by two referees, one involved in molecular modeling and another in artificial neural networks.

Neural Networks in QSAR and Drug Design is the second volume in the series *Principles of QSAR and Drug Design*. It is hoped that the readers will find this book to be a valuable source of reference, as well as a source of stimulation of further research, on artificial neural network applications in molecular modeling.

It is a pleasure to acknowledge the help I have received from my team at CTIS in the preparation of this book. I also wish to thank Hugo Devillers for the design of the cover artwork.

James Devillers

Preface

1 Strengths and Weaknesses of the Backpropagation Neural Network in QSAR and QSPR Studies

J. DEVILLERS
CTIS, 21 rue de la Bannière, 69003 Lyon, France

An overview is given of the current usage of the backpropagation neural network (BNN) in quantitative structure–activity relationship (QSAR) and quantitative structure–property relationship (QSPR) studies. Emphasis is placed on practical aspects related to the selection of the training and testing sets, the preprocessing of the data, the choice of an architecture with adequate parameters, and the comparison of models. Advantages and limitations of BNN are stressed, as well as the usefulness of hybrid systems which mix and match a BNN with other intelligent techniques for solving complex modeling problems.

KEY WORDS: *backpropagation neural network; overtraining; overfitting; model selection; validation; hybrid system.*

INTRODUCTION

Artificial neural networks are computer models derived from a simplified concept of the brain (Kohonen, 1988; Anderson, 1990; Boddy *et al.*, 1990). Research into neural networks has been active since the 1940s but underwent a decline in the late 1960s following the work of Minsky and Papert which showed that a simple neural network, called perceptron, was not able to solve relatively trivial problems, including the function performed by a simple exclusive-or (XOR) gate (Wasserman, 1989; Aleksander and Morton, 1990). In the 1980s, neural network research has received fresh impetus as new network paradigms have been developed (Hopfield, 1982; Rumelhart *et al.*, 1986; Le Cun, 1987; Lippmann, 1987; Grossberg, 1988; Kohonen, 1988,

In, *Neural Networks in QSAR and Drug Design* (J. Devillers, Ed.)
Academic Press, London, 1996, pp. 1–46.
ISBN 0-12-213815-5

1990). The most popular and widely used of these is the backpropagation algorithm (Rumelhart *et al.*, 1986; Le Cun, 1987; North, 1987; Crick, 1989; Rumelhart, 1995). A backpropagation neural network (BNN) is constructed from simple processing units called 'neurons' or 'nodes' which are arranged in a series of layers bounded by input and output layers encompassing a variable number of hidden layers. Each neuron is connected to other neurons in the network by connections of different strengths or weights. BNN belongs to the class of supervised learning techniques because the method consists in comparing the response of the output units to the desired responses *via* an iterative process in which an error term is calculated and used to readjust the weights in the network in order to obtain BNN responses close to the desired responses. The most important feature of a BNN is its ability to learn from examples and to generalize, since the learnt information is stored across the network weights. It is also able to make decisions and draw conclusions when presented with complex, noisy and partial information (Borman, 1989; de Saint Laumer *et al.*, 1991; Salt *et al.*, 1992; Medsker, 1994). Due to these different advantages, it is not surprising that BNN has gained momentum in numerous industrial and scientific areas. It is estimated that more than 90 per cent of the neural network applications in use today employ BNN or some variant of it (Hammerstrom, 1995, page 349). Successful applications can be found in finance and business (e.g., Refenes *et al.*, 1994; Furness, 1995), process control operation (e.g., Guillon and Crommelynck, 1991; Linko and Zhu, 1991; Pollard *et al.*, 1992), astronomy (e.g., Angel *et al.*, 1990; Sandler *et al.*, 1991), physics (e.g., Gernoth and Clark, 1995), handwriting, text and speech recognition (e.g., Basak *et al.*, 1995; Skoneczny and Szostakowski, 1995; Bengio, 1996), image processing (e.g., Silverman and Noetzel, 1990), medicine (e.g., Zipser and Andersen, 1988; Miller *et al.*, 1992; Weinstein *et al.*, 1992, 1994; Schaltenbrand *et al.*, 1993; Zheng *et al.*, 1994), biotechnology and biochemistry (e.g., Hirst and Sternberg, 1992; Montague and Morris, 1994; Neal *et al.*, 1994), geology (e.g., Romeo *et al.*, 1995), and chemistry (e.g., Zupan and Gasteiger, 1991; Tusar *et al.*, 1992; Sumpter *et al.*, 1994). In the same way, the introduction of BNN in (quantitative) structure–activity relationship ((Q)SAR) studies and quantitative structure–property relationship (QSPR) studies can be traced back to the early 1990s (Aoyama *et al.*, 1990a,b; Aoyama and Ichikawa, 1991a,b,c; 1992; de Saint Laumer *et al.*, 1991). Indeed, BNN has been successfully used in pharmacology (Andrea and Kalayeh, 1991a,b; So and Richards, 1992; Ichikawa and Aoyama, 1993; King *et al.*, 1993; Hirst *et al.*, 1994a,b; Tetko *et al.*, 1994; Achanta *et al.*, 1995) and (eco)toxicology (Devillers and Cambon, 1993; Wiese and Schaper, 1993; Calleja *et al.*, 1994; Devillers *et al.*, 1995) and for modeling odors (Chastrette and de Saint Laumer, 1991; Chastrette *et al.*, 1993; Song *et al.*, 1993; Chastrette and El Aidi, 1996; Devillers *et al.*, 1996b), biodegradation (Zitko, 1991; Cambon and Devillers, 1993; Devillers, 1993; Tabak and Govind, 1993; Devillers *et al.*, 1996a; Devillers and Putavy, 1996), atmospheric degradation (Rorije *et al.*, 1995), half-lives (Domine *et al.*, 1993a), carcinogenicity (Devillers and Cambon, 1993; Villemin *et al.*, 1994) and mutagenicity (Brinn *et al.*, 1993; Ghoshal *et al.*, 1993; Villemin *et al.*, 1993) of chemicals. In the same way, BNN and derived algorithms have been used for the prediction of the *n*-octanol/water partition coefficient (log P) (Grunenberg and Herges, 1995; Domine *et al.*, 1996a), aqueous solubility (Bodor *et al.*, 1991) and aqueous activity coefficients (Chow *et al.*, 1995), interatomic distance (Jordan *et al.*, 1995), heat capacity (Gakh *et al.*, 1994), critical temperature (Cherqaoui *et al.*, 1994a), boiling point (Balaban *et al.*, 1994; Cherqaoui *et al.*, 1994a,b; Egolf *et al.*, 1994; Gakh *et al.*, 1994), melting point (Cherqaoui *et al.*, 1994a), density (Gakh *et al.*, 1994), Gibbs energy (Gakh *et al.*, 1994), standard enthalpy of formation (Gakh *et al.*, 1994), molar volume (Cherqaoui *et al.*, 1994a), refractive index (Gakh *et al.*, 1994), NMR shifts (Doucet *et al.*, 1993; Panaye *et al.*, 1994; Sharma *et al.*, 1994; Svozil *et al.*, 1995), ionization potentials (Sigman and Rives, 1994), and impact sensitivity (Nefati *et al.*, 1993) of chemicals. The above large list of bibliographical references dealing with applications of BNN in (Q)SAR and QSPR studies clearly demonstrates the increasing interest of BNN in the domain. This is principally due to the fact that a BNN allows one to find complex relationships between the structure of molecules and their physicochemical properties or biological activities. In addition, a BNN provides a way of modeling the patterns in a data set without the need to make any assumptions about the underlying reasons for the patterns. Under these conditions, the scope of this paper is only to underline some problems and some solutions related to the use of a BNN in the particular case of the analysis of QSPR and (Q)SAR data matrices. The aim of this article is also to provide a quick guide into the labyrinth of BNN research.

STANDARD BNN ALGORITHM

The theory of BNN has been widely described in numerous textbooks and theoretical articles (e.g., Werbos, 1988; Pao, 1989; Wasserman, 1989, 1993; Aleksander and Morton, 1990; Davalo and Naïm, 1990; Eberhart and Dobbins, 1990a; Zeidenberg, 1990; Bourret *et al.*, 1991; Levine, 1991; McCord Nelson and Illingworth, 1991; Ripley, 1992; Wythoff, 1993; Lee and Hajela, 1994a,b).

Therefore, the basic concepts of the BNN algorithm are only recalled below in order to review all the elements to take into account when using a BNN in practice for (Q)SAR and QSPR modeling. The basic entity of a BNN is the formal neuron (Figure 1). Its action consists in summing weighted inputs and producing an output signal through an activation function. Although the activation functions can have several forms (Wasserman, 1989; McCord Nelson and Illingworth, 1991), the most commonly used is the

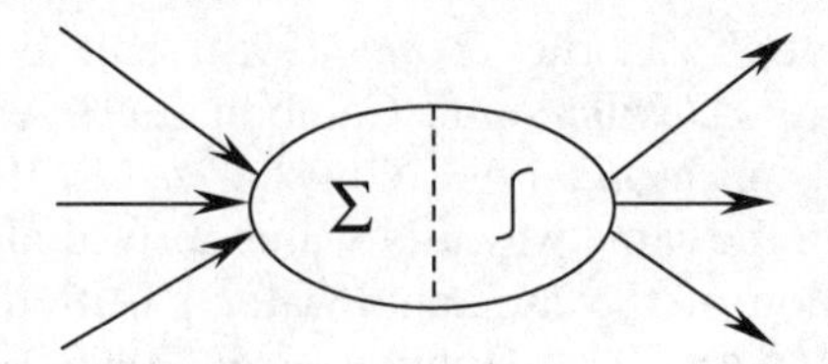

Figure 1 Formal neuron.

sigmoid function, which is characterized by the following equation:

$$f(x) = 1/[1 + \exp(-x)] \quad (1)$$

The nonlinear nature of the sigmoid function (also called squashing function) plays an important role in the performances of the BNN.

Topologically a BNN presents three types of layers (Figure 2):

- one input layer (with a number of neurons corresponding to the number of molecular descriptors);
- one (or more) hidden layer(s) with adjustable numbers of neurons;
- one output layer with a number of neurons depending on the modeled activity or property. This layer generates the calculated outputs.

The neurons of each layer are connected in the forward direction (i.e., input to output). Each of the input and hidden layers can have an additional unit, called bias, connected as shown in Figure 2. Biases allow a more rapid convergence of the training process (Wasserman, 1989, page 53).

Before starting the training (learning) process, all the weights associated with the connections between the neurons within the network must be initialized to small random numbers (e.g., [–0.3, 0.3]). This ensures that the network is not saturated by large values of the weights, and prevents some training pathologies (Wasserman, 1989, page 47). During the training phase, each input pattern of the training set is presented to the network, which

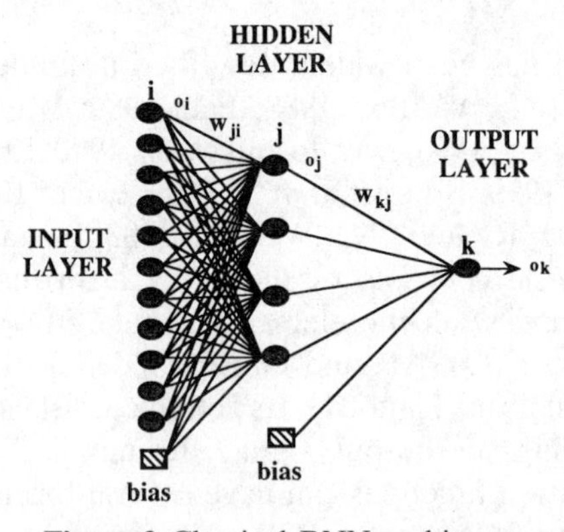

Figure 2 Classical BNN architecture.

generates a calculated output. At this stage, the network has performed the feedforward step. An error (Eq. (2)) is computed from the calculated (o_{pk}) and target (t_{pk}) outputs for a pattern p.

$$E_p = \frac{1}{2} \sum_k (t_{pk} - o_{pk})^2 \quad (2)$$

For all patterns, we obtain:

$$E = \frac{1}{2} \sum_p \sum_k (t_{pk} - o_{pk})^2 \quad (3)$$

Weights are adjusted by backpropagating the error from the output to the input layer. This is performed after presentation of each training pattern (on-line or single pattern training). However, note that calculation times can be reduced by adjusting the weights once all patterns have been presented (batch or epoch training) (Eberhart and Dobbins, 1990b, page 44).

After presentation of a pattern p, the adjustment of the weights located between the output layer k and the hidden layer j is performed by means of the following equation:

$$\Delta_p w_{kj} = \eta \delta_{pk} o_{pj} \quad (4)$$

where η is the learning rate and $\delta_{pk} = (t_{pk} - o_{pk}) o_{pk} (1 - o_{pk})$ for a sigmoid function. Note that a large η value corresponds to a rapid learning but might also result in oscillations. If η is set too low, the convergence is difficult and the risk of falling into and remaining in local minima is high (Pao, 1989, page 128; de Saint Laumer *et al.*, 1991).

The adjustment of the weights located between the hidden layer j and the input layer i are calculated in a similar manner. Rumelhart and coworkers (Rumelhart *et al.*, 1986) have modified Eq. (4) by including a momentum term (α) which prevents oscillations (Pao, 1989, page 128; Wasserman, 1989, page 54). The adjustment of the weights between the (n)th and the (n + 1)th steps then becomes:

$$\Delta_p w_{ji} (n + 1) = \eta \delta_{pj} o_{pi} + \alpha \Delta_p w_{ji} (n) \quad (5)$$

The BNN algorithm has been studied and improved since its emergence. Numerous algorithms have been proposed to help speed convergence, to avoid local minima, and to enhance generalization accuracy. Modifications of the original BNN algorithm deal mainly with architecture, momentum term (α), error metric, transfer function, and optimization process (Fahlman, 1988; Jacobs, 1988; Baba, 1989; Biswas and Kumar, 1990; Leonard and Kramer, 1990; Shawe-Taylor and Cohen, 1990; Specht, 1990; Tollenaere, 1990; Samad, 1991; Weir, 1991; Weymaere and Martens, 1991; Bhat and McAvoy, 1992; Shah *et al.*, 1992; van Ooyen and Nienhuis, 1992; Zollner *et al.*, 1992; Anguita *et al.*, 1993; Biegler-König and Bärmann, 1993; Janakiraman and Honavar, 1993; Karras *et al.*, 1993; Kim and May, 1993; Lee, 1993; Møller, 1993; Sperduti and Starita, 1993; Brunelli, 1994; Harrington, 1994; Humpert,

1994; Kasparian *et al.*, 1994; Kendrick, 1994; Klawun and Wilkins, 1994; Lee, 1994; Liang and Xia, 1994; Maillard and Gueriot, 1994; Mastriani, 1994; Ochiai *et al.*, 1994; van der Smagt, 1994; Hunt and Deller, 1995; Perantonis and Karras, 1995; Vitthal *et al.*, 1995). Almost every new BNN algorithm introduced in the neural network literature comes equipped with some performance comparisons to validate its efficiency. However, these comparisons are generally based on artificial data. Thus, these algorithms are frequently tested on the XOR problem, by fitting a polynomial relationship (PR), and so on. This is problematic because simple simulation examples are generally unrealistic in the sense that they generate insufficient data set complexity to provide a reasonable test of the usefulness of a new BNN algorithm, especially for estimating its generalization ability. In addition, problems can also be directly related to the BNN paradigm. Thus, for example, it has been shown (Cardell *et al.*, 1994) that training algorithms have trouble converging when trained on artificial data generated by a PR in which the order of the polynomial is less than or equal to the number of hidden units in the BNN. In addition, comparisons between algorithms should only be made when their various parameters are optimally tuned. The difficulty of fine tuning can itself become a point of comparison (Alpsan *et al.*, 1995). The number of comparative studies based on real data is limited. Thus, for example, Skurikhin and Surkan (1995) have compared the performances of the backpropagation, quickpropagation, and cascade correlation algorithms in energy control problems. In the same way, Alpsan and coworkers (Alpsan *et al.*, 1995) have compared numerous modifications of the standard BNN algorithm on a real-world medical problem.

DESIGNING THE MODEL

Selection of the training and testing sets

The amount of data required for training a BNN is dependent on the network architecture and the problem being addressed. In the neural network literature, some theoretical attempts have been made for addressing the problem of training set selection when the available data are numerous (Reeves, 1995a). In the same way, some practical rules of thumb have been proposed. Thus, for example, for pattern analysis, according to Eberhart and Dobbins (Eberhart and Dobbins, 1990c, page 60) '*a minimum of about 10 to a few dozen pattern vectors for each possible output classification are often required*'. In QSPR and (Q)SAR, the use of a BNN requires the constitution of a training set providing a full and accurate representation of the problem domain. If possible, statistical design has to be used (Hansch *et al.*, 1973; Pleiss and Unger, 1990; Simon *et al.*, 1993; Domine *et al.*, 1994a,b, 1996b; Devillers, 1995; Putavy *et al.*, 1996; Wienke *et al.*, 1996).

It is well known that the choice of a BNN configuration must be made after numerous runs and has to be a compromise between the learning and generalization abilities of the network (Finnoff *et al.*, 1993; Musavi *et al.*, 1994). Basically, the performances of a BNN during the training phase are evaluated from the error (Eq. (3)). The generalization performances are assessed by means of a testing phase. During this phase, unknown patterns are presented to the BNN to check whether the model is able to predict their actual outputs. To use the potentialities of a BNN correctly, it is required to employ a large testing set. Furthermore, it is well admitted that the interpolation performances of a BNN are better than those of extrapolation. The use of an in-sample testing set (ISTS) and an out-of-sample testing set (OSTS) allows one to overcome this drawback (Devillers *et al.*, 1995). Indeed, the ISTS contains patterns similar to those belonging to the training set of the BNN model and therefore allows one to assess its interpolation performances. The OSTS includes patterns presenting an appreciable difference with those of the training set and can be used for estimating the extrapolation performances of the BNN model. In practice, experimental data of lesser quality can also be included in the OSTS. This allows their rational exploitation. The use of different testing sets has been experienced in chemometrics by Borggaard and Thodberg (1992), in QSAR by Devillers and coworkers (Devillers *et al.*, 1995), and in QSPR by Domine and coworkers (Domine *et al.*, 1996a).

It is obvious that to use the potentialities of a BNN correctly, one is required to employ large data sets. However, in practice, this condition is difficult to satisfy in QSAR studies where biological activities are often measured on limited series of molecules. Even if inspection of the QSAR literature reveals that leave-*n*-out procedures (Cramer *et al.*, 1988) have been recommended with BNN (Schüürmann and Müller, 1994), we have to mention that these techniques are not the most appropriate for estimating the accuracy of a BNN model. Indeed, the removal of the patterns can greatly influence the training, especially when the number of individuals is reduced. It is also important to keep in mind that, for the same data set, different solutions (i.e., models) exist. Thus, the results obtained with a leave-one-out (or leave-*n*-out) procedure cannot be interpreted as easily as when used with classical linear methods. Last, it is important to note that the leave-one-out procedure has also been criticized in linear statistics (Shao, 1993). Under these conditions, in the case of lack of data, with a BNN and related algorithms, it is preferable to use permutation tests or bootstrap (Efron and Tibshirani, 1993; Katz *et al.*, 1994; Russo *et al.*, 1994; Reeves, 1995b) which are more robust.

Data preparation

Once a suitable training set has been designed, preprocessing the input data can often play a key role in the ability of the BNN to carry out the desired mapping (Bos *et al.*, 1993). Preprocessing can be used to satisfy different constraints. Thus, for example, it has to satisfy requirements imposed by the selected transfer function(s). Preprocessing also depends on the nature and quality of the raw data used for designing the model.

Boolean data can be transformed to 0.1 and 0.9 or scaled *via* correspondence factor analysis (CFA) (Cambon and Devillers, 1993). CFA is particularly suitable since it also allows one to reduce the number of inputs and provides interesting graphical displays (Figure 3).

When dealing with frequency or quantitative data, it has been shown that in most cases, learning from raw data does not allow one to obtain convergence within a reasonable number of cycles. Thus, several data transformations can be used. The most commonly employed is the min/max procedure (e.g., Eq. (6)).

$$x'_i = [a(x_i - x_{min})/(x_{max} - x_{min})] + b \quad (6)$$

In Eq. (6), x'_i is the scaled input, x_i the original input, x_{min} and x_{max} are respectively the minimum and maximum values in each column of the data matrix, and a and b are constants allowing to fix the limits of the interval for the scaled values. For example, a = 0.9 and b = 0.05 yield scaled data ranging between 0.05 and 0.95. Principal components analysis (PCA), partial least squares (PLS), and CFA have also been successfully used for preprocessing (Bhandare *et al.*, 1993; Bos *et al.*, 1993; Devillers and Cambon, 1993; Domine *et al.*, 1993a; Dong and Chan, 1993; Maxwell *et al.*, 1993; Otto, 1993; Andrews and Lieberman, 1994; Devillers *et al.*, 1995, 1996a; Moreno *et al.*, 1995). These methods also provide a reduction of dimensionality. This reduction is very interesting since the number of connections within a BNN should not be too high compared with the number of learning examples. Thus, by reducing the number of inputs, one also reduces the number of connections within the network. The graphical displays obtained with PCA and CFA can also be of interest. The use of stochastic analysis as preprocessing for a BNN also allows one to reduce the input representational space (Basti *et al.*, 1993).

Nonlinear transformations of the inputs can also increase the performances of a BNN. This concept has led to the development of a particular neural network paradigm called functional-link net (Pao, 1989; Pao and Takefuji, 1992; Pao and Igelnik, 1994; Hosseini-Nezhad *et al.*, 1995). The functional-link net has been used in QSAR and its performances have been compared with those of BNN, regression analysis, adaptive least squares, and fuzzy adaptive least squares (Liu *et al.*, 1992a,b,c). In the studies of Liu and coworkers (Liu *et al.*, 1992a,b,c), the inputs x_i (e.g., F, R) were transformed into $\sin(\pi x_i)$, $\cos(\pi x_i)$, and so on. Manallack and Livingstone (1994a) have stressed the dangers of chance correlations with this type of transformation which allows the generation of a plethora of descriptors which are difficult to interpret in terms of chemical structure.

Addition of noise to the training inputs (Elman and Zipser, 1988; Elrod *et al.*, 1990; Sietsma and Dow, 1991; Bos *et al.*, 1992; Hartnett *et al.*, 1993)

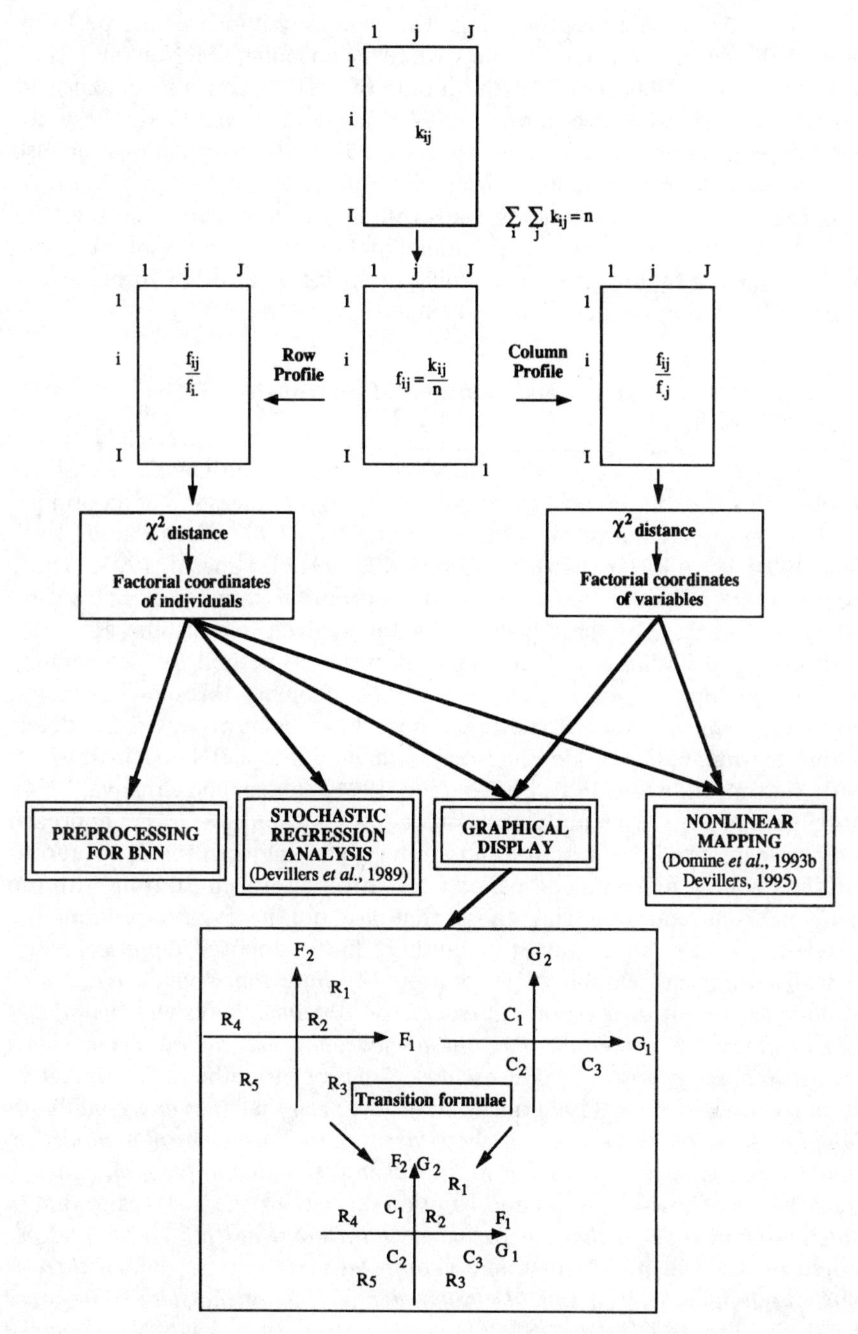

Figure 3 Principle of CFA and possible hybridizations.

can be considered as a preprocessing. This generally improves the performances of the neural net. In the same way, fuzzy encoding (McNeill and Thro, 1994; Yager and Filev, 1994) of the inputs of a BNN can also enhance its performances (Pizzi and Somorjai, 1994; Maggiora *et al.*, 1996). Last, for specific applications, Fourier transforms and wavelet transformations can also be employed (MacIntyre and O'Brien, 1995).

As regards the outputs, scaling transformations allow one to account for the 0–1 activity interval of the sigmoid function of the BNN output. This can be achieved by means of a min/max procedure, but other formulas can be used depending on the problem at hand (Bos *et al.*, 1993).

Number of hidden layers and number of neurons in the hidden layer(s)

The number of hidden layers in a BNN depends on the complexity of the problem to solve, but in most cases one hidden layer consisting of an optimal number of sigmoid neurons is sufficient (Funahashi, 1989; Hornik *et al.*, 1989, 1990, 1994; Hornik, 1991; Ito, 1991a,b, 1992, 1994; Takahashi, 1993). There are no theoretical rules to determine the appropriate number of neurons on the hidden layer of a three-layer BNN for a given application. Recently, Amirikian and Nishimura (1994) have proposed a method for determining the optimal topology of a three-layer BNN of binary threshold elements. However, even if some automatic methods have been proposed for determining the number of hidden neurons to include into a BNN (Hirose *et al.*, 1991; Sietsma and Dow, 1991; Lim and Ho, 1994; Nabhan and Zomaya, 1994), more generally, a trial and error procedure is employed. The approach consists in using either a bottom-up strategy, starting with too few neurons and then adding more if need be, or a top-down approach, starting with too many neurons and removing those that are not necessary (pruning the network). To save time, numerous authors have proposed empirical rules. Thus, Eberhart and Dobbins (1990b, page 42) '*have found that a reasonable number to start with in many cases can be obtained by taking the square root of the number of input plus output neurodes, and adding a few-which sounds like a recipe for baking a cake...*'. Using a mathematical result of Kolmogorov, Zaremba (1990, page 268) advances that '*for any continuous mapping function there exists a three-layer neural network of* n *inputs,* m *output units, and* 2n + *1 hidden units that implements the function exactly*'. Weiss and Kulikowski (Weiss and Kulikowski, 1991, page 104) claim that '*a widely used heuristic is that a network should average at least ten samples per weight in the network*'. Bailey and Thompson (1990) stress that '*if there is only one hidden layer, a suitable initial size is 75% of the size of the input layer*'. Sigillito and Hutton (Sigillito and Hutton, 1990, page 241) advance that '*other applications had shown that the number of hidden nodes required was usually less than half of the number of input nodes*'. Leigh (1995, page 66) mentions that '*the key issue is determining the size of the hidden layer(s). We invested a large amount of time in testing various topologies. The published information in the field indicates that 90% of applications using backpropagation can be solved using one hidden layer and that, as a guide, the size of the hidden layer should not exceed 30% of the size of the input layer. Our experience was in accordance with these guidelines*'. Last, we have to mention that some authors try to use statistical criteria. Thus, for example, Hamad and Betrouni (1995) employ the Akaike's information criterion (AIC) (Akaike, 1974) for selecting the number of hidden neurons in a BNN.

In practice, according to the Ockham's Razor principle (also called principle of economy), the number of neurons on the hidden layer of a three-layer BNN must be as reduced as possible. It must take into account the ratio between the number of individuals in the training set and the number of connections in the BNN in order to ensure its generalization capabilities (Andrea and Kalayeh, 1991a; Livingstone and Salt, 1992; Manallack and Livingstone, 1992, 1995; Livingstone and Manallack, 1993; Manallack *et al.*, 1994). Indeed, too many free parameters in the BNN will allow the network to fit the training data arbitrarily closely (overfitting), but will not lead to an optimal generalization. This kind of BNN is called a 'grandmother' network in the neural network literature. Different methods have been proposed for addressing the problem of overfitting. They are presented later, in the section dedicated to the selection of a BNN model.

Sometimes, the architecture of a BNN is fixed in order to satisfy specific tasks. Thus, ReNDeR (reversible nonlinear dimensionality reduction) network (Livingstone *et al.*, 1991) is a particular BNN allowing one to obtain nonlinear graphical displays. In such a network, we find (Figure 4) an input layer containing as many neurons as there are variables in the data set, an encoding hidden layer containing a smaller number of neurons than the input layer, a parameter hidden layer with two (or three) neurons, and a decoding hidden layer and an output layer which are symmetrical with the encoding and input layers, respectively. Input patterns are presented to ReNDeR network one at a time and the connection weights are adjusted until the output values match the input values (Livingstone and Salt, 1995). Once trained, the ReNDeR network squeezes the data through the central bottleneck of two neurons. For each input pattern, this produces two values which can be used as the x and y coordinates for the construction of a two-dimensional plot. Even if some criticisms have been made (Reibnegger *et al.*, 1993), the published results on the ReNDeR network (Good *et al.*, 1993; Livingstone and Salt, 1995; Livingstone, 1996; Manallack *et al.*, 1996) are encouraging. They complement those obtained with classical linear and nonlinear multivariate approaches. It is interesting to note that the same kind of network has been described in the literature. Thus, Kramer (1992) used the so-called autoassociative networks in process engineering for noise filtering, missing

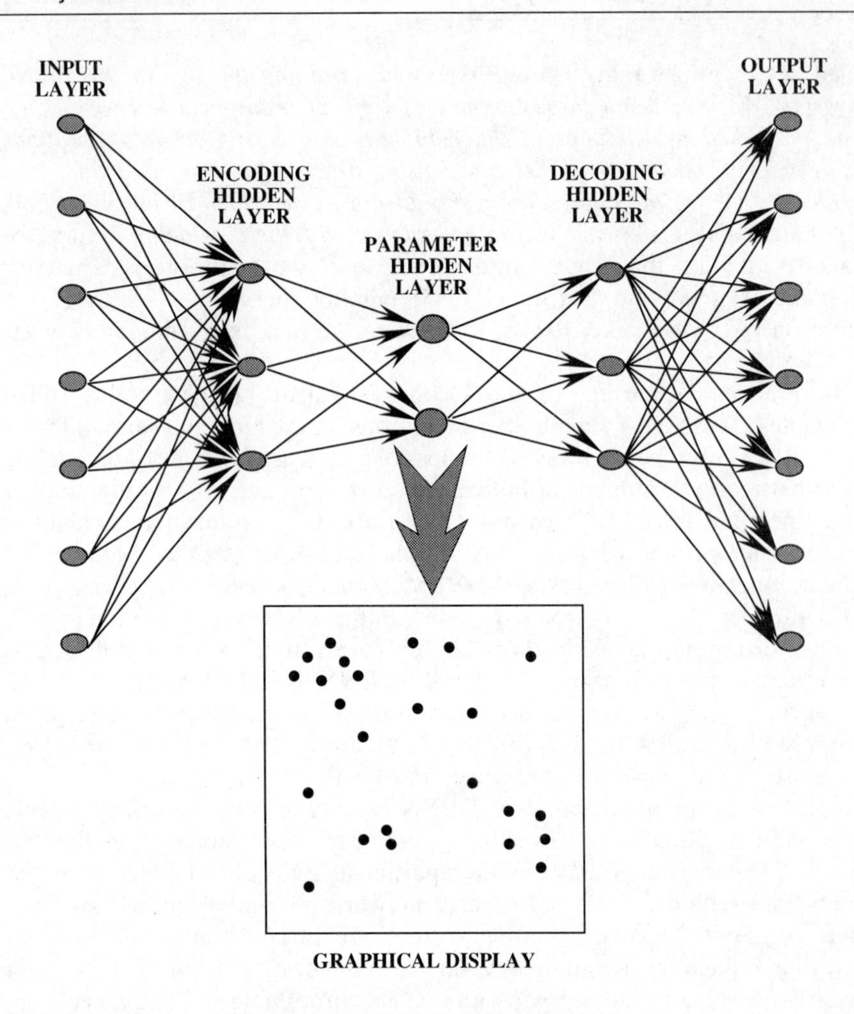

Figure 4 ReNDeR BNN architecture.

sensor replacement, and gross error detection. Garrido and coworkers (Garrido *et al.*, 1993) termed their system NNA (neural net analysis) and presented it from a theoretical problem related to Gaussian distributions.

Selection of the learning rate (η) and the momentum (α)

The systematic search of the best values for the learning rate (η) and the momentum (α) is generally performed by means of a trial and error procedure. However, Kamiyama and coworkers (Kamiyama *et al.*, 1993) have shown that the relation between η and α could be expressed as a linear function by $\eta = K(1 - \alpha)$ where K was decided by the ratio between the number of output and hidden units. In the same way, Eaton and Olivier (1992) have proposed the following rule of thumb:

$$\eta = 1.5/\sqrt{(N_1^2 + N_2^2 + \cdots + N_m^2)} \text{ with } \alpha = 0.9 \quad (7)$$

where N_1 is the number of patterns of type '1' and m is the number of different pattern types.

An empirical rule for dynamically adjusting the learning rate (η) was also proposed by Song and Yu (1993). Last, it is interesting to note that the use of the scaled conjugate gradient algorithm (Møller, 1993) allows one to overcome the problem of the learning and momentum coefficients adjustments.

When to stop the training?

A crucial step in the training of a BNN is the decision to stop the iterative learning at the right time. Prolonged training (overtraining) to reduce an error beyond a realistic value is time-consuming and, more important, deteriorates the generalization performances of the BNN model. Different techniques have been proposed for avoiding overtraining. They are presented later, in the section dedicated to the selection of a BNN model.

Interpretation of the BNN model

One of the major disadvantages of the BNN models is that they are generally difficult to interpret. Although this limitation may not matter for problems in which only the predictive power of a model is important, in some other cases, it is important to understand the rationale of a modeling solution. In order to gain any insight into a BNN solution, one should look at the weights. The best way to represent and analyze the weights of a BNN is by using graphics. Among them, the Hinton diagram is widely used (Gorman and Sejnowski, 1988; Qian and Sejnowski, 1988; Bottou *et al.*, 1990; Kneller *et al.*, 1990; Lang *et al.*, 1990; Morabito, 1993, 1994; Wolfer *et al.*, 1994). It shows the weights of connections to or from a layer of a BNN. Classically, white and black squares (or rectangles) represent positive and negative weights, respectively. The size of a square (or a rectangle) is proportional to the magnitude of a weight. There are several variations on this scheme (Dobbins and Eberhart, 1990, page 99). An example of Hinton diagram is given in Figure 5. Another way to interpret the weights is to plot the values of the weights versus different information related to BNN (Muskal and Kim, 1992; O'Neill, 1991, 1992; Guo *et al.*, 1994) or to use classical multivariate methods such as cluster analysis (Gorman and Sejnowski, 1988). Last, Sumpter and coworkers (Sumpter *et al.*, 1994) propose using a second network for analyzing the weights.

The analysis of a BNN model is also provided by estimation of the strengths of the relationships between its inputs and outputs. The output value(s) of the BNN model is(are) calculated when the value of a given input

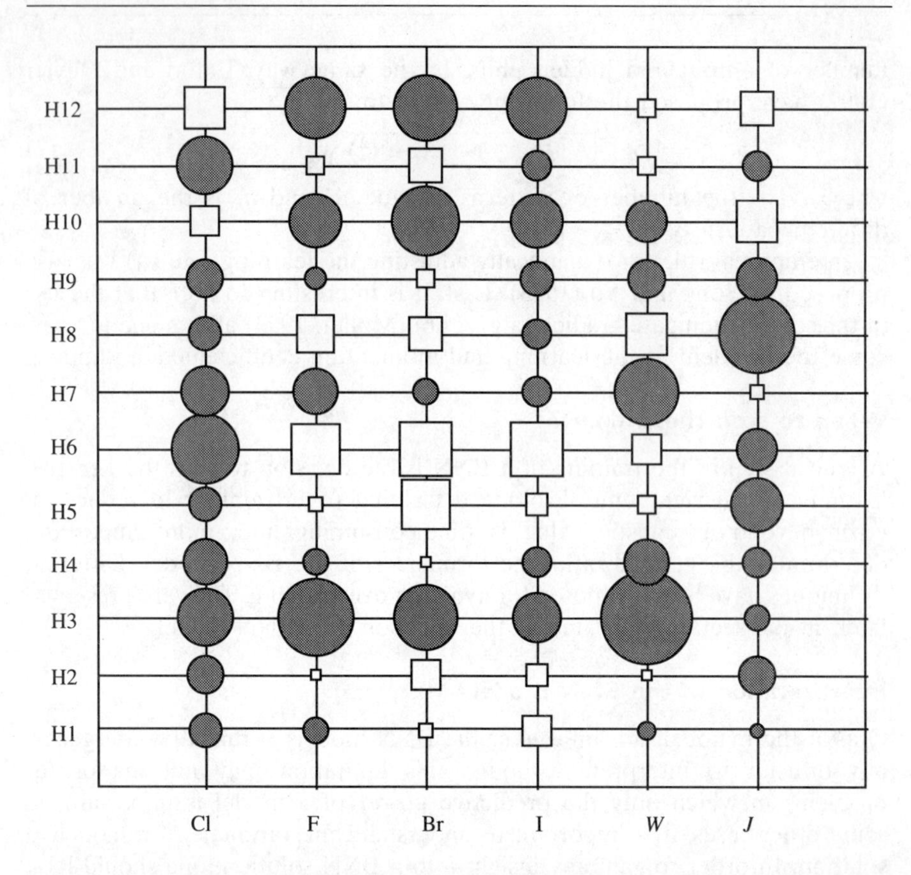

Figure 5 Example of Hinton diagram showing the weights from the input neurons (Cl, F, Br, I, *W*, and *J*) to the twelve hidden neurons (H1 to H12) of a BNN model. The weight values were obtained from Table 6 of Balaban *et al.* (1994). The sizes of the circles (negative values) and squares (positive values) represent the magnitudes of the weights.

is changed while all other inputs remain unchanged. The results of this sensitivity analysis are generally analyzed from graphics (Refenes *et al.*, 1994). Song and Yu (1993) propose the calculation of a partial correlation index for measuring the degree of influence of individual input variables on a BNN output. Dow (1993) uses a 'voting' mechanism to rank the importance of the inputs. Last, recently, Takeda and coworkers (Takeda *et al.*, 1994) have proposed a method allowing estimation of the effects of de-activations of hidden layer units on the activity of the outputs.

SELECTION OF THE BEST BNN MODEL

Estimating BNN performances

Neural network performances can be estimated from different measurements. For classification systems, the simplest way is the estimation of the percentage of correct predictions. Receiver operating characteristic (ROC) curves (Swets, 1973, 1988) provide useful tools for measuring the performances and for quantifying the accuracy of classification systems (Lusted, 1971a,b; McNeil *et al.*, 1975a,b; Komaroff, 1979; Adlassnig and Scheithauer, 1989; Yang *et al.*, 1994). They are particularly suitable when they are employed with neural networks because the results are not sensitive to the probability distribution of the training/testing set patterns or decision bias (Eberhart *et al.*, 1990, page 169). The ROC curves are principally used to reflect the performance of the BNN output(s). Thus, for a given activity (e.g., mutagenicity), indicated by a given output neuron, four possible alternatives exist: true positive (TP), false positive (FP), false negative (FN), and true negative (TN) (Table I). The ROC curve makes use of two ratios involving the four possible alternatives listed in Table I. The first ratio, called true positive ratio or sensitivity, is TP/(TP + FN). The second ratio, called false positive ratio, is FP/(FP + TN). The ROC curve is a plot of the true positive ratio versus the false positive ratio. When applied to the performances of different BNN models, the curve is usually obtained by plotting points for various values of the studied parameters (e.g., number of cycles, number of hidden neurons), then connecting the points with either line segments (Binder *et al.*, 1994) or a smooth curve (Eberhart *et al.*, 1990, page 171). Figure 6 shows hypothetical ROC curves for two BNN configurations. The design of the ROC curves (Figure 6) allows identification of a model presenting the best decision performances. It is located in the upper-left corner of the graph where the sensitivity is maximized and the false positive ratio is minimized. The information encoded by the area under a ROC curve (Hanley and McNeil, 1982, 1983) can also be used to estimate the performances of BNN models. From Table I, it is also possible to calculate the false negative ratio (FN/(FN + TP)) and the true negative ratio also called the specificity (TN/(FP + TN)) for providing additional information on the performances of a BNN model (Eberhart *et al.*, 1990, page 173; Pattichis *et al.*, 1995). A chi-square test can

Table I *Contingency table used in ROC curve definition.*

	Class positive	Class negative
Prediction positive	TP (true positive)	FP (false positive)
Prediction negative	FN (false negative)	TN (true negative)

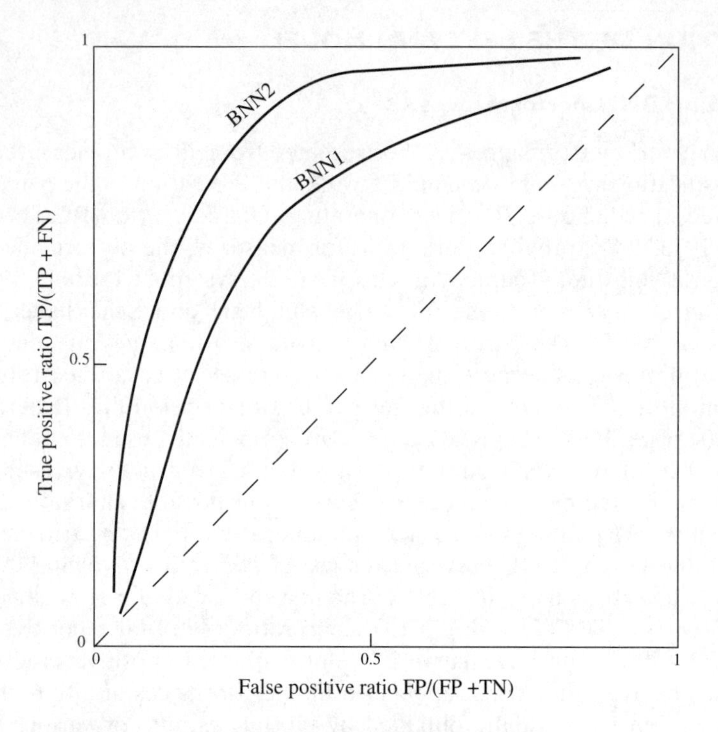

Figure 6 Examples of ROC curves.

be used to determine whether a given set of output categories (or answers or classifications), when compared with an expected distribution, presents a variance from probability or a predefined expectation greater than would be expected by chance alone (Eberhart *et al.*, 1990, page 175). Last, the calculation of a Hamming distance between targets and outputs can also provide interesting results for comparison purposes (Kamimura and Nakanishi, 1994a).

When the BNN model produces continuous output values, the BNN performances are generally estimated from the calculation of the root mean square (RMS) error (Hair *et al.*, 1992) which is calculated as follows:

$$\text{RMS} = \sqrt{(\sum (\text{target} - \text{output})^2)/\text{number of patterns}} \tag{8}$$

It is obvious that a single indicator of performance like the RMS error cannot provide a complete appreciation of the various aspects of the BNN performances. Therefore, it is often required to consider other statistical and/or graphical information (Bhandare *et al.*, 1993; Ricard *et al.*, 1993; Refenes *et al.*, 1994, 1995).

Model selection

One of the central problems in the field of neural networks is that of model selection. As regards the classical BNN algorithm, this problem has been widely investigated and a vast array of techniques and results have been suggested to perform this task. Basically, the choice of a BNN model must be a compromise between its learning and its generalization performances. Thus, the complexity of the network should be high enough to learn the structures hidden in the training set and at the same time avoid overfitting. It should also be necessary to stop the training process in due time to avoid overtraining. These conditions allow the BNN model to learn the training data correctly without losing its ability to generalize. Different methods have been proposed for selecting an optimal BNN model. A first class of techniques deals with architecture modifications. A common method for directing architecture modification and selection is to start with an 'oversized' network and remove (prune) elements (i.e., neurons and/or weights) which produce poor test results. Neuron pruning has been examined in a previous section. Weight pruning is usually performed interactively according to different procedures (Deco *et al.*, 1993; Finnoff *et al.*, 1993; Ying *et al.*, 1993; Depenau and Møller, 1994; Kamimura and Nakanishi, 1994b). The BNN is trained either to localize a minimum of the error function or until generalization performance begins to deteriorate on a validation set of examples. Then, an appropriate number of weights is removed and the training process is restarted. However, according to Finnoff and coworkers (Finnoff *et al.*, 1993) '*the problem lies in determining what an appropriate number might be, and can be influenced by the number of pruning steps already performed, . . ., and eventually other topology modifications that are performed in parallel (for example input neuron pruning)*'.

A second class of techniques for selecting an optimal BNN model consists in adding penalty terms to the error function near its extreme values for reducing the ability of the BNN to learn useless information (Curry and Rumelhart, 1990; Finnoff *et al.*, 1993; Klawun and Wilkins, 1994).

The last class of methods consists in using a training set for adjusting the values of the weights and another set (called a validation set) for deciding when to stop the training phase (Weigend *et al.*, 1990; Finnoff *et al.*, 1993; Tsaptsinos and Leigh, 1993; Kreesuradej *et al.*, 1994). The role of the validation set is to track the generalization abilities of the BNN model. Note that the performance of the validation set is also monitored by means of the calculation of an error. When it ceases to improve, the training is stopped. This method is sometimes termed crossvalidation (Curry and Rumelhart, 1990; Bos *et al.*, 1992; Jodouin, 1994) but is different from the classical cross-validation procedure used in QSAR studies which is generally related to the leave-one-out procedure (Cramer *et al.*, 1988). The method of stopped training presents different shortcomings. When the validation set error is narrow, sampling error in the validation set itself creates uncertainty about the right time to stop (Curry and Rumelhart, 1990). The method cannot be used in real situations where the true answer is truly unknown (Zhang *et al.*, 1992). It also requires the use of another external testing set to estimate the real generalization capabilities of the BNN (Jodouin, 1994). This can be problematic in QSAR studies since the number of tested chemicals is often reduced. Last, if the training and validation sets are relatively small or only partially cover the pattern space to model, the generalization capabilities of the model will be very limited. All the above methods (and others) for improving model selection have been tested by Finnoff and coworkers (Finnoff *et al.*, 1993) in a comparative study. Their results clearly show that the effectiveness of these methods is domain dependent, but in all cases, they are able to improve the generalization performances of the network significantly.

It is interesting to note that sometimes, instead of selecting a unique BNN model, authors prefer averaging the results obtained with different runs (Hansen and Salamon, 1990; Tetko *et al.*, 1995).

COMPARISON OF THE PERFORMANCES OF A BNN MODEL WITH THOSE OBTAINED WITH OTHER APPROACHES

The neural network literature contains numerous studies dealing with the comparison of the tasks performed by BNNs with those carried out by multivariate methods such as principal components analysis (Bourlard and Kamp, 1988; Baldi and Hornik, 1989; Oja, 1989, 1992; Koutsougeras, 1994), discriminant analysis (Gallinari *et al.*, 1991), and so on (Bounds *et al.*, 1990). These works often enhance mathematical considerations allowing one to find filiations between methods but generally never use real data. Therefore, their interest for a (Q)SAR or QSPR practitioner is limited. In the same way, it is obvious that the majority of the BNN models which are published in the (Q)SAR literature are compared with classical models designed from the other statistical tools generally used in the discipline (e.g., regression analysis, PLS, discriminant analysis). However, these comparisons are also sometimes of limited interest. Indeed, they are generally performed on limited sets of molecules. In addition, numerous factors and practical aspects are not taken into account. Hence they are discussed below.

With a BNN, the goal of the training phase is to minimize an error. This error (Eq. (3)) is obtained by computing the difference between the output value that an output neuron is supposed to have, and the value the neuron actually has as a result of the feedforward calculation. This difference is squared and then the sum of the squares is taken over all output nodes. Finally, the calculation is repeated for each pattern in the training set. The grand total sum over all nodes and all patterns, multiplied by 0.5, provides the total error

E (Eq. (3)). When the total error (Eq. (3)) is divided by the number of patterns, we obtain the average sum-squared error which can be used for comparing results. There are some shortcomings when employing this error for comparison purposes. Indeed, many implementations ignore the factor of 0.5 in the calculation of the error (Eberhart *et al.*, 1990, page 165). Therefore, if you are not aware of how the error term is implemented in your software (or in the software used in a published article), this can lead to erroneous conclusions. Second, the error term is summed over all output neurons. Therefore, if the comparison deals with various configurations with different numbers of output neurons, the average sum-squared error may not reflect the performances of the BNNs. Last, Eberhart and coworkers (Eberhart *et al.*, 1990, pages 166–167) have clearly shown that the average sum-squared error may not adequately measure the classification performances of a BNN since it is linked to the value selected for the threshold. From the above, it appears that when we want to compare the performances of different BNN models, it is preferable to use another parameter. For the comparison of SAR models (pattern recognition) obtained with a BNN and other statistical tools, the percentages of good and/or bad classifications can be employed. The design of ROC curves and the calculation of the area under the curve (AUC) can also provide interesting results (Burke *et al.*, 1994) which can be easily interpreted. Last, it is interesting to note that to compare the results obtained with different classifiers, information-theoretical techniques can be employed (Reibnegger *et al.*, 1991). When QSAR or QSPR models designed from BNNs and classical linear methods are compared, the mean square (MS) or RMS errors are generally used. Other statistical parameters can be also employed (e.g., Refenes *et al.*, 1994, 1995). However, the key problem dealing with the comparison of the performances of a BNN with those obtained by the use of a discriminant analysis, regression analysis, PCR or PLS is that we try to compare a nonlinear method with linear method(s). This is a difficult task which must take into account statistical and practical aspects. Indeed, it is obvious that the comparison of statistical methods has to be based on the use of valuable statistical parameters like those presented above. In addition, when the comparison includes a BNN, some parameters can be calculated for estimating its nonlinearity effect. Thus, for example, Weigend and coworkers (Weigend *et al.*, 1990) propose a parameter (Eq. (9)) for assessing the overall improvement due to sigmoids in the BNN compared with linear regression.

$$\mathcal{R} = \text{residual variance (nonlinear model)} \,/\, \text{residual variance (linear model)} \tag{9}$$

The value of $\mathcal{R}$ lies between 0 and 1, since the performance can improve only if additional nonlinearities are used (Weigend *et al.*, 1990).

However, besides these statistical aspects, the comparison of the performances of a BNN with those of classical linear methods must also take into account some practical features. Thus, it is important to stress that a BNN cannot be used as easily as classical linear approaches, such as regression analysis or PLS. Indeed, the proper use of a BNN requires experience and expertise to select the best model since many elements must be adjusted, monitored and/or selected. In addition, other parameters like computation time and memory requirements have also to be taken into account. These constraints have a cost which can be evaluated and integrated in the comparison process.

SOFTWARE AVAILABILITY

There is a huge amount of software available commercially and in the public domain for the construction of BNN models. A recent review has been provided by Manallack and Livingstone (1994b). Additional information on BNN and other neural network software packages can also be obtained from the annex. These programs run on most commonly available hardware platforms such as IBM® PC and compatibles, Macintosh®, and UNIX® workstations (e.g., SUN®, Silicon Graphics®). The main problem with the commercial packages concerns the source code which is generally not available. Therefore, without a well designed user's guide, these packages can appear as black boxes and lead to badly designed models. They can be very expensive and difficult to implement on different hardware platforms in the same laboratory without the purchase of a multisite license. The ability to create composite models by bolting together a BNN model and other modeling tool(s) can be difficult. Last, the transformation of a designed model into an autonomous commercial application can be problematic. Numerous computer tools available in the public domain are more open and flexible since generally the source code is available. However, they often present bugs which must be corrected for proper use. This is always a difficult and time-consuming task. Under these conditions, the optimal choice consists in using programs provided with reference books (e.g., Pao, 1989; Eberhart and Dobbins, 1990a; Caudill and Butler, 1992; Freeman, 1994; Welstead, 1994; Müller *et al.*, 1995). The source code, generally in C, is included in the books and the programs can be easily transformed and adapted for specific requirements which are numerous in (Q)SAR and QSPR studies.

HYBRID SYSTEMS WITH BNN

As previously stressed, for improving the performances of a BNN, one can try to modify the standard algorithm. Another way to overcome the shortcomings of the classical BNN is to include it into a hybrid system. The notion of hybrid system is based on the idea that each technique has particular strengths and weaknesses and cannot be applied to solve all problems

Table II *Comparison of intelligent techniques (ITs) which can be hybridized (adapted from Goonatilake (1995)).*

ITs	Learning	Flexibility	Adaptation	Explanation	Discovery	Use in (Q)SAR
BNNs	*****	*****	*****	*	**	****
GAs	*****	****	****	***	*****	**
FSs	*	*****	*	***	*	*
ESs	*	*	*	*****	*	*

BNNs, Backpropagation neural networks; GAs, Genetic algorithms; FSs, Fuzzy systems; ESs, Expert systems.

(Goonatilake, 1995; Goonatilake and Khebbal, 1995). Therefore, the judicious hybridization of different statistical and mathematical approaches can overcome their individual limitations for performing a particular task. Table II summarizes some properties of different intelligent techniques which can be hybridized with a BNN. According to Goonatilake and Khebbal (1995), hybrid systems can be differentiated on the basis of their functionality. In function-replacing hybrids, a principal function of a given technique is replaced by another intelligent processing technique. Here, the aim is to take an approach presenting weaknesses in a particular property and combine it with a technique that has strengths in the same property. Thus, for example, genetic algorithms (GAs) are a class of search methods rooted on the principles of the natural evolution (Goldberg, 1989; Davis, 1991; Michalewicz, 1992; Forrest, 1993; Devillers, 1996). In practice, they have been shown to be very effective at function optimization, efficiently searching large and complex spaces to find nearly global optima. Under these conditions, GAs have been largely employed for designing and/or training neural networks (Harp *et al.*, 1989; Miller *et al.*, 1989; Jones, 1993; Shamir *et al.*, 1993; Montana, 1995; Roberts and Turega, 1995). In the same way, note that evolutionary programming (Fogel, 1995) has been used by McDonnell and Waagen (1993) for simultaneously determining the weights and number of hidden units in a multilayer neural network.

According to Goonatilake and Khebbal (1995), the second class of hybrid systems corresponds to the polymorphic hybrids which use a single processing architecture to achieve the functionality of different intelligent processing techniques. The cellular encoding methodology proposed by Gruau (1995) which allows the synthesis of neural networks belongs to this category of hybrid systems. Indeed, according to Gruau (1995, page 269) cellular encoding enables genetic programming to decompose a problem into subproblems automatically and dynamically, discover subneural networks that compute the subproblems, and find a way to allow assembly of the subneural networks into neural networks computing the overall problem.

Under these conditions, cellular encoding can be used to overcome some methodological problems in the design of BNN architectures.

The last class of hybrid systems corresponds to the intercommunicating hybrids which are independent, self-contained, processing modules that exchange information and perform separate functions in order to generate optimal solutions (Goonatilake, 1995; Goonatilake and Khebbal, 1995). Thus, if a problem can be subdivided into distinct processing tasks, then different independent modules can be used to solve the parts of the problem at which they are the best. They work in synergy for performing a particular task. The different modules can process subtasks either in a sequential manner or in a competitive/cooperative framework. Intercommunicating hybrids using GAs and BNNs have been used for designing oil pumping units (Vahidov *et al.*, 1995) but also for optimizing QSAR studies (Devillers and Putavy, 1996; van Helden *et al.*, 1996; Venkatasubramanian *et al.*, 1996). The encapsulation of an expert system and a neural network into an intercommunicating hybrid system can also provide interesting results (Hillman, 1990). Indeed, expert systems perform reasoning based on previously established rules for a well-defined and narrow domain (Medsker, 1994). The strength of expert systems comes from the fact that shallow knowledge of experts can be represented by simple IF–THEN rules (Scherer and Schlageter, 1995). In addition, the knowledge bases can be easily modified when new rules and facts become available (Medsker, 1994). However, expert systems present numerous shortcomings. Among them, we can cite the knowledgc acquisition, the development time, the maintenance, the verification and the validation of large knowledge bases (Medsker, 1994; Powell *et al.*, 1995; Scherer and Schlageter, 1995). These limitations can be compensated by neural networks. A survey of the literature shows numerous successful hybridizations between expert systems and BNNs for performing various tasks (Lübbert and Simutis, 1994; Medsker, 1994; Theocharidou and Burianec, 1995). Sima (1995) also indicates an example of application in QSAR dealing with the use of a neural expert system called EXPSYS for deriving relationships between the structure of catechol analogs and malignant melanoma. Systems based on fuzzy logic exhibit a flexibility and have proven to be successful in numerous tasks (Dubois *et al.*, 1993; Brown and Harris, 1994; Zhang *et al.*, 1995). Central to the flexibility that fuzzy logic provides, is the notion of fuzzy sets. A crisp set consists of objects that either belong or not to the set, while a fuzzy set presents objects that may partially belong. Therefore, the set boundary is not crisp and appears blurred. Usually the shapes and ranges of the fuzzy memberships are designed by an expert. Data which have been converted into fuzzy membership functions are referred to as having been 'fuzzified'. Defuzzification is the reverse process. Fuzzy inference rules provide the relationships among the fuzzy variables. They have to be specified by an expert. One of the main strengths of fuzzy logic is that their knowledge bases are easy to examine and understand. However, it is a cumbersome and time-consuming procedure to decide the IF–THEN type of linguistic rules and membership functions. These limitations can be compensated by the aid of neural networks (Shi and Shimizu, 1992). This provides an intercommunicating hybrid system where the advantages of each technique are combined (Shi and Shimizu, 1992; Hu and Hertz, 1993; Khan, 1993; Maillard and Gresser, 1993). Note that the notion of an intercommunicating hybrid is very large and that the degree of intercommunication between different approaches can be variable in terms of complexity, number of systems involved, and so on. Thus, for example, preprocessing of the inputs in a BNN by means of the PLS method (Bhandare *et al.*, 1993) or mixing the PLS and BNN algorithms (Holcomb and Morari, 1992; Qin and McAvoy, 1992) can be considered as intercommunicating hybridizations. The use of a learning vector quantization network or a Kohonen self-organizing map (Kohonen, 1990) for filtering the input data of a BNN (Cammarata *et al.*, 1995) can be also considered as an intercommunicating hybridization.

CONCLUSION

The backpropagation neural network (BNN) is a very powerful tool in (Q)SAR and QSPR studies. Indeed, a BNN can find nonlinear relationships between the structure of the organic molecules and their activity or property. Basically, a BNN is able to learn from examples and therefore can acquire its own knowledge. It is able to solve new problems by exhibiting generalization. BNN is also fault tolerant and can handle noisy and incomplete data. However, it presents some shortcomings. The training phase is a time-consuming task. Using BNN in practice implies adjustment of a variety of parameters (e.g., data preprocessing, architecture, learning rate) which influence its behavior. In addition, in some cases there is uncertainty as to whether the final learning goal is achieved. To validate that the BNN has learnt, testing sets are necessary. The obtained models are often difficult to interpret and comparisons with classical methods are not straightforward. Large amounts of works have been made to overcome these problems. They principally deal with modifications of the original BNN algorithm and/or the creation of hybrid systems. The flowering of interesting results regularly obtained in these two axes of research implies that even if BNN is not a cure-all recipe in (Q)SAR and QSPR studies, this powerful tool cannot be ignored in the domain and has to be used when the classical linear methods fail.

ANNEX: ARTIFICIAL NEURAL NETWORKS (ANNs) ON INTERNET

Surfing the ANN-WEB information on the Internet is time-consuming and expensive. Indeed, it is amazing to follow how day per day the amount of available information about ANNs increases on the network. Under these conditions, we have only compiled a selected list of Internet resources that provide access to key information on ANNs. From this list, the readers who are interested in some of the latest developments on ANNs, software availability (BNN but also other paradigms), conferences, and other specific topics on ANNs will explore this valuable source of information on their own.

http://www.neuronet.ph.kcl.ac.uk
http://www.eeb.ele.tue.nl/neural/neural.html
http://nucleus.hut.fi/~jari/research.html
http://wwwipd.ira.uka.de/~prechelt/FAQ/neural-net-faq.html
http://glimpse.cs.arizona.edu:1994/bib/
http://liinwww.ira.uka.de/bibliography/Neural/index.html
http://msia02.msi.se/~lindsey/elba2html/elba2html.html

REFERENCES

Achanta, A.S., Kowalski, J.G., and Rhodes, C.T. (1995). Artificial neural networks: Implications for pharmaceutical sciences. *Drug Dev. Ind. Pharm.* **21,** 119–155.

Adlassnig, K.P. and Scheithauer, W. (1989). Performance evaluation of medical expert systems using ROC curves. *Comput. Biomed. Res.* **22,** 297–313.

Akaike, H. (1974). A new look at the statistical model identification. *IEEE Trans. Appl. Comp.* **AC19,** 716–723.

Aleksander, I. and Morton, H. (1990). *An Introduction to Neural Computing.* Chapman & Hall, London, p. 240.

Alpsan, D., Towsey, M., Ozdamar, O., Chung Tsoi, A., and Ghista, D.N. (1995). Efficacy of modified backpropagation and optimisation methods on a real-world medical problem. *Neural Networks* **8,** 945–962.

Amirikian, B. and Nishimura, H. (1994). What size network is good for generalization of a specific task of interest? *Neural Networks* **7,** 321–329.

Anderson, J.A. (1990). Data representation in neural networks. *AI Expert* **June,** 30–37.

Andrea, T.A. and Kalayeh, H. (1991a). Applications of neural networks in quantitative structure–activity relationships of dihydrofolate reductase inhibitors. *J. Med. Chem.* **34,** 2824–2836.

Andrea, T.A. and Kalayeh, H. (1991b). Applications of neural networks in quantitative structure–activity relationships. In, *QSAR: Rational Approaches to the Design of Bioactive Compounds* (C. Silipo and A. Vittoria, Eds.). Elsevier Science Publishers, Amsterdam, pp. 209–212.

Andrews, J.M. and Lieberman, S.H. (1994). Neural network approach to qualitative identification of fuels and oils from laser induced fluorescence spectra. *Anal. Chim. Acta* **285,** 237–246.

Angel, J.R.P., Wizinowich, P., Lloyd-Hart, M., and Sandler, D. (1990). Adaptive optics for array telescopes using neural-network techniques. *Nature* **348,** 221–224.

Anguita, D., Parodi, G., and Zunino, R. (1993). Speed improvement of the back-propagation on current-generation workstations. In, *World Congress on Neural Networks.* Oregon Convention Center, Portland, Oregon, July 11–15, 1993, Lawrence Erlbaum Associates and INNS Press, pp. I-165–I-168.

Aoyama, T. and Ichikawa, H. (1991a). Basic operating characteristics of neural networks when applied to structure–activity studies. *Chem. Pharm. Bull.* **39,** 358–366.

Aoyama, T. and Ichikawa, H. (1991b). Obtaining the correlation indices between drug activity and structural parameters using a neural network. *Chem. Pharm. Bull.* **39,** 372–378.

Aoyama, T. and Ichikawa, H. (1991c). Reconstruction of weight matrices in neural networks – A method of correlating outputs with inputs. *Chem. Pharm. Bull.* **39,** 1222–1228.

Aoyama, T. and Ichikawa, H. (1992). Neural networks as nonlinear structure–activity relationship analyzers. Useful functions of the partial derivative method in multilayer neural networks. *J. Chem. Inf. Comput. Sci.* **32,** 492–500.

Aoyama, T., Suzuki, Y., and Ichikawa, H. (1990a). Neural networks applied to structure–activity relationships. *J. Med. Chem.* **33,** 905–908.

Aoyama, T., Suzuki, Y., and Ichikawa, H. (1990b). Neural networks applied to quantitative structure–activity relationship analysis. *J. Med. Chem.* **33,** 2583–2590.

Baba, N. (1989). A new approach for finding the global minimun of error function of neural networks. *Neural Networks* **2,** 367–373.

Bailey, D. and Thompson, D. (1990). How to develop neural-network. *AI Expert* **June,** 38–47.

Balaban, A.T., Basak, S.C., Colburn, T., and Grunwald, G.D. (1994). Correlation between structure and normal boiling points of haloalkanes C_1–C_4 using neural networks. *J. Chem. Inf. Comput. Sci.* **34,** 1118–1121.

Baldi, P. and Hornik, K. (1989). Neural networks and principal components analysis: Learning from examples without local minima. *Neural Networks* **2,** 53–58.

Basak, J., Pal, N.R., and Pal, S.K. (1995). A connectionist system for learning and recognition of structures: Application to handwritten characters. *Neural Networks* **8,** 643–657.

Basti, G., Casolino, M., Messi, R., Perrone, A., and Picozza, P. (1993). Back-propagation with stochastic pre-processing for particle recognition in a Silicon calorimeter. In, *World Congress on Neural Networks.* Oregon Convention Center, Portland, Oregon, July 11–15, 1993, Lawrence Erlbaum Associates and INNS Press, pp. I-320–I-328.

Bengio, Y. (1996). *Neural Networks for Speech and Sequence Recognition.* International Thomson Computer Press, London, p. 167.

Bhandare, P., Mendelson, Y., Peura, R.A., Janatsch, G., Kruse-Jarres, J.D., Marbach, R., and Heise, H.M. (1993). Multivariate determination of glucose in whole blood using partial least-squares and artificial neural networks based on mid-infrared spectroscopy. *Appl. Spectrosc.* **47,** 1214–1221.

Bhat, N.V. and McAvoy, T.J. (1992). Determining model structure for neural models by network stripping. *Comput. Chem. Engng* **16,** 271–281.

Biegler-König, F. and Bärmann, F. (1993). A learning algorithm for multilayered neural networks based on linear least squares problems. *Neural Networks* **6,** 127–131.

Binder, M., Steiner, A., Schwarz, M., Knollmayer, S., Wolff, K., and Pehamberger, H. (1994). Application of an artificial neural network in epiluminescence microscopy pattern analysis of pigmented skin lesions: A pilot study. *British J. Dermatol.* **130,** 460–465.

Biswas, N.N. and Kumar, R. (1990). A new algorithm for learning representations in Boolean neural networks. *Current Sci.* **59,** 595–600.

Boddy, L., Morris, C.W., and Wimpenny, J.W.T. (1990). Introduction to neural networks. *Binary* **2,** 179–185.

Bodor, N., Harget, A., and Huang, M.J. (1991). Neural network studies. I. Estimation of the aqueous solubility of organic compounds. *J. Am. Chem. Soc.* **113,** 9480–9483.

Borggaard, C. and Thodberg, H.H. (1992). Optimal minimal neural interpretation of spectra. *Anal. Chem.* **64,** 545–551.

Borman, S. (1989). Neural network applications in chemistry begin to appear. *Chem. Engng News* **April,** 24–28.

Bos, A., Bos, M., and van der Linden, W.E. (1992). Artificial neural networks as a tool for soft-modelling in quantitative analytical chemistry: The prediction of the water content of cheese. *Anal. Chim. Acta* **256,** 133–144.

Bos, M., Bos, A., and van der Linden, W.E. (1993). Data processing by neural networks in quantitative chemical analysis. *Analyst* **118,** 323–328.

Bottou, L., Fogelman Soulié, F., Blanchet, P., and Liénard, J.S. (1990). Speaker-independent isolated digit recognition: Multilayer perceptrons vs. dynamic time warping. *Neural Networks* **3,** 453–465.

Bounds, D.G., Lloyd, P.J., and Matthew, B.G. (1990). A comparison of neural network and other pattern recognition approaches to the diagnosis of low back disorders. *Neural Networks* **3,** 583–591.

Bourlard, H. and Kamp, Y. (1988). Auto-association by multilayer perceptrons and singular value decomposition. *Biol. Cyber.* **59,** 291–294.

Bourret, P., Reggia, J., and Samuelides, M. (1991). *Réseaux Neuronaux: Une Approche Connexionniste de l'Intelligence Artificielle*. Teknea, Toulouse, p. 269.

Brinn, M., Walsh, P.T., Payne, M.P., and Bott, B. (1993). Neural network classification of mutagens using structural fragment data. *SAR QSAR Environ. Res.* **1,** 169–211.

Brown, M. and Harris, C. (1994). *Neurofuzzy Adaptive Modelling and Control.* Prentice Hall, New York, p. 508.

Brunelli, R. (1994). Training neural nets through stochastic minimization. *Neural Networks* **7,** 1405–1412.

Burke, H.B., Goodman, P.H., and Rosen, D.B. (1994). Artificial neural networks for outcome prediction in cancer. In, *World Congress on Neural Networks*. Town & Country Hotel, San Diego, California, USA, June 5–9, 1994, Lawrence Erlbaum Associates and INNS Press, pp. I-53–I-56.

Calleja, M.C., Geladi, P., and Persoone, G. (1994). QSAR models for predicting the acute toxicity of selected organic chemicals with diverse structures to aquatic non-vertebrates and humans. *SAR QSAR Environ. Res.* **2,** 193–234.

Cambon, B. and Devillers, J. (1993). New trends in structure–biodegradability relationships. *Quant. Struct.–Act. Relat.* **12,** 49–56.

Cammarata, G., Cavalieri, S., and Fichera, A. (1995). A neural network architecture for noise prediction. *Neural Networks* **8,** 963–973.

Cardell, N.S., Joerding, W., and Li, Y. (1994). Why some feedforward networks cannot learn some polynomials. *Neural Comput.* **6,** 761–766.

Caudill, M. and Butler, C. (1992). *Understanding Neural Networks*. Vol. I and II. MIT Press, Cambridge.

Chastrette, M. and de Saint Laumer, J.Y. (1991). Structure–odor relationships using neural networks. *Eur. J. Med. Chem.* **26,** 829–833.

Chastrette, M., de Saint Laumer, J.Y., and Peyraud, J.F. (1993). Adapting the structure of a neural network to extract chemical information. Application to structure–odour relationships. *SAR QSAR Environ. Res.* **1,** 221–231.

Chastrette, M. and El Aidi, C. (1996). Structure–bell-pepper odor relationships for pyrazines and pyridines using neural networks. In, *Neural Networks in QSAR and Drug Design* (J. Devillers, Ed.). Academic Press, London, pp. 83–96.

Cherqaoui, D., Villemin, D., and Kvasnicka, V. (1994a). Application of neural network approach for prediction of some thermochemical properties of alkanes. *Chemom. Intell. Lab. Syst.* **24,** 117–128.

Cherqaoui, D., Villemin, D., Mesbah, A., Cense, J.M., and Kvasnicka, V. (1994b). Use of a neural network to determine the normal boiling points of acyclic ethers, peroxides, acetals and their sulfur analogues. *J. Chem. Soc. Faraday Trans.* **90,** 2015–2019.

Chow, H., Chen, H., Ng, T., Myrdal, P., and Yalkowsky, S.H. (1995). Using back-propagation networks for the estimation of aqueous activity coefficients of aromatic organic compounds. *J. Chem. Inf. Comput. Sci.* **35,** 723–728.

Cramer, R.D., Bunce, J.D., Patterson, D.E., and Frank, I.E. (1988). Crossvalidation, bootstrapping, and partial least squares compared with multiple regression in conventional QSAR studies. *Quant. Struct.–Act. Relat.* **7,** 18–25.

Crick, F. (1989). The recent excitement about neural networks. *Nature* **337,** 129–132.

Curry, B. and Rumelhart, D.E. (1990). MSnet: A neural network which classifies mass spectra. *Tetrahedron Comput. Methodol.* **3,** 213–237.

Davalo, E. and Naïm, P. (1990). *Des Réseaux de Neurones*. Eyrolles, Paris, p. 232.

Davis, L. (1991). *Handbook of Genetic Algorithms*. Van Nostrand Reinhold, New York, p. 385.

Deco, G., Finnoff, W., and Zimmermann, H.G. (1993). Unsupervised elimination of overtraining by a mutual information criterion. In, *World Congress on Neural Networks.* Oregon Convention Center, Portland, Oregon, July 11–15, 1993, Lawrence Erlbaum Associates and INNS Press, pp. III-433–III-436.

Depenau, J. and Møller, M. (1994). Aspects of generalization and pruning. In, *World Congress on Neural Networks*. Town & Country Hotel, San Diego, California, USA, June 5–9, 1994, Lawrence Erlbaum Associates and INNS Press, pp. III-504–III-509.

de Saint Laumer, J.Y., Chastrette, M., and Devillers, J. (1991). Multilayer neural networks applied to structure–activity relationships. In, *Applied Multivariate Analysis in SAR and Environmental Studies* (J. Devillers and W. Karcher, Eds.). Kluwer Academic Publishers, Dordrecht, pp. 479–521.

Devillers, J. (1993). Neural modelling of the biodegradability of benzene derivatives. *SAR QSAR Environ. Res.* **1,** 161–167.

Devillers, J. (1995). Display of multivariate data using non-linear mapping. In, *Chemometric Methods in Molecular Design* (H. van de Waterbeemd, Ed.). VCH, Weinheim, pp. 255–263.

Devillers, J. (1996). Genetic algorithms in computer-aided molecular design. In, *Genetic Algorithms in Molecular Modeling* (J. Devillers, Ed.). Academic Press, London, pp. 1–34.

Devillers, J., Bintein, S., Domine, D., and Karcher, W. (1995). A general QSAR model for predicting the toxicity of organic chemicals to luminescent bacteria (Microtox® Test). *SAR QSAR Environ. Res.* **4,** 29–38.

Devillers, J. and Cambon, B. (1993). Modeling the biological activity of PAH by neural networks. *Polycyclic Aromatic Compounds* **3 (supp),** 257–265.

Devillers, J., Domine, D., and Boethling, R.S. (1996a). Use of a backpropagation neural network and autocorrelation descriptors for predicting the biodegradation of organic chemicals. In, *Neural Networks in QSAR and Drug Design* (J. Devillers, Ed.). Academic Press, London, pp. 65–82.

Devillers, J., Guillon, C., and Domine, D. (1996b). A neural structure–odor threshold model for chemicals of environmental and industrial concern. In, *Neural Networks in QSAR and Drug Design* (J. Devillers, Ed.). Academic Press, London, pp. 97–117.

Devillers, J. and Putavy, C. (1996). Designing biodegradable molecules from the combined use of a backpropagation neural network and a genetic algorithm. In, *Genetic Algorithms in Molecular Modeling* (J. Devillers, Ed.). Academic Press, London, pp. 303–314.

Devillers, J., Zakarya, D., Chastrette, M., and Doré, J.C. (1989). The stochastic regression analysis as a tool in ecotoxicological QSAR studies. *Biomed. Environ. Sci.* **2,** 385–393.

Dobbins, R.W. and Eberhart, R.C. (1990). Software tools. In, *Neural Network PC Tools. A Practical Guide* (R.C. Eberhart and R.W. Dobbins, Eds.). Academic Press, San Diego, pp. 81–110.

Domine, D., Devillers, J., and Chastrette, M. (1994a). A nonlinear map of substituent constants for selecting test series and deriving structure–activity relationships. 1. Aromatic series. *J. Med. Chem.* **37,** 973–980.

Domine, D., Devillers, J., and Chastrette, M. (1994b). A nonlinear map of substituent constants for selecting test series and deriving structure–activity relationships. 2. Aliphatic series. *J. Med. Chem.* **37,** 981–987.

Domine, D., Devillers, J., Chastrette, M., and Karcher, W. (1993a). Estimating pesticide field half-lives from a backpropagation neural network. *SAR QSAR Environ. Res.* **1,** 211–219.

Domine, D., Devillers, J., Chastrette, M., and Karcher, W. (1993b). Non-linear mapping for structure–activity and structure–property modelling. *J. Chemometrics* **7,** 227–242.

Domine, D., Devillers, J. and Karcher, W. (1996a). AUTOLOGP versus neural network estimation of *n*-octanol/water partition coefficients. In, *Neural Networks in QSAR and Drug Design* (J. Devillers, Ed.). Academic Press, London, pp. 47–63.

Domine, D., Wienke, D., Devillers, J., and Buydens, L. (1996b). A new nonlinear neural mapping technique for visual exploration of QSAR data. In, *Neural Networks in QSAR and Drug Design* (J. Devillers, Ed.). Academic Press, London, pp. 223–253.

Dong, D.W. and Chan, Y.D. (1993). Three layer network for identifying Cherenkov radiation patterns. In, *World Congress on Neural Networks.* Oregon Convention Center, Portland, Oregon, July 11–15, 1993, Lawrence Erlbaum Associates and INNS Press, pp. I-312–I-315.

Doucet, J.P., Panaye, A., Feuilleaubois, E., and Ladd, P. (1993). Neural networks and ^{13}C NMR shift prediction. *J. Chem. Inf. Comput. Sci.* **33,** 320–324.

Dow, E.R. (1993). Self-training neural network for generalization and data reduction. In, *World Congress on Neural Networks.* Oregon Convention Center, Portland, Oregon, July 11–15, 1993, Lawrence Erlbaum Associates and INNS Press, pp. I-416–I-419.

Dubois, D., Prade, H., and Yager, R.R. (1993). *Readings in Fuzzy Sets for Intelligent Systems*. Morgan Kaufmann Publishers, San Mateo, p. 916.

Eaton, H.A.C. and Olivier, T.L. (1992). Learning coefficient dependence on training set size. *Neural Networks* **5,** 283–288.

Eberhart, R.C. and Dobbins, R.W. (1990a). *Neural Network PC Tools. A Practical Guide*. Academic Press, San Diego, p. 414.

Eberhart, R.C. and Dobbins, R.W. (1990b). Implementations. In, *Neural Network PC Tools. A Practical Guide* (R.C. Eberhart and R.W. Dobbins, Eds.). Academic Press, San Diego, pp. 35–58.

Eberhart, R.C. and Dobbins, R.W. (1990c). Systems considerations. In, *Neural Network PC Tools. A Practical Guide* (R.C. Eberhart and R.W. Dobbins, Eds.). Academic Press, San Diego, pp. 59–79.

Eberhart, R.C., Dobbins, R.W., and Hutton, L.V. (1990). Performance metrics. In, *Neural Network PC Tools. A Practical Guide* (R.C. Eberhart and R.W. Dobbins, Eds.). Academic Press, San Diego, pp. 161–176.

Efron, B. and Tibshirani, R.J. (1993). *An Introduction to the Bootstrap*. Chapman & Hall, New York, p. 436.

Egolf, L.M., Wessel, M.D., and Jurs, P.C. (1994). Prediction of boiling points and critical temperatures of industrially important organic compounds from molecular structure. *J. Chem. Inf. Comput. Sci.* **34,** 947–956.

Elman, J.L. and Zipser, D. (1988). Learning the hidden structure of speech. *J. Acoust. Soc. Am.* **83,** 1615–1626.

Elrod, D.W., Maggiora, G.M., and Trenary, R.G. (1990). Applications of neural networks in chemistry. 2. A general connectivity representation for the prediction of regiochemistry. *Tetrahedron Comput. Methodol.* **3,** 163–174.

Fahlman, S.E. (1988). *An Empirical Study of Learning Speed in Backpropagation Networks*. Technical Report CMU-CS-88-162, Department of Computer Science, Carnegie Mellon University, Pittsburgh.

Finnoff, W., Hergert, F., and Zimmermann, H.G. (1993). Improving model selection by nonconvergent methods. *Neural Networks* **6,** 771–783.

Fogel, D. (1995). *Evolutionary Computation. Toward a New Philosophy of Machine Intelligence*. IEEE Press, p. 288.

Forrest, S. (1993). Genetic algorithms: Principles of natural selection applied to computation. *Science* **261,** 872–878.

Freeman, J.A. (1994). *Simulating Neural Networks with Mathematica*. Addison-Wesley Publishing Company, Reading, p. 341.

Funahashi, K.I. (1989). On the approximate realization of continuous mappings by neural networks. *Neural Networks* **2,** 183–192.

Furness, P. (1995). Neural networks for data-driven marketing. In, *Intelligent Systems for Finance and Business* (S. Goonatilake and P. Treleaven, Eds.). John Wiley & Sons, Chichester, pp. 73–96.

Gakh, A.A., Gakh, E.G., Sumpter, B.G., and Noid, D.W. (1994). Neural network-graph theory approach to the prediction of the physical properties of organic compounds. *J. Chem. Inf. Comput. Sci.* **34,** 832–839.

Gallinari, P., Thiria, S., Badran, F., and Fogelman-Soulie, F. (1991). On the relations between discriminant analysis and multilayer perceptrons. *Neural Networks* **4,** 349–360.

Garrido, L., Gaitan, V., Serra-Ricart, M., and Calbet, X. (1993). Use of layered neural nets as a display method for *n*-dimensional distributions. In, *World Congress on Neural Networks.* Oregon Convention Center, Portland, Oregon, July 11–15, 1993, Lawrence Erlbaum Associates and INNS Press, pp. II-477–II-480.

Gernoth, K.A. and Clark, J.W. (1995). Neural networks that learn to predict probabilities: Global models of nuclear stability and decay. *Neural Networks* **8,** 291–311.

Ghoshal, N., Mulkhopadhyay, S.N., Ghoshal, T.K., and Achari, B. (1993). Quantitative structure–activity relationship studies of aromatic and heteroaromatic nitro compounds using neural network. *Bioorg. Med. Chem. Lett.* **3,** 329–332.

Goldberg, D.E. (1989). *Genetic Algorithms in Search, Optimization & Machine Learning*. Addison-Wesley Publishing Company, Reading, p. 412.

Good, A.C., So, S.S., and Richards, W.G. (1993). Structure–activity relationships from molecular similarity matrices. *J. Med. Chem.* **36,** 433–438.

Goonatilake, S. (1995). Intelligent systems for finance and business: An overview. In, *Intelligent Systems for Finance and Business* (S. Goonatilake and P. Treleaven, Eds.). John Wiley & Sons, Chichester, pp. 1–28.

Goonatilake, S. and Khebbal, S. (1995). Intelligent hybrid systems: Issues, classifications and future directions. In, *Intelligent Hybrid Systems* (S. Goonatilake and S. Khebbal, Eds.). John Wiley & Sons, Chichester, pp. 1–20.

Gorman, R.P. and Sejnowski, T.J. (1988). Analysis of hidden units in a layered network trained to classify sonar targets. *Neural Networks* **1,** 75–89.

Grossberg, S. (1988). Nonlinear neural networks: Principles, mechanisms, and architectures. *Neural Networks* **1,** 17–61.

Gruau, F. (1995). Genetic programming of neural networks: Theory and practice. In, *Intelligent Hybrid Systems* (S. Goonatilake and S. Khebbal, Eds.). John Wiley & Sons, Chichester, pp. 245–271.

Grunenberg, J. and Herges, R. (1995). Prediction of chromatographic retention values (R_M) and partition coefficients (log P_{oct}) using a combination of semiempirical self-consistent reaction field calculations and neural networks. *J. Chem. Inf. Comput. Sci.* **35,** 905–911.

Guillon, M. and Crommelynck, V. (1991). Les applications de l'intelligence artificielle à la gestion des réseaux de distribution. *T.S.M.-L'Eau* **11,** 519–526.

Guo, Z., Durand, L.G., Lee, H.C., Allard, L., Grenier, M.C., and Stein, P.D. (1994). Artificial neural networks in computer-assisted classification of heart sounds in patients with porcine bioprosthetic valves. *Med. Biol. Eng. Comput.* **32,** 311–316.

Hair, J.F., Anderson, R.E., Tatham, R.L., and Black, W.C. (1992). *Multivariate Data Analysis with Readings.* Macmillan Publishing Company, New York, p. 544.

Hamad, D. and Betrouni, M. (1995). Artificial neural networks for nonlinear projection and exploratory data analysis. In, *Artificial Neural Nets and Genetic Algorithms.* Proceedings of the International Conference, Alès, France, 1995 (D.W. Pearson, N.C. Steele, and R.F. Albrecht, Eds.). Springer-Verlag, Wien, pp. 164–167.

Hammerstrom, D. (1995). A digital VLSI architecture for real-world applications. In, *An Introduction to Neural and Electronic Networks,* Second Edition (S.F. Zornetzer, J.L. Davis, C. Lau, and T. McKenna, Eds.). Academic Press, San Diego, pp. 335–358.

Hanley, J.A. and McNeil, B.J. (1982). The meaning and use of the area under a receiver operating characteristic (ROC) curve. *Radiology* **143,** 29–36.

Hanley, J.A. and McNeil, B.J. (1983). A method of comparing the areas under receiver operating characteristic curves derived from the same cases. *Radiology* **148,** 839–843.

Hansch, C., Unger, S.H., and Forsythe, A.B. (1973). Strategy in drug design. Cluster analysis as an aid in the selection of substituents. *J. Med. Chem.* **16,** 1217–1222.

Hansen, L.K. and Salamon, P. (1990). Neural network ensembles. *IEEE Trans. Pattern Anal. Machine Intell.* **12,** 993–1001.

Harp, S.A., Samad, T., and Guha, A. (1989). Towards the genetic synthesis of neural networks. In, *Proceedings of the Third International Conference on Genetic Algorithms* (J.D. Schaffer, Ed.). Morgan Kaufmann Publishers, San Mateo, pp. 360–369.

Harrington, P.B. (1994). Temperature-constrained backpropagation neural networks. *Anal. Chem.* **66,** 802–807.

Hartnett, M., Diamond, D., and Barker, P.G. (1993). Neural network based recognition of flow injection patterns. *Analyst* **118,** 347–354.

Hillman, D.V. (1990). Integrating neural nets and expert systems. *AI Expert* **June,** 54–59.

Hirose, Y., Yamashita, K., and Hijiya, S. (1991). Back-propagation algorithm which varies the number of hidden units. *Neural Networks* **4,** 61–66.

Hirst, J.D., King, R.D., and Sternberg, M.J.E. (1994a). Quantitative structure–activity relationships by neural networks and inductive logic programming. I. The inhibition of dihydrofolate reductase by pyrimidines. *J. Comp. Aided Mol. Des.* **8,** 405–420.

Hirst, J.D., King, R.D., and Sternberg, M.J.E. (1994b). Quantitative structure–activity relationships by neural networks and inductive logic programming. II. The inhibition of dihydrofolate reductase by triazines. *J. Comp. Aided Mol. Des.* **8,** 421–432.

Hirst, J.D. and Sternberg, M.J.E. (1992). Prediction of structural and functional features of protein and nucleic acid sequences by artificial neural networks. *Biochemistry* **31,** 7211–7218.

Holcomb, T.R. and Morari, M. (1992). PLS/neural networks. *Comput. Chem. Engng* **16,** 393–411.

Hopfield, J.J. (1982). Neural networks and physical systems with emergent collective computational abilities. *Proc. Natl Acad. Sci. USA* **79,** 2554–2558.

Hornik, K. (1991). Approximation capabilities of multilayer feedforward networks. *Neural Networks* **4,** 251–257.

Hornik, K., Stinchcombe, M., and White, H. (1989). Multilayer feedforward networks are universal approximators. *Neural Networks* **2,** 359–366.

Hornik, K., Stinchcombe, M., and White, H. (1990). Universal approximation of an unknown mapping and its derivatives using multilayer feedforward networks. *Neural Networks* **3,** 551–560.

Hornik, K., Stinchcombe, M., White, H., and Auer, P. (1994). Degree of approximation results for feedforward networks approximating unknown mappings and their derivatives. *Neural Comput.* **6,** 1262–1275.

Hosseini-Nezhad, S.M., Yamashita, T.S., Bielefeld, R.A., Krug, S.E., and Pao, Y.H. (1995). A neural network approach for the determination of interhospital transport mode. *Comput. Biomed. Res.* **28,** 319–334.

Hu, Q. and Hertz, D.B. (1993). Improving the convergence of backpropagation using fuzzy logic control. In, *World Congress on Neural Networks.* Oregon Convention Center, Portland, Oregon, July 11–15, 1993, Lawrence Erlbaum Associates and INNS Press, pp. II-47–II-51.

Humpert, B.K. (1994). Improving back propagation with a new error function. *Neural Networks* **7,** 1191–1192.

Hunt, S.D. and Deller, J.R. (1995). Selective training of feedforward artificial neural networks using matrix perturbation theory. *Neural Networks* **8,** 931–944.

Ichikawa, H. and Aoyama, T. (1993). How to see characteristics of structural parameters in QSAR analysis: Descriptor mapping using neural networks. *SAR QSAR Environ. Res.* **1,** 115–130.

Ito, Y. (1991a). Representation of functions by superpositions of a step or sigmoid function and their applications to neural network theory. *Neural Networks* **4,** 385–394.

Ito, Y. (1991b). Approximation of functions on a compact set by finite sums of a sigmoid function without scaling. *Neural Networks* **4,** 817–826.

Ito, Y. (1992). Approximation of continuous functions on R^d by linear combinations of shifted rotations of a sigmoid function with and without scaling. *Neural Networks* **5,** 105–115.

Ito, Y. (1994). Approximation capability of layered neural networks with sigmoid units on two layers. *Neural Comput.* **6,** 1233–1243.

Jacobs, R.A. (1988). Increased rates of convergence through learning rate adaptation. *Neural Networks* **1,** 295–307.

Janakiraman, J. and Honavar, V. (1993). Adaptive learning rate for increasing learning speed in backpropagation networks. In, *World Congress on Neural Networks.* Oregon Convention Center, Portland, Oregon, July 11–15, 1993, Lawrence Erlbaum Associates and INNS Press, pp. IV-378–IV-381.

Jodouin, J.F. (1994). *Les Réseaux de Neurones. Principes et Définitions*. Hermès, Paris, p. 124.

Jones, A.J. (1993). Genetic algorithms and their applications to the design of neural networks. *Neural Comput. Applic.* **1,** 32–45.

Jordan, S.N., Leach, A.R., and Bradshaw, J. (1995). The application of neural networks in conformational analysis. 1. Prediction of minimum and maximum interatomic distances. *J. Chem. Inf. Comput. Sci.* **35,** 640–650.

Kamimura, R. and Nakanishi, S. (1994a). Improving generalization performance by entropy maximization. In, *World Congress on Neural Networks*. Town & Country Hotel, San Diego, California, USA, June 5–9, 1994, Lawrence Erlbaum Associates and INNS Press, pp. III-352–III-357.

Kamimura, R. and Nakanishi, S. (1994b). Weight decay as a process of redundancy reduction. In, *World Congress on Neural Networks*. Town & Country Hotel, San Diego, California, USA, June 5–9, 1994, Lawrence Erlbaum Associates and INNS Press, pp. III-486–III-491.

Kamiyama, N., Taguchi, A., Watanabe, H., Pechranin, N., Yoshida, Y., and Sone, M. (1993). Tuning of learning rate and momentum on back-propagation. In, *World Congress on Neural Networks.* Oregon Convention Center, Portland, Oregon, July 11–15, 1993, Lawrence Erlbaum Associates and INNS Press, pp. IV-119–IV-122.

Karras, D.A., Perantonis, S.J., and Varoufakis, S.J. (1993). Constrained learning: A new approach to pattern classification using feedforward networks. In, *World Congress on Neural Networks.* Oregon Convention Center, Portland, Oregon, July 11–15, 1993, Lawrence Erlbaum Associates and INNS Press, pp. IV-235–IV-238.

Kasparian, V., Batur, C., Zhang, H., and Padovan, J. (1994). Davidon least squares-based learning algorithm for feedforward neural networks. *Neural Networks* **7,** 661–670.

Katz, A.S., Katz, S., and Lowe, N. (1994). Fundamentals of the bootstrap based analysis of neural network's accuracy. In, *World Congress on Neural Networks*. Town & Country Hotel, San Diego, California, USA, June 5–9, 1994, Lawrence Erlbaum Associates and INNS Press, pp. III-673–III-678.

Kendrick, C.T. (1994). Backpropagation algorithm modification reducing number of local minima and convergence time. In, *World Congress on Neural Networks*. Town & Country Hotel, San Diego, California, USA, June 5–9, 1994, Lawrence Erlbaum Associates and INNS Press, pp. III-561–III-567.

Khan, E. (1993). Neural network based algorithms for rule evaluation & defuzzification in fuzzy logic design. In, *World Congress on Neural Networks.* Oregon Convention Center, Portland, Oregon, July 11–15, 1993, Lawrence Erlbaum Associates and INNS Press, pp. II-31–II-38.

Kim, B. and May, G.S. (1993). A new learning rule for neural process modeling. In, *World Congress on Neural Networks.* Oregon Convention Center, Portland, Oregon, July 11–15, 1993, Lawrence Erlbaum Associates and INNS Press, pp. I-233–I-236.

King, R.D., Hirst, J.D., and Sternberg, M.J.E. (1993). New approaches to QSAR:

Neural networks and machine learning. *Perspect. Drug Discovery Design* **1,** 279–290.

Klawun, C. and Wilkins, C.L. (1994). A novel algorithm for local minimum escape in backpropagation neural networks: Application to the interpretation of matrix isolation infrared spectra. *J. Chem. Inf. Comput. Sci.* **34,** 984–993.

Kneller, D.G., Cohen, F.E., and Langridge, R. (1990). Improvements in protein secondary structure prediction by an enhanced neural network. *J. Mol. Biol.* **214,** 171–182.

Kohonen, T. (1988). An introduction to neural computing. *Neural Networks* **1,** 3–16.

Kohonen, T. (1990). The self-organizing map. *Proc. IEEE* **78,** 1464–1480.

Komaroff, A.L. (1979). The variability and inaccuracy of medical data. *Proc. IEEE* **67,** 1196–1207.

Koutsougeras, C. (1994). Principal components and neural nets. In, *World Congress on Neural Networks*. Town & Country Hotel, San Diego, California, USA, June 5–9, 1994, Lawrence Erlbaum Associates and INNS Press, pp. III-307–III-311.

Kramer, M.A. (1992). Autoassociative neural networks. *Comput. Chem. Engng* **16,** 313–328.

Kreesuradej, W., Wunsch II, D.C., and Lane, M. (1994). Time delay neural network for small time series data sets. In, *World Congress on Neural Networks*. Town & Country Hotel, San Diego, California, USA, June 5–9, 1994, Lawrence Erlbaum Associates and INNS Press, pp. II-248–II-253.

Lang, K.J., Waibel, A.H., and Hinton, G.E. (1990). A time-delay neural network architecture for isolated word recognition. *Neural Networks* **3,** 23–43.

Le Cun, Y. (1987). *Modèles Connexionnistes de l'Apprentissage*. Thesis, Paris.

Lee, C.W. (1993). Learning in neural networks by using tangent planes to constraint surfaces. *Neural Networks* **6,** 385–392.

Lee, J. (1994). A novel design method for multilayer feedforward neural networks. *Neural Comput.* **6,** 885–901.

Lee, H. and Hajela, P. (1994a). On a unified geometrical interpretation of multilayer feedforward networks. Part I: The interpretation. In, *World Congress on Neural Networks*. Town & Country Hotel, San Diego, California, USA, June 5–9, 1994, Lawrence Erlbaum Associates and INNS Press, pp. III-625–III-630.

Lee, H. and Hajela, P. (1994b). On a unified geometrical interpretation of multilayer feedforward networks. Part II: Quantifying mapping nonlinearity. In, *World Congress on Neural Networks*. Town & Country Hotel, San Diego, California, USA, June 5–9, 1994, Lawrence Erlbaum Associates and INNS Press, pp. III-631–III-636.

Leigh, D. (1995). Neural networks for credit scoring. In, *Intelligent Systems for Finance and Business* (S. Goonatilake and P. Treleaven, Eds.). John Wiley & Sons, Chichester, pp. 61–69.

Leonard, J. and Kramer, M.A. (1990). Improvement of the backpropagation algorithm for training neural networks. *Comput. Chem. Engng* **14**, 337–341.

Levine, D.S. (1991). *Introduction to Neural & Cognitive Modeling.* Lawrence Erlbaum Associates, Hillsdale, p. 439.

Liang, X. and Xia, S. (1994). Principle and methods of structure variation in feed-forward neural networks. In, *World Congress on Neural Networks*. Town & Country Hotel, San Diego, California, USA, June 5–9, 1994, Lawrence Erlbaum Associates and INNS Press, pp. III-522–III-527.

Lim, S.F. and Ho, S.B. (1994). Dynamic creation of hidden units with selective pruning in backpropagation. In, *World Congress on Neural Networks*. Town & Country Hotel, San Diego, California, USA, June 5–9, 1994, Lawrence Erlbaum Associates and INNS Press, pp. III-492–III-497.

Linko, P. and Zhu, Y. (1991). Neural network programming in bioprocess variable estimation and state prediction. *J. Biotechnol.* **21,** 253–270.

Lippmann, R.P. (1987). An introduction to computing with neural nets. *IEEE ASSP Mag.* **April,** 4–22.

Liu, Q., Hirono, S., and Moriguchi, I. (1992a). Application of functional-link net in QSAR. 1. QSAR for activity data given by continuous variate. *Quant. Struct.–Act. Relat.* **11,** 135–141.

Liu, Q., Hirono, S., and Moriguchi, I. (1992b). Application of functional-link net in QSAR. 2. QSAR for activity data given by ratings. *Quant. Struct.–Act. Relat.* **11,** 318–324.

Liu, Q., Hirono, S., and Moriguchi, I. (1992c). Comparison of the functional-link net and the generalized delta rule net in quantitative structure–activity relationship studies. *Chem. Pharm. Bull.* **40,** 2962–2969.

Livingstone, D.J. (1996). Multivariate data display using neural networks. In, *Neural Networks in QSAR and Drug Design* (J. Devillers, Ed.). Academic Press, London, pp. 157–176.

Livingstone, D.J., Hesketh, G., and Clayworth, D. (1991). Novel method for the display of multivariate data using neural networks. *J. Mol. Graphics* **9,** 115–118.

Livingstone, D.J. and Manallack, D.T. (1993). Statistics using neural networks: Chance effects. *J. Med. Chem.* **36,** 1295–1297.

Livingstone, D.J. and Salt, D.W. (1992). Regression analysis for QSAR using neural networks. *Bioorg. Med. Chem. Lett.* **2,** 213–218.

Livingstone, D.J. and Salt, D.W. (1995). Neural networks in the search for similarity and structure–activity. In, *Molecular Similarity in Drug Design* (P.M. Dean, Ed.). Chapman & Hall, Glasgow, pp. 187–214.

Lübbert, A. and Simutis, R. (1994). Using measurement data in bioprocess modelling and control. *TIBTECH* **12,** 304–311.

Lusted, L.B. (1971a). Decision-making studies in patient management. *New Engl. J. Med.* **284,** 416–424.

Lusted, L.B. (1971b). Signal detectability and medical decision-making. Signal detectability studies help radiologists evaluate equipment systems and performance of assistants. *Science* **171,** 1217–1219.

MacIntyre, J. and O'Brien, J.C. (1995). Investigations into the use of wavelet transformations as input into neural networks for condition monitoring. In, *Artificial Neural Nets and Genetic Algorithms.* Proceedings of the International Conference, Alès, France, 1995 (D.W. Pearson, N.C. Steele, and R.F. Albrecht, Eds.). Springer-Verlag, Wien, pp. 116–119.

Maggiora, G.M., Zhang, C.T., Chou, K.C., and Elrod, D.W. (1996). Combining fuzzy clustering and neural networks to predict protein structural classes. In, *Neural Networks in QSAR and Drug Design* (J. Devillers, Ed.). Academic Press, London, pp. 255–279.

Maillard, E. and Gresser, J. (1993). A new fuzzy perceptron classifier. In, *World Congress on Neural Networks.* Oregon Convention Center, Portland, Oregon, July 11–15, 1993, Lawrence Erlbaum Associates and INNS Press, pp. II-78–II-81.

Maillard, E. and Gueriot, D. (1994). Learning the sigmoid slopes to increase the convergence speed of the multilayer perceptron. In, *World Congress on Neural Networks.* Town & Country Hotel, San Diego, California, USA, June 5–9, 1994, Lawrence Erlbaum Associates and INNS Press, pp. III-539–III-544.

Manallack, D.T., Ellis, D.D., and Livingstone, D.J. (1994). Analysis of linear and nonlinear QSAR data using neural networks. *J. Med. Chem.* **37,** 3758–3767.

Manallack, D.T., Gallagher, T., and Livingstone, D.J. (1996). Quantitative structure–activity relationships of nicotinic agonists. In, *Neural Networks in QSAR and Drug Design* (J. Devillers, Ed.). Academic Press, London, pp. 177–208.

Manallack, D.T. and Livingstone, D.J. (1992). Artificial neural networks: Application and chance effects for QSAR data analysis. *Med. Chem. Res.* **2,** 181–190.

Manallack, D.T. and Livingstone, D.J. (1994a). Limitations of functional-link nets as applied to QSAR data analysis. *Quant. Struct.–Act. Relat.* **13,** 18–21.

Manallack, D.T. and Livingstone, D.J. (1994b). Neural networks – A tool for drug design. In, *Advanced Computer-Assisted Techniques in Drug Discovery* (H. van de Waterbeemd, Ed.). VCH, Weinheim, pp. 293–318.

Manallack, D.T. and Livingstone, D.J. (1995). Relating biological activity to chemical structure using neural networks. *Pestic. Sci.* **45,** 167–170.

Mastriani, M. (1994). Pattern recognition using a faster new algorithm for training feed-forward neural networks. In, *World Congress on Neural Networks.* Town & Country Hotel, San Diego, California, USA, June 5–9, 1994, Lawrence Erlbaum Associates and INNS Press, pp. III-601–III-606.

Maxwell, R.J., Howells, S.L., Peet, A.C., and Griffiths, J.R. (1993). Tumour classification using ^{1}H nuclear magnetic resonance spectroscopy, principal component analysis and a neural network. In, *Techniques and Applications of Neural Networks* (M.J. Taylor and P.J.G. Lisboa, Eds.). Ellis Horwood, New York, pp. 63–75.

McCord Nelson, M. and Illingworth, W.T. (1991). *A Practical Guide to Neural Nets.* Addison-Wesley Publishing Company, Reading, p. 344.

McDonnell, J.R. and Waagen, D. (1993). Determining neural network hidden layer size using evolutionary programming. In, *World Congress on Neural Networks.* Oregon Convention Center, Portland, Oregon, July 11–15, 1993, Lawrence Erlbaum Associates and INNS Press, pp. III-564–III-567.

McNeil, B.J., Keeler, E., and Adelstein, S.J. (1975a). Primer on certain elements of medical decision making. *New Engl. J. Med.* **293,** 211–215.

McNeil, B.J., Varady, P.D., Burrows, B.A., and Adelstein, S.J. (1975b). Measures of clinical efficacy. Cost-effectiveness calculations in the diagnosis and treatment of hypertensive renovascular disease. *New Engl. J. Med.* **293,** 216–221.

McNeill, F.M. and Thro, E. (1994). *Fuzzy Logic: A Practical Approach.* Academic Press Professional, Cambridge, p. 292.

Medsker, L. (1994). Neural network connections to expert systems. In, *World Congress on Neural Networks.* Town & Country Hotel, San Diego, California, USA, June 5–9, 1994, Lawrence Erlbaum Associates and INNS Press, pp. IV-411–IV-417.

Michalewicz, Z. (1992). *Genetic Algorithms + Data Structures = Evolution Programs.* Springer-Verlag, Berlin, p. 250.

Miller, A.S., Blott, B.H., and Hames, T.K. (1992). Review of neural network applications in medical imaging and signal processing. *Med. Biol. Eng. Comput.* **30,** 449–464.

Miller, G.F., Todd, P.M., and Hegde, S.U. (1989). Designing neural networks using genetic algorithms. In, *Proceedings of the Third International Conference on Genetic Algorithms* (J.D. Schaffer, Ed.). Morgan Kaufmann Publishers, San Mateo, pp. 379–384.

Møller, M.F. (1993). A scaled conjugate gradient algorithm for fast supervised learning. *Neural Networks* **6,** 525–533.

Montague, G. and Morris, J. (1994). Neural-network contributions in biotechnology. *TIBTECH* **12,** 312–324.

Montana, D.J. (1995). Neural network weight selection using genetic algorithms. In, *Intelligent Hybrid Systems* (S. Goonatilake and S. Khebbal, Eds.). John Wiley & Sons, Chichester, pp. 85–104.

Morabito, F.C. (1993). Multilayer neural network for identification of non-linear electromagnetic systems. In, *World Congress on Neural Networks.* Oregon Convention Center, Portland, Oregon, July 11–15, 1993, Lawrence Erlbaum Associates and INNS Press, pp. I-428–I-432.

Morabito, F.C. (1994). Plasma shape recognition in a tokamak machine. In, *World Congress on Neural Networks*. Town & Country Hotel, San Diego, California, USA, June 5–9, 1994, Lawrence Erlbaum Associates and INNS Press, pp. I-254–I-259.

Moreno, L., Pineiro, J.D., Sanchez, J.L., Manas, S., Merino, J.J., Acosta, L., and Hamilton, A. (1995). Using neural networks to improve classification: Application to brain maturation. *Neural Networks* **8,** 815–820.

Müller, B., Reinhardt, J., and Strickland, M.T. (1995). *Neural Networks: An Introduction.* Springer-Verlag, Berlin, p. 329.

Musavi, M.T., Chan, K.H., Hummels, D.M., and Kalantri, K. (1994). On the generalization ability of neural network classifiers. *IEEE Trans. Pattern Anal. Machine Intell.* **16,** 659–663.

Muskal, S.M. and Kim, S.H. (1992). Predicting protein secondary structure content. A tandem neural network approach. *J. Mol. Biol.* **225,** 713–727.

Nabhan, T.M. and Zomaya, A.Y. (1994). Toward generating neural network structures for function approximation. *Neural Networks* **7,** 89–99.

Neal, M.J., Goodacre, R., and Kell, D.B. (1994). On the analysis of pyrolysis mass spectra using artificial neural networks. Individual input scaling leads to rapid learning. In, *World Congress on Neural Networks*. Town & Country Hotel, San Diego, California, USA, June 5–9, 1994, Lawrence Erlbaum Associates and INNS Press, pp. I-318–I-323.

Nefati, H., Diawara, B., and Legendre, J.J. (1993). Predicting the impact sensitivity of explosive molecules using neuromimetic networks. *SAR QSAR Environ. Res.* **1,** 131–136.

North, G. (1987). A celebration of connectionism. *Nature* **328,** 107.

Ochiai, K., Toda, N., and Usui, S. (1994). Kick-out learning algorithm to reduce the oscillation of weights. *Neural Networks* **7,** 797–807.

Oja, E. (1989). Neural networks, principal components, and subspaces. *Int. J. Neural Syst.* **1,** 61–68.

Oja, E. (1992). Principal components, minor components, and linear neural networks. *Neural Networks* **5,** 927–935.

O'Neill, M.C. (1991). Training back-propagation neural networks to define and detect DNA-binding sites. *Nucleic Acids Res.* **19,** 313–318.

O'Neill, M.C. (1992). *Escherichia coli* promoters: Neural networks develop distinct descriptions in learning to search for promoters of different spacing classes. *Nucleic Acids Res.* **20,** 3471–3477.

Otto, M. (1993). Neural networks for calibration of unspecific chemical sensors. In, *World Congress on Neural Networks.* Oregon Convention Center, Portland, Oregon, July 11–15, 1993, Lawrence Erlbaum Associates and INNS Press, pp. I-346–I-349.

Panaye, A., Doucet, J.P., Fan, B.T., Feuilleaubois, E., and Rahali El Azzouzi, S. (1994). Artificial neural network simulation of ^{13}C NMR shifts for methyl substituted cyclohexanes. *Chemom. Intell. Lab. Syst.* **24,** 129–135.

Pao, Y.H. (1989). *Adaptive Pattern Recognition and Neural Networks.* Addison-Wesley Publishing Company, Reading, p. 309.

Pao, Y.H. and Igelnik, B. (1994). Mathematical concepts underlying the functional-link approach. In, *World Congress on Neural Networks*. Town & Country Hotel, San Diego, California, USA, June 5–9, 1994, Lawrence Erlbaum Associates and INNS Press, pp. I-236–I-246.

Pao, Y.H. and Takefuji, Y. (1992). Functional-link net computing: Theory, system architecture, and functionalities. *Computer* **May,** 76–79.

Pattichis, C.S., Charalambous, C., and Middleton, L.T. (1995). Efficient training of neural network models in classification of electromyographic data. *Med. Biol. Eng. Comput.* **33,** 499–503.

Perantonis, S.J. and Karras, D.A. (1995). An efficient constrained learning algorithm with momentum acceleration. *Neural Networks* **8,** 237–249.

Pizzi, N. and Somorjai, R.L. (1994). Fuzzy encoding as a preprocessing method for artificial neural networks. In, *World Congress on Neural Networks*. Town & Country Hotel, San Diego, California, USA, June 5–9, 1994, Lawrence Erlbaum Associates and INNS Press, pp. III-643–III-648.

Pleiss, M.A. and Unger, S.H. (1990). The design of test series and the significance of QSAR relationships. In, *Comprehensive Medicinal Chemistry,* Vol. 4 (C.A. Ramsden, Ed.). Pergamon Press, Oxford, pp. 561–587.

Pollard, J.F., Broussard, M.R., Garrison, D.B., and San, K.Y. (1992). Process identification using neural networks. *Comput. Chem. Engng* **16,** 253–270.

Powell, D.J., Skolnick, M.M., and Tong, S.S. (1995). A unified approach for engineering design. In, *Intelligent Hybrid Systems* (S. Goonatilake and S. Khebbal, Eds.). John Wiley & Sons, Chichester, pp. 107–119.

Putavy, C., Devillers, J., and Domine, D. (1996). Genetic selection of aromatic substituents for designing test series. In, *Genetic Algorithms in Molecular Modeling* (J. Devillers, Ed.). Academic Press, London, pp. 243–269.

Qian, N. and Sejnowski, T.J. (1988). Predicting the secondary structure of globular proteins using neural network models. *J. Mol. Biol.* **202,** 865–884.

Qin, S.J. and McAvoy, T.J. (1992). Nonlinear PLS using neural networks. *Comput. Chem. Engng* **16,** 379–391.

Reeves, C.R. (1995a). Training set selection in neural network applications. In, *Artificial Neural Nets and Genetic Algorithms.* Proceedings of the International Conference, Alès, France, 1995 (D.W. Pearson, N.C. Steele, and R.F. Albrecht, Eds.). Springer-Verlag, Wien, pp. 476–478.

Reeves, C.R. (1995b). Bias estimation for neural network predictions. In, *Artificial Neural Nets and Genetic Algorithms.* Proceedings of the International Conference, Alès, France, 1995 (D.W. Pearson, N.C. Steele, and R.F. Albrecht, Eds.). Springer-Verlag, Wien, pp. 242–244.

Refenes, A.N., Zapranis, A.D., Connor, J.T., and Bunn, D.W. (1995). Neural networks in investment management. In, *Intelligent Systems for Finance and Business* (S. Goonatilake and P. Treleaven, Eds.). John Wiley & Sons, Chichester, pp. 177–208.

Refenes, A.N., Zapranis, A., and Francis, G. (1994). Stock performance modeling using neural networks: A comparative study with regression models. *Neural Networks* **7,** 375–388.

Reibnegger, G., Weiss, G., Werner-Felmayer, G., Judmaier, G., and Wachter, H. (1991). Neural networks as a tool for utilizing laboratory information: Comparison with linear discriminant analysis and with classification and regression trees. *Proc. Natl. Acad. Sci. USA* **88,** 11426–11430.

Reibnegger, G., Werner-Felmayer, G., and Wachter, H. (1993). A note on the low-dimensional display of multivariate data using neural networks. *J. Mol. Graphics* **11,** 129–133.

Ricard, D., Cachet, C., and Cabrol-Bass, D. (1993). Neural network approach to structural feature recognition from infrared spectra. *J. Chem. Inf. Comput. Sci.* **33,** 202–210.

Ripley, B.D. (1992). *Statistical Aspects of Neural Networks.* Invited Lecture for SemStat, Sandjerg, Denmark, April 25–30, 1992, p. 70.

Roberts, S.G. and Turega, M. (1995). Evolving neural network structures: An evaluation of encoding techniques. In, *Artificial Neural Nets and Genetic Algorithms.* Proceedings of the International Conference, Alès, France, 1995 (D.W. Pearson, N.C. Steele, and R.F. Albrecht, Eds.). Springer-Verlag, Wien, pp. 96–99.

Romeo, G., Mele, F., and Morelli, A. (1995) Neural networks and discrimination of seismic signals. *Comput. Geosci.* **21,** 279–288.

Rorije, E., van Wezel, M.C., and Peijnenburg, W.J.G.M. (1995). On the use of back-propagation neural networks in modeling environmental degradation. *SAR QSAR Environ. Res.* **4,** 219–235.

Rumelhart, D.E. (1995). Brain style computation: Learning and generalization. In, *An Introduction to Neural and Electronic Networks,* Second Edition (S.F. Zornetzer, J.L. Davis, C. Lau, and T. McKenna, Eds.). Academic Press, San Diego, pp. 431–446.

Rumelhart, D.E., Hinton, G.E., and Williams, R.J. (1986). Learning representations by back-propagating errors. *Nature* **323,** 533–536.

Russo, M.F., Huff, A.C., Heckler, C.E., and Evans, A.C. (1994). An improved DNA encoding scheme for neural network modeling. In, *World Congress on Neural Networks.* Town & Country Hotel, San Diego, California, USA, June 5–9, 1994, Lawrence Erlbaum Associates and INNS Press, pp. I-354–I-359.

Salt, D.W., Yildiz, N., Livingstone, D.J., and Tinsley, C.J. (1992). The use of artificial neural networks in QSAR. *Pestic. Sci.* **36,** 161–170.

Samad, T. (1991). Back propagation with expected source values. *Neural Networks* **4,** 615–618.

Sandler, D.G., Barrett, T.K., Palmer, D.A., Fugate, R.Q., and Wild, W.J. (1991). Use of a neural network to control an adaptive optics system for an astronomical telescope. *Nature* **351,** 300–302.

Schaltenbrand, N., Lengelle, R., and Macher, J.P. (1993). Neural network model: Application to automatic analysis of human sleep. *Comput. Biomed. Res.* **26,** 157–171.

Scherer, A. and Schlageter, G. (1995). A multi-agent approach for the integration of neural networks and expert systems. In, *Intelligent Hybrid Systems* (S. Goonatilake and S. Khebbal, Eds.). John Wiley & Sons, Chichester, pp. 153–173.

Schüürmann, G. and Müller, E. (1994). Back-propagation neural networks-recognition vs. prediction capability. *Environ. Toxicol. Chem.* **13,** 743–747.

Shah, S., Palmieri, F., and Datum, M. (1992). Optimal filtering algorithms for fast learning in feedforward neural networks. *Neural Networks* **5,** 779–787.

Shamir, N., Saad, D., and Marom, E. (1993). Using the functional behavior of neurons for genetic recombination in neural nets training. *Complex Syst.* **7,** 445–467.

Shao, J. (1993). Linear model selection by cross-validation. *J. Am. Stat. Assoc.* **88,** 486–494.

Sharma, A.K., Sheikh, S., Pelczer, I., and Levy, G.C. (1994). Classification and clustering: Using neural networks. *J. Chem. Inf. Comput. Sci.* **34,** 1130–1139.

Shawe-Taylor, J.S. and Cohen, D.A. (1990). Linear programming algorithm for neural networks. *Neural Networks* **3,** 575–582.

Shi, Z. and Shimizu, K. (1992). Neuro-fuzzy control of bioreactor systems with pattern recognition. *J. Ferment. Bioeng.* **74,** 39–45.

Sietsma, J. and Dow, R.J.F. (1991). Creating artificial neural networks that generalize. *Neural Networks* **4,** 67–79.

Sigillito, V.G. and Hutton, L.V. (1990). Case study II: Radar signal processing. In, *Neural Network PC Tools. A Practical Guide* (R.C. Eberhart and R.W. Dobbins, Eds.). Academic Press, San Diego, pp. 235–250.

Sigman, M.E. and Rives, S.S. (1994). Prediction of atomic ionization potentials I-III using an artificial neural network. *J. Chem. Inf. Comput. Sci.* **34,** 617–620.

Silverman, R.H. and Noetzel, A.S. (1990). Image processing and pattern recognition in ultrasonograms by backpropagation. *Neural Networks* **3,** 593–603.

Sima, J. (1995). Neural expert systems. *Neural Networks* **8,** 261–271.

Simon, V., Gasteiger, J., and Zupan, J. (1993). A combined application of two different neural network types for the prediction of chemical reactivity. *J. Am. Chem. Soc.* **115,** 9148–9159.

Skoneczny, S. and Szostakowski, J. (1995). Advanced neural networks methods for recognition of handwritten characters. In, *Artificial Neural Nets and Genetic Algorithms.* Proceedings of the International Conference, Alès, France, 1995 (D.W. Pearson, N.C. Steele, and R.F. Albrecht, Eds.). Springer-Verlag, Wien, pp. 140–143.

Skurikhin, A.N. and Surkan, A.J. (1995). Neural network training compared for BACKPROP QUICKPROP and CASCOR in energy control problems. In, *Artificial Neural Nets and Genetic Algorithms.* Proceedings of the International Conference, Alès, France, 1995 (D.W. Pearson, N.C. Steele, and R.F. Albrecht, Eds.). Springer-Verlag, Wien, pp. 444–447.

So, S.S. and Richards, W.G. (1992). Application of neural networks: Quantitative structure–activity relationships of the derivatives of 2,4-diamino-5-(substituted-benzyl)pyrimidines as DHFR inhibitors. *J. Med. Chem.* **35,** 3201–3207.

Song, X.H., Chen, Z., and Yu, R.Q. (1993). Artificial neural networks applied to odor classification for chemical compounds. *Comput. Chem.* **17,** 303–308.

Song, X.H. and Yu, R.Q. (1993). Artificial neural networks applied to the quantitative structure–activity relationship study of dihydropteridine reductase inhibitors. *Chemom. Intell. Lab. Syst.* **19,** 101–109.

Specht, D.F. (1990). Probabilistic neural networks. *Neural Networks* **3,** 109–118.

Sperduti, A. and Starita, A. (1993). Speed up learning and network optimization with extended back propagation. *Neural Networks* **6,** 365–383.

Sumpter, B.G., Getino, C., and Noid, D.W. (1994). Theory and applications of neural computing in chemical science. *Ann. Rev. Phys. Chem.* **45,** 439–481.

Svozil, D., Pospichal, J., and Kvasnicka, V. (1995). Neural network prediction of carbon-13 NMR chemical shifts of alkanes. *J. Chem. Inf. Comput. Sci.* **35,** 924–928.

Swets, J.A. (1973). The relative operating characteristic in psychology. A technique for isolating effects of response bias finds wide use in the study of perception and cognition. *Science* **182,** 990–1000.

Swets, J.A. (1988). Measuring the accuracy of diagnostic systems. *Science* **240,** 1285–1293.

Tabak, H.H. and Govind, R. (1993). Prediction of biodegradation kinetics using a nonlinear group contribution method. *Environ. Toxicol. Chem.* **12,** 251–260.

Takahashi, Y. (1993). Generalization and approximation capabilities of multilayer networks. *Neural Comput.* **5,** 132–139.

Takeda, T., Kishi, K., Yamanouchi, T., Mizoe, H., and Matsuoka, T. (1994). Significance of distributed representation in the output layer of a neural network in a pattern recognition task. *Med. Biol. Eng. Comput.* **32,** 77–84.

Tetko, I.V., Livingstone, D.J., and Luik, A.I. (1995). Neural network studies. I. Comparison of overfitting and overtraining. *J. Chem. Inf. Comput. Sci.* **35,** 826–833.

Tetko, I.V., Tanchuk, V.Y., Chentsova, N.P., Antonenko, S.V., Poda, G.I., Kukhar, V.P., and Luik, A.I. (1994). HIV-1. Reverse transcriptase inhibitor design using artificial neural networks. *J. Med. Chem.* **37,** 2520–2526.

Theocharidou, E. and Burianec, Z. (1995). Using of neural-network and expert system for imissions prediction. In, *Artificial Neural Nets and Genetic Algorithms.* Proceedings of the International Conference, Alès, France, 1995 (D.W. Pearson, N.C. Steele, and R.F. Albrecht, Eds.). Springer-Verlag, Wien, pp. 65–68.

Tollenaere, T. (1990). SuperSAB: Fast adaptive back propagation with good scaling properties. *Neural Networks* **3,** 561–573.

Tsaptsinos, D. and Leigh, J.R. (1993). A step by step approach for the construction of a fermentation process estimator. In, *World Congress on Neural Networks.* Oregon Convention Center, Portland, Oregon, July 11–15, 1993, Lawrence Erlbaum Associates and INNS Press, pp. I-216–I-219.

Tusar, M., Zupan, J., and Gasteiger, J. (1992). Neural networks and modelling in chemistry. *J. Chim. Phys.* **89,** 1517–1529.

Vahidov, R.M., Vahidov, M.A., and Eyvazova, Z.E. (1995). Use of genetic and neural technologies in oil equipment computer-aided design. In, *Artificial Neural Nets and*

Genetic Algorithms (D.W. Pearson, N.C. Steele, and R.F. Albrecht, Eds.). Springer-Verlag, Wien, pp. 317–320.

van der Smagt, P.P. (1994). Minimisation methods for training feedforward neural networks. *Neural Networks* **7,** 1–11.

van Helden, S.P., Hamersma, H., and van Geerestein V.J. (1996). Prediction of the progesterone receptor binding of steroids using a combination of genetic algorithms and neural networks. In, *Genetic Algorithms in Molecular Modeling* (J. Devillers, Ed.). Academic Press, London, pp. 159–192.

van Ooyen, A. and Nienhuis, B. (1992). Improving the convergence of the back-propagation algorithm. *Neural Networks* **5,** 465–471.

Venkatasubramanian, V., Sundaram, A., Chan, K., and Caruthers, J.M. (1996). Computer-aided molecular design using neural networks and genetic algorithms. In, *Genetic Algorithms in Molecular Modeling* (J. Devillers, Ed.). Academic Press, London, pp. 271–302.

Villemin, D., Cherqaoui, D., and Cense, J.M. (1993). Neural networks studies: quantitative structure–activity relationship of mutagenic aromatic nitro compounds. *J. Chim. Phys.* **90,** 1505–1519.

Villemin, D., Cherqaoui, D., and Mesbah, A. (1994). Predicting carcinogenicity of polycyclic aromatic hydrocarbons from back-propagation neural network. *J. Chem. Inf. Comput. Sci.* **34,** 1288–1293.

Vitthal, R., Sunthar, P., and Durgaprasada Rao, C. (1995). The generalized proportional-integral-derivative (PID) gradient descent back propagation algorithm. *Neural Networks* **8,** 563–569.

Wasserman, P.D. (1989). *Neural Computing: Theory and Practice.* Van Nostrand Reinhold, New York, p. 230.

Wasserman, P.D. (1993). *Advanced Methods in Neural Computing.* Van Nostrand Reinhold, New York, p. 255.

Weigend, A.S., Huberman, B.A., and Rumelhart, D.E. (1990). Predicting the future: A connectionist approach. *Int. J. Neural Systems* **1,** 193–209.

Weinstein, J.N., Kohn, K.W., Grever, M.R., Viswanadhan, V.N., Rubinstein, L.V., Monks, A.P., Scudiero, D.A., Welch, L., Koutsoukos, A.D., Chiausa, A.J., and Paull, K.D. (1992). Neural computing in cancer drug development: Predicting mechanism of action. *Science* **258,** 447–451.

Weinstein, J.N., Myers, T., Casciari, J.J., Buolamwini, J., and Raghavan, K. (1994). Neural networks in the biomedical sciences: A survey of 386 publications since the beginning of 1991. In, *World Congress on Neural Networks*. Town & Country Hotel, San Diego, California, USA, June 5–9, 1994, Lawrence Erlbaum Associates and INNS Press, pp. I-121–I-126.

Weir, M.K. (1991). A method for self-determination of adaptive learning rates in back propagation. *Neural Networks* **4,** 371–379.

Weiss, S.M. and Kulikowski, C.A. (1991). *Computer Systems that Learn: Classification and Prediction Methods from Statistics, Neural Nets, Machine Learning, and Expert Systems.* Morgan Kaufmann Publishers, San Francisco, p. 223.

Welstead, S.T. (1994). *Neural Network and Fuzzy Logic Applications in C/C++.* John Wiley & Sons, New York, p. 494.

Werbos, P.J. (1988). Generalization of backpropagation with application to a recurrent gas market model. *Neural Networks* **1,** 339–356.

Weymaere, N. and Martens, J.P. (1991). A fast and robust learning algorithm for feed-forward neural networks. *Neural Networks* **4,** 361–369.

Wienke, D., Domine, D., Buydens, L., and Devillers, J. (1996). Adaptive resonance theory based neural networks explored for pattern recognition analysis of QSAR data. In, *Neural Networks in QSAR and Drug Design* (J. Devillers, Ed.). Academic Press, London, pp. 119–156.

Wiese, M. and Schaper, K.J. (1993). Application of neural networks in the QSAR analysis of percent effect biological data: Comparison with adaptive least squares and nonlinear regression analysis. *SAR QSAR Environ. Res.* **1,** 137–152.

Wolfer, J., Robergé, J., and Grace, T. (1994). Robust multispectral road classification in landsat thematic mapper imagery. In, *World Congress on Neural Networks*. Town & Country Hotel, San Diego, California, USA, June 5–9, 1994, Lawrence Erlbaum Associates and INNS Press, pp. I-260–I-268.

Wythoff, B.J. (1993). Backpropagation neural networks. A tutorial. *Chemom. Intell. Lab. Syst.* **18,** 115–155.

Yager, R.R. and Filev, D.P. (1994). *Essentials of Fuzzy Modeling and Control.* John Wiley & Sons, New York, p. 388.

Yang, T.F., Devine, B., and Macfarlane, P.W. (1994). Artificial neural networks for the diagnosis of atrial fibrillation. *Med. Biol. Eng. Comput.* **32,** 615–619.

Ying, X., Surkan, A.J., and Guan, Q. (1993). Simplifying neural networks by pruning alternated with backpropagation training. In, *World Congress on Neural Networks.* Oregon Convention Center, Portland, Oregon, July 11–15, 1993, Lawrence Erlbaum Associates and INNS Press, pp. III-364–III-367.

Zaremba, T. (1990). Case study III: Technology in search of a buck. In, *Neural Network PC Tools. A Practical Guide* (R.C. Eberhart and R.W. Dobbins, Eds.). Academic Press, San Diego, pp. 251–283.

Zeidenberg, M. (1990). *Neural Network Models in Artificial Intelligence*. Ellis Horwood, New York, p. 268.

Zhang, Q., Litchfield, J.B., Reid, J.F., Ren, J., and Chang, S.W. (1995). Coupling a machine vision sensor and a neural net supervised controller: Controlling microbial cultivations. *J. Biotechnol.* **38,** 219–228.

Zhang, X., Mesirov, J.P., and Waltz, D.L. (1992). Hybrid system for protein secondary structure prediction. *J. Mol. Biol.* **225,** 1049–1063.

Zheng, B., Qian, W., and Clarke, L.P. (1994). Artificial neural network for pattern recognition in mammography. In, *World Congress on Neural Networks*. Town &

Country Hotel, San Diego, California, USA, June 5–9, 1994, Lawrence Erlbaum Associates and INNS Press, pp. I-57–I-62.

Zipser, D. and Andersen, R.A. (1988). A back-propagation programmed network that simulates response properties of a subset of posterior parietal neurons. *Nature* **331,** 679–684.

Zitko, V. (1991). Prediction of biodegradability of organic chemicals by an artificial neural network. *Chemosphere* **23,** 305–312.

Zollner, R., Schmitz, H.J., Wünsch, F., and Krey, U. (1992). Fast generating algorithm for a general three-layer perceptron. *Neural Networks* **5,** 771–777.

Zupan, J. and Gasteiger, J. (1991). Neural networks: A new method for solving chemical problems or just a passing phase? *Anal. Chim. Acta* **248,** 1–30.

2 AUTOLOGP Versus Neural Network Estimation of *n*-Octanol/Water Partition Coefficients

D. DOMINE[1]*, J. DEVILLERS[1], and W. KARCHER[2]
[1]*CTIS, 21 rue de la Bannière, 69003 Lyon, France*
[2]*CEC, JRC, Ispra Establishment, I-21020 Ispra, Varese, Italy*

The performances of AUTOLOGP (version 2.11) for estimating the n-octanol/water partition coefficient of organic molecules are compared with those of a three-layer backpropagation neural network (BNN) model on the basis of the same training set (n = 800) and several heterogeneous testing sets (n = 182 + 380 + 21 + 72 + 16). Analysis of the residuals obtained during the learning and testing phases shows that the BNN model outperforms AUTOLOGP. This suggests that it may be useful to develop a new version of AUTOLOGP from a BNN as modeling tool.

KEY WORDS: *AUTOLOGP; n-octanol/water partition coefficient; autocorrelation method; regression analysis; backpropagation neural network.*

INTRODUCTION

The *n*-octanol/water partition coefficient (log P) provides a useful quantitative parameter for representing the lipophilic/hydrophobic nature of organic chemicals. It describes the tendency for a compound to partition preferentially in aqueous or lipophilic media. Under these conditions, it is widely used in many modeling areas such as design of drugs and pesticides (Dearden, 1990; Hansch and Leo, 1995), prediction of bioconcentration (Veith *et al.*, 1979, 1980; Mackay, 1982; Isnard and Lambert, 1988; Bintein *et al.*, 1993),

* Author to whom all correspondence should be addressed.

In, *Neural Networks in QSAR and Drug Design* (J. Devillers, Ed.)
Academic Press, London, 1996, pp. 47–63.
ISBN 0-12-213815-5

(eco)toxicity (Karcher and Devillers, 1990; Hansch and Leo, 1995), and soil and sediment sorption (Karickhoff *et al.*, 1979; Kenaga and Goring, 1980; Bintein and Devillers, 1994). The *n*-octanol/water partition coefficient (log P) is also a key parameter in the models designed for estimating the environmental fate of organic chemicals (Mackay, 1991; Devillers *et al.*, 1995b). For numerous chemicals, log P values are readily available. However, its experimental determination is time-consuming, expensive, and can yield various technical problems (Gago *et al.*, 1987; Dearden and Bresnen, 1988). In addition, it is practically impossible to measure the log P of all existing chemicals, and the ability to predict the log P of yet unsynthesized compounds can be very useful. Under these conditions, additive–constitutive calculation schemes based on molecular structures have been proposed and used (Fujita *et al.*, 1964; Leo *et al.*, 1971; Nys and Rekker, 1974; Hine and Mookerjee, 1975; Rekker and de Kort, 1979; Broto *et al.*, 1984; Ghose and Crippen, 1986; Ghose *et al.*, 1988; Leo, 1990; Suzuki and Kudo, 1990; Klopman and Wang, 1991). Among them, the most widely used are the Leo and Hansch approach (Leo, 1990; Hansch and Leo, 1995) and that of Rekker (Rekker and Mannhold, 1992). These two approaches work well for simple chemicals but have limitations for complex molecules since correction factors have to be introduced (Dunn *et al.*, 1983; Klopman *et al.*, 1985, 1994; Devillers *et al.*, 1995c). In the Rekker's method for example, 'magic constants' are added to produce a correction factor allowing one to adjust the log P value by taking into account proximity effects, condensations in aromatic systems, steric effects, and so on (Rekker and Mannhold, 1992). This prompted us to develop a QSAR model for log P allowing one to overcome this problem (Devillers *et al.*, 1995c). For this purpose, we proposed using a modified algorithm of the autocorrelation method reducing the redundancy of the descriptors and increasing their physicochemical significance in conjunction with the Rekker's fragmental constants (Devillers *et al.*, 1995c). This allowed us to derive a simple and powerful log P model for organic chemicals based on regression analysis from the components of the autocorrelation vectors. AUTOLOGP (version 2.11), a user-friendly program, has been developed from this regression model. However, it is obvious that this type of software can always be adequately refined for increasing its accuracy and domain of application. One way consists of refining the contributions allowing one to describe the molecules (work in preparation). It is also possible to envision the use of different modeling techniques. Thus, it has been shown recently that the back-propagation neural network (BNN) and related algorithms were useful modeling tools in QSPR studies (Bodor *et al.*, 1991; Doucet *et al.*, 1993; Kvasnicka *et al.*, 1993; Nefati *et al.*, 1993; Balaban *et al.*, 1994; Cherqaoui *et al.*, 1994a,b; Egolf *et al.*, 1994; Gakh *et al.*, 1994; Panaye *et al.*, 1994; Sharma *et al.*, 1994; Sigman and Rives, 1994; Chow *et al.*, 1995; Grunenberg and Herges, 1995; Jordan *et al.*, 1995; Svozil *et al.*, 1995; Tschinke, 1995; Wessel and Jurs, 1995a,b; West, 1995a,b). Under these conditions, this paper is aimed at deriving a BNN model for predicting the log P of chemicals and comparing its performances with those of AUTOLOGP (version 2.11).

MATERIALS AND METHODS

AUTOLOGP (version 2.11)

AUTOLOGP (version 2.11) (Devillers *et al.*, 1995c) is a user-friendly program implementing a QSAR model (Eq. (1)) based on a new autocorrelation method for the estimation of the log P values of chemicals.

$$\log P = 0.827h_0 - 0.053h_3 + 0.117h_8 + 0.112h_9 + 0.672 \quad (1)$$

$$n = 800;\ r = 0.981;\ s = 0.387;\ F = 5034.64;\ p < 0.01\%$$

In Eq. (1), h_0, h_3, h_8, and h_9 represent components of the autocorrelation vector **H** encoding the hydrophobicity of the molecules for distances 0, 3, 8, and 9, respectively. To derive this model, the autocorrelation method which combines a topological and physicochemical description of the molecules was modified in order to reduce the redundancy of the autocorrelation vectors and to use atomic contributions presenting positive or negative values. Briefly, in the original autocorrelation method (Moreau and Broto, 1980a,b; Broto and Devillers, 1990), the structural formula of an organic compound is viewed as a hydrogen-suppressed molecular graph where the atoms are the vertices and the bonds the edges. In this graph, the distance between two atoms is defined as the smallest number of bonds between them. If we consider that a property G of a molecule can be calculated from specific contributions $g(i)$ of its atoms, it is possible to calculate the products $g(i) \times g(i)$, $g(i) \times g(j)$, ..., corresponding to the same distance between the different atoms of this molecule. The sum of these products for the same distance gives a component of the autocorrelation vector for the property G. By using the fragmental constants of Rekker and Mannhold (1992) and by strictly following their methodological approach (Nys and Rekker, 1974; Rekker and de Kort, 1979; Rekker and Mannhold, 1992; Rekker *et al.*, 1993), it was possible to compute an autocorrelation vector **H** encoding the hydrophobicity of organic molecules (Devillers *et al.*, 1995c). Figure 1 gives an example of coding for phenylpropyl ether. To optimize the descriptive power and the weak redundancy of the autocorrelation vector **H**, a new algorithm was designed (Devillers *et al.*, 1992). This algorithm uses a modified autocorrelation function allowing one to obtain the first component of the autocorrelation vector by simply summing the positive and negative contributions attributed to the atoms and functional groups constituting a molecule. In AUTOLOGP (version 2.11), the first component, h_0, of the autocorrelation vector **H** encoding the hydrophobicity of chemicals, simply corresponds to the sum of the contributions attributed to the atoms or

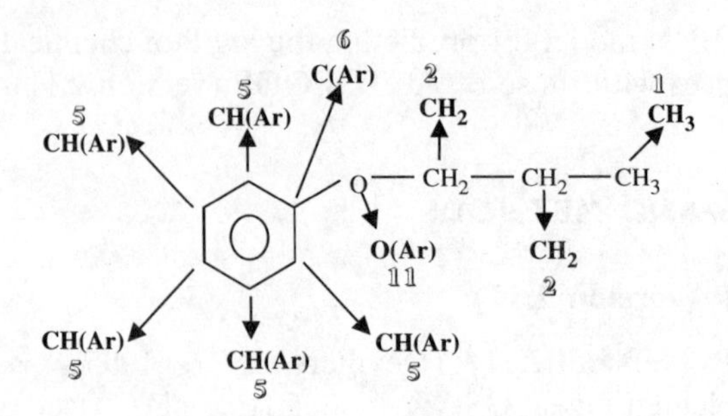

Figure 1 Autocorrelation coding of phenylpropyl ether. The coding of this molecule requires the use of the contributions for the five following types: CH(Ar), C(Ar), O(Ar), CH_2, and CH_3 (types no. 5, 6, 11, 2, and 1, respectively in our coding) where Ar stands for aromatic.

functional groups constituting the studied molecule. As a result, in Eq. (1), h_0 is equivalent to a simple summation of the fragmental constants of Rekker while h_3, h_8, and h_9 act as correction factors.

Backpropagation neural network

A classical three-layer BNN consisting of four input neurons corresponding to the four variables used in Eq. (1), an adjustable number of hidden units and one log P output neuron was used. Biases were included in the structure of the BNN. The theoretical foundations of this technique are beyond the scope of this paper and can be found in several articles and textbooks (e.g., Rumelhart *et al.*, 1986; Pao, 1989; Eberhart and Dobbins, 1990; Cambon and Devillers, 1993). Recent extensive reviews in chemistry and QSAR are also available (Sumpter *et al.*, 1994; Devillers, 1996). Numerous trials were performed to determine the optimal set of parameters for the network (i.e., number of hidden neurons, learning rate (η), momentum term (α), number of learning cycles). The training was monitored with a validation set (term used by Weigend *et al.* (1990)) consisting of molecules presenting a high structural diversity but not deviating too much from the structures included in the training set. This technique, termed crossvalidation in the BNN literature, is known to avoid problems of generalization (Curry and Rumelhart, 1990; Weigend *et al.*, 1990; Borggaard and Thodberg, 1992; Bos *et al.*, 1992; Finnoff *et al.*, 1993; Tsaptsinos and Leigh, 1993; Jodouin, 1994; Kreesuradej *et al.*, 1994). Indeed, in order to obtain a good convergence during learning while retaining the generalization ability of a BNN model, when applicable, it is recommended to monitor the training with a validation set which does not intervene in the learning itself, but determines when the training should be stopped. In the present study, after having observed, during a large number of trials, the evolution of the errors in the training and validation sets as a function of the number of cycles, it appeared that both curves converged within approximately 5000 cycles but exhibited oscillations. As a result, to obtain the best compromise between the learning and generalization ability, the network was run until a sufficiently low error in the training set was obtained and the validation set was then used to stop the training when the minimal errors were obtained in both the training and validation sets. As stressed above, for comparison purposes, the variables used for training the BNN were the same as those used in Eq. (1). Before introduction in the BNN, these autocorrelation components were scaled by means of Eq. (2). The log P values were scaled by means of the same equation.

$$x'_i = [\mathrm{a}(x_i - x_{\min})/(x_{\max} - x_{\min})] + \mathrm{b} \quad (2)$$

In Eq. (2), x'_i was the scaled value, x_i the original data, and $x_{\min}$ and $x_{\max}$ were the minimum and maximum values of the column, respectively. In order to have scaled data ranging between 0.05 and 0.95 the values of a and b were taken as 0.9 and 0.05, respectively.

The BNN calculations were carried out with the STATQSAR package (STATQSAR, 1994).

Sampling the training and testing sets

The selection of suitable training and testing sets is crucial when using BNNs, and in addition when comparing the performances of this nonlinear tool with those of a classical linear method. Indeed, when developing a BNN model, it is important to have a sufficient number of known examples for training and validation. As regards the training set (TR), in the frame of the project dealing with the design of AUTOLOGP, it had been possible to compile 800 chemicals for which reliable experimental log P values were available. This set had been used for deriving the regression model included in AUTOLOGP (version 2.11) and therefore constituted the training set for deriving the BNN model. In the same way, when comparing methods, large heterogeneous sets of test chemicals are required. Indeed, even if numerous authors only focused their attention on the constitution of a training set (TR) and considered that the use of a crossvalidation procedure (Cramer *et al.*, 1988) was sufficient for validating QSAR or QSPR models, recently, Shao (1993) clearly demonstrated that the leave-one-out crossvalidation method was inconsistent for validating a linear model. However, this author claimed that a leave-*n*-out procedure was preferable. In nonlinear multivariate analysis, the use of any type of crossvalidation procedures (in its QSAR sense) must not be recommended. Indeed, the removal of patterns in the training set can greatly influence the learning phase, especially when the number of individuals is reduced. In addition, it should be borne in mind that a BNN can yield different solutions (i.e., models) for the same data set. This complicates the interpretation of the results obtained from a crossvalidation procedure and it can be problematic when comparing the performances of regression analysis and BNN by means of this technique. Under these conditions, in all cases, it is preferable to constitute separate testing sets. In addition, it is obvious that data of various natures and presenting various levels of accuracy can be found in the literature, especially when dealing with physico-chemical and/or complex biological data. This shortcoming also justifies the creation of different testing sets. With BNNs, another problem occurs due to the scaling of the output data in the range [0, 1]. Indeed, even if margins outside the range of log P values of the training set can be used by taking various values of a and b in Eq. (2), it is obvious that the BNN cannot predict the log P values of chemicals outside a predefined range of values which generally correspond (after rescaling) to the minimum and maximum values of log P in the training set. From these considerations, it should therefore be recommended to use, when possible, several testing sets with data of different natures, in order to estimate the interpolation and extrapolation performances of the models and their limits.

In order to have a comparable basis for evaluating the performances of our models, the log P data used in the present study to constitute our different testing sets were retrieved from the 'star list' (Hansch *et al.*, 1995) or from a critical compilation of Mackay and coworkers (Mackay *et al.*, 1992). Chemicals containing three or more aliphatic halogen atoms were removed since it has been stressed in a previous publication (Devillers *et al.*, 1995c) that the atomic contributions for these atoms should be refined. In the same way, the amino acids were not taken into account since they presented very low log P values which were not correctly predicted. Last, in this study we did not keep chemicals such as uracils, which can be coded either as aromatic or non-aromatic.

As a result, our sampling strategy was the following. First, as stressed earlier, a validation set (TS1) was constituted. It contained 182 molecules taken from the 'star list' (Hansch *et al.*, 1995) or recommended by Mackay and coworkers (Mackay *et al.*, 1992). These chemicals were structurally representative of the molecules of the training set and had log P values ranging between –0.80 and 7.57. A prediction set (TS2) (term used by Weigend *et al.* (1990)) of 380 molecules with log P values ranging between –1.51 and 7.6 was also designed in order to compare on a same basis the performances of AUTOLOGP and the BNN models. Most of these 380 molecules were also structurally representative of the chemicals included in the training set. Indeed, as TS1 was used as the validation set during the training of the BNN, the recorded performances for the BNN model with this set are biased. Beside these sets, which could be called in-sample testing sets (Devillers *et al.*, 1995a; Devillers, 1996) due to their constitution, three other sets bearing the properties of out-of-sample testing sets (Devillers *et al.*, 1995a; Devillers,

1996) were designed. The first (TS3) contained 21 interesting chemicals but for which Hansch and coworkers (Hansch *et al.*, 1995) indicated that the log P had been measured under specific experimental conditions (e.g., high temperature), calculated from π values or by means of a regression based on HPLC retention times. The second (TS4) contained 72 chemicals with structures not well represented in the training set and the last (TS5), 16 chemicals for which the log P value was inferior to the minimal log P value found in the training set. This last testing set was designed to have an insight into the problem of output scaling.

Performance evaluation

The general performances of models were estimated from the calculation of the root mean square (RMS) errors for the training and testing sets. This parameter is computed as shown in Eq. (3).

$$\text{RMS} = \sqrt{(\sum (\text{target} - \text{output})^2)/\text{number of patterns}} \quad (3)$$

However, it is obvious that a single parameter like the RMS error cannot provide a complete appreciation of the various aspects of the performances of models. Therefore, it is often required to consider other statistical and/or graphical information (Devillers, 1996). Under these conditions, detailed inspection and comparison of the residuals were also performed in order to detect more peculiar trends in the prediction behavior of the AUTOLOGP and BNN models. The number of chemicals for which the residual (absolute value) falls within certain ranges (e.g., <0.5; ≥ 0.8) were tabulated. Last, plots of the observed versus calculated log P values were drawn.

RESULTS AND DISCUSSION

After a trial and error procedure involving more than 500 runs, it appeared that the best results were obtained with five neurons in the hidden layer. With such an architecture, the ratio between the number of input samples and the number of samples equaled 25.8. With a momentum term (α) of 0.9, the learning rate (η) did not seem to influence the results since with values ranging between 0.1 and 0.9, good results could be obtained. These runs also showed that 5000 cycles were sufficient to reach convergence. Under these conditions, the BNN was run again more than 500 times with the above architecture, a momentum term of 0.9 and learning rates ranging between 0.1 and 0.9. As explained earlier, the choice of the best BNN model was founded on a compromise between the performances with the training set and those obtained with the validation set (TS1). This latter set was used to stop the training. The very best results were obtained with a learning rate of 0.5 after 4599 cycles. The RMS errors obtained with the BNN model for the training set (TR) and the two in-sample testing sets (i.e., TS1 and TS2) are reported in

Table I *Root mean square (RMS) errors for the training set (TR) and the in-sample testing sets (i.e., TS1 and TS2) obtained with AUTOLOGP (version 2.11) and the best BNN model.*

Data set	RMS	
	AUTOLOGP	BNN
TR	0.39	0.33
TS1	0.45	0.38
TS2	0.57	0.49

Table I, together with those obtained from AUTOLOGP (version 2.11). Although the RMS errors suggest that both models yield accurate simulations, the reported values underline the superiority of the BNN model over AUTOLOGP. This is confirmed by inspection of the plots of the observed versus calculated log P values for AUTOLOGP (Figure 2) and the BNN model (Figure 3). If other parameters such as the number of chemicals for which the residual values (absolute value) are lower than 0.5 or the number of chemicals with a residual value superior or equal to 0.8 are considered (Table II), the better performances of the BNN model become more evident. The results obtained with TR are not surprising since BNNs are known to perform better than regression analysis during the training, due to their learning abilities. Therefore, to compare the results properly, one must inspect their generalization ability. In the same way, with TS1, which was used as validation set during the training of the BNN, the lower RMS value (Table I) and the better trends in the prediction behavior (Table II) of the BNN model could be expected. As regards TS2, however, the better results of the BNN model are more significant. They indicate that within the log P domain defined by the training set (i.e., between –1.51 and 9.29) and for structures well represented in the training set, the BNN model is able to propose valuable estimations. It should be noted that due to its size (i.e., 800 molecules), the training set is highly representative in terms of chemical structures, functionalities, and so on (Devillers *et al.*, 1995c). Although the comparison parameters for AUTOLOGP are unfavorable, it is a reliable simple model also providing useful predictions. Indeed, from a practical point of view, the usefulness of a model is not only judged from its performances but also, for example, from its ease of use or implementation. Thus, we have to take into account that a regression model is very easy to use and implement while BNNs require expertise and more complicated computations.

For the three out-of-sample testing sets (i.e., TS3, TS4, and TS5) which were principally designed to compare the extrapolation performances of AUTOLOGP and the BNN model, the main results are summarized in Table III. With TS3, both models provide satisfying results. Table III shows that

Table II *Number of chemicals yielding residual values (absolute value) inferior to 0.5 and superior or equal to 0.8 for the training set (TR) and the in-sample testing sets (i.e., TS1 and TS2).*

Data set	Total*	Residual <0.5		Residual ≥0.8	
		AUTOLOGP	BNN	AUTOLOGP	BNN
TR	800	636	704	35	19
TS1	182	123	149	10	5
TS2	380	224	267	58	37

*Total number of chemicals in the set.

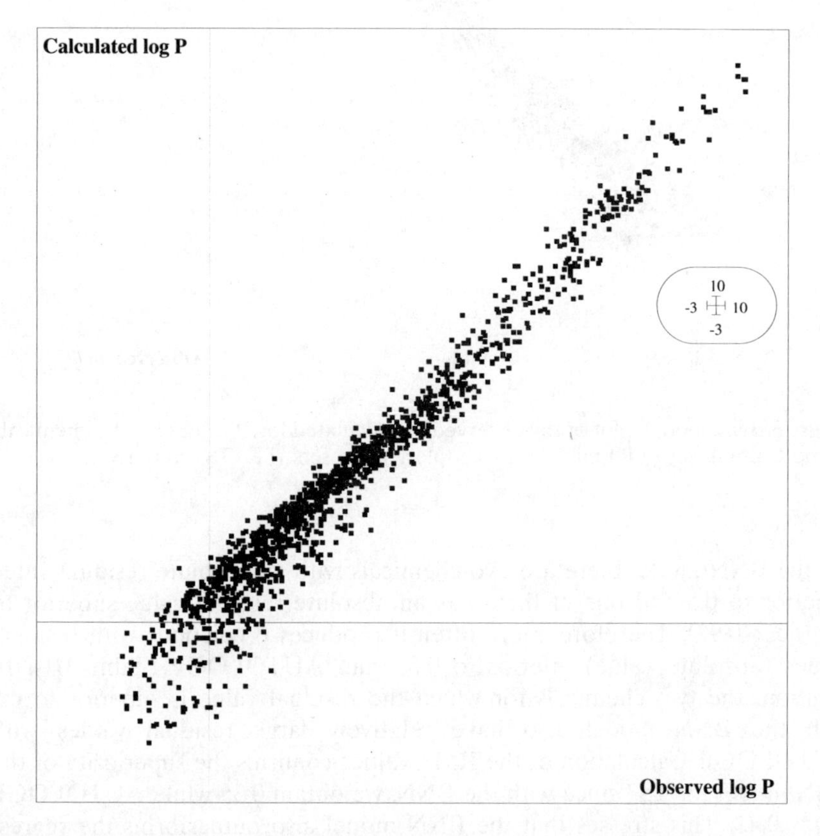

Figure 2 AUTOLOGP model: plot of the observed vs. calculated log P values for the chemicals of the training set (TR) and the in-sample testing sets (i.e., TS1 and TS2).

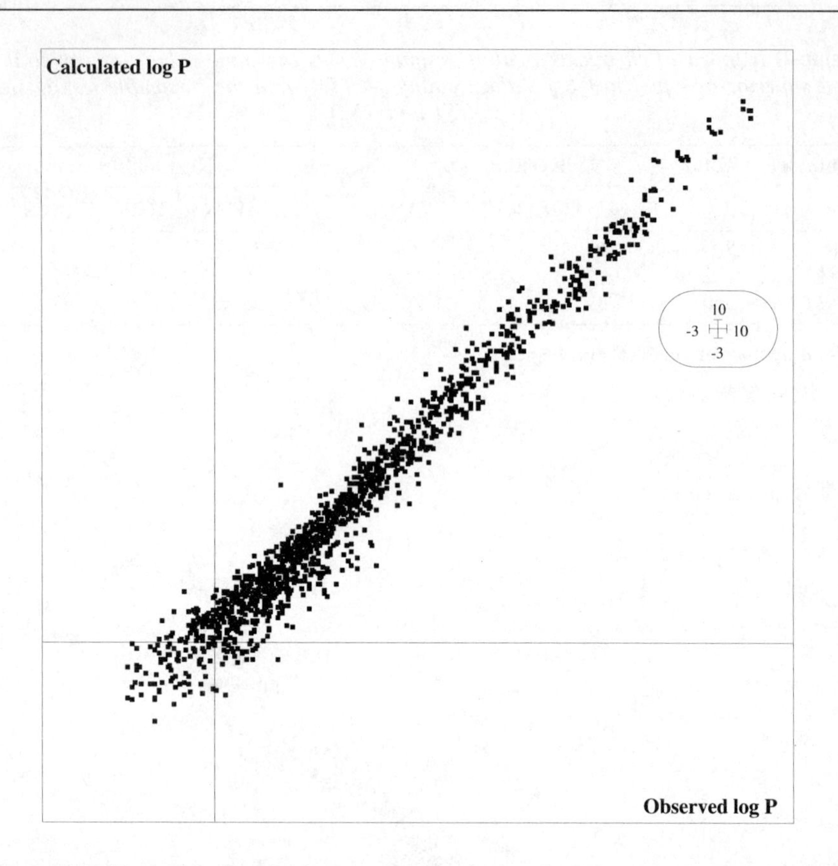

Figure 3 BNN model: plot of the observed vs. calculated log P values for the chemicals of the training set (TR) and the in-sample testing sets (i.e., TS1 and TS2).

for the BNN model there are two chemicals with an absolute residual value superior to 0.5 and one of them has an absolute residual value superior to 0.8 (i.e., –0.92). Therefore, more often it produces predictions with residual values (absolute value) inferior to 0.5, than AUTOLOGP (Table III). In addition, the two chemicals for which the residual value is superior to 0.5 with the BNN model also have relatively large residual values with AUTOLOGP. Calculation of the RMS values confirms the superiority of the BNN model for TS3 since with the BNN, we obtain 0.35 while AUTOLOGP yields 0.41. This stresses that the BNN model also outperforms the regression equation in extrapolation. For TS4, constituted of chemicals presenting structures not widely represented in the training set, Table III indicates that the results are of lesser quality than with the previous sets and the BNN model compares favorably with AUTOLOGP. Last, for TS5, which contains chemicals with very low log P values (i.e., < –1.51), the results obtained by

Table III *Number of chemicals yielding residual values (absolute value) inferior to 0.5 and superior or equal to 0.8 for the out-of-sample testing sets (i.e., TS3, TS4, and TS5).*

Data set	Total*	Residual <0.5		Residual ≥0.8	
		AUTOLOGP	BNN	AUTOLOGP	BNN
TS3	21	14	19	0	1
TS4	72	22	38	44	24
TS5	16	2	5	13	11

*Total number of chemicals in the set.

both the linear and nonlinear approaches are poor. With both methods, for the majority of the chemicals, the absolute value of the residuals exceeds 1 and sometimes 2 or 3. For the BNN model, there are only five chemicals with a residual value (absolute value) inferior to 0.5 and only two for the regression model (Table III). It is also noteworthy that these chemicals are not the same. In many cases, these highly hydrophilic chemicals possess several functional groups tending to increase their lipophilicity. The incorrect predictions might therefore be explained by the fact that this kind of situation is not sufficiently represented in the training set. In the future versions of AUTOLOGP, which will contain more chemicals, it will be possible to take into account more components of the autocorrelation vectors allowing one to encode the interactions between the highly hydrophilic groups.

CONCLUDING REMARKS

On the basis of the same training set and several heterogeneous testing sets, AUTOLOGP (version 2.11) compares unfavorably with a three-layer back-propagation neural network (BNN) model using the same inputs. These results seem to confirm the superiority of the nonlinear techniques for modeling physicochemical properties. In addition, we can expect that the results will be further improved by the addition of other components of the autocorrelation vector **H**. In a recent study (Devillers *et al.*, 1995a), we underlined the usefulness of in-sample and out-of-sample testing sets for evidencing the interpolation and extrapolation performances of BNN models. In the present study, we refine this approach and when possible we propose to use several testing sets, each allowing one to uncover some particular properties of the models to be compared.

Acknowledgment

This study was granted by the EU (contract no. 5570–93–11 ED ISP F).

REFERENCES

Balaban, A.T., Basak, S.C., Colburn, T., and Grunwald, G.D. (1994). Correlation between structure and normal boiling points of haloalkanes C_1–C_4 using neural networks. *J. Chem. Inf. Comput. Sci.* **34,** 1118–1121.

Bintein, S. and Devillers, J. (1994). QSAR for organic chemical sorption in soils and sediments. *Chemosphere* **28,** 1171–1188.

Bintein, S., Devillers, J., and Karcher, W. (1993). Nonlinear dependence of fish bioconcentration on *n*-octanol/water partition coefficient. *SAR QSAR Environ. Res.* **1,** 29–39.

Bodor, N., Harget, A., and Huang, M.-J. (1991). Neural network studies. I. Estimation of the aqueous solubility of organic compounds. *J. Am. Chem. Soc.* **113,** 9480–9483.

Borggaard, C. and Thodberg, H.H. (1992). Optimal minimal neural interpretation of spectra. *Anal. Chem.* **64,** 545–551.

Bos, A., Bos, M., and van der Linden, W.E. (1992). Artificial neural networks as a tool for soft-modelling in quantitative analytical chemistry: The prediction of the water content of cheese. *Anal. Chim. Acta* **256,** 133–144.

Broto, P. and Devillers, J. (1990). Autocorrelation of properties distributed on molecular graphs. In, *Practical Applications of Quantitative Structure–Activity Relationships (QSAR) in Environmental Chemistry and Toxicology* (W. Karcher and J. Devillers, Eds.). Kluwer Academic Publishers, Dordrecht, pp. 105–127.

Broto, P., Moreau, G., and Vandycke, C. (1984). Molecular structures: Perception, autocorrelation descriptor and sar studies. System of atomic contributions for the calculation of the *n*-octanol/water partition coefficients. *Eur. J. Med. Chem. – Chim. Ther.* **19,** 71–78.

Cambon, B. and Devillers, J. (1993). New trends in structure–biodegradability relationships. *Quant. Struct.-Act. Relat.* **12,** 49–56.

Cherqaoui, D., Villemin, D., and Kvasnicka, V. (1994a). Application of neural network approach for prediction of some thermochemical properties of alkanes. *Chemom. Intell. Lab. Syst.* **24,** 117–128.

Cherqaoui, D., Villemin, D., Mesbah, A., Cense, J.M., and Kvasnicka, V. (1994b). Use of a neural network to determine the normal boiling points of acyclic ethers, peroxides, acetals and their sulfur analogues. *J. Chem. Soc. Faraday Trans.* **90,** 2015–2019.

Chow, H., Chen, H., Ng, T., Myrdal, P., and Yalkowsky, S.H. (1995). Using backpropagation networks for the estimation of aqueous activity coefficients of aromatic organic compounds. *J. Chem. Inf. Comput. Sci.* **35,** 723–728.

Cramer, R.D., Bunce, J.D., Patterson, D.E., and Frank, I.E. (1988). Crossvalidation, bootstrapping, and partial least squares compared with multiple regression in conventional QSAR studies. *Quant. Struct.–Act. Relat.* **7,** 18–25.

Curry, B. and Rumelhart, D.E. (1990). MSnet: A neural network which classifies mass spectra. *Tetrahedron Comput. Methodol.* **3,** 213–237.

Dearden, J.C. (1990). Physico-chemical descriptors. In, *Practical Applications of Quantitative Structure–Activity Relationships (QSAR) in Environmental Chemistry and Toxicology* (W. Karcher and J. Devillers, Eds.). Kluwer Academic Publishers, Dordrecht, pp. 25–59.

Dearden, J.C. and Bresnen, G.M. (1988). The measurement of partition coefficients. *Quant. Struct.–Act. Relat.* **7,** 133–144.

Devillers, J. (1996). Strengths and weaknesses of the backpropagation neural network in QSAR and QSPR studies. In, *Neural Networks in QSAR and Drug Design* (J. Devillers, Ed.). Academic Press, London, pp. 1–46.

Devillers, J., Domine, D., and Chastrette, M. (1992). A new method of computing the octanol/water partition coefficient. In, *Proceedings of QSAR92*. The University of Minnesota Duluth, Duluth, pp. 12.

Devillers, J., Bintein, S., Domine, D., and Karcher, W. (1995a). A general QSAR model for predicting the toxicity of organic chemicals to luminescent bacteria (Microtox® test). *SAR QSAR Environ. Res.* **4,** 29–38.

Devillers, J., Bintein, S., and Karcher, W. (1995b). CHEMFRANCE: A regional level III fugacity model applied to France. *Chemosphere* **30,** 457–476.

Devillers, J., Domine, D., and Karcher, W. (1995c). Estimating *n*-octanol/water partition coefficients from the autocorrelation method. *SAR QSAR Environ. Res.* **3,** 301–306.

Doucet, J.P., Panaye, A., Feuilleaubois, E., and Ladd, P. (1993). Neural networks and ^{13}C NMR shift prediction. *J. Chem. Inf. Comput. Sci.* **33,** 320–324.

Dunn III, W.J., Johansson, E., and Wold, S. (1983). Nonadditivity of 1-octanol/water partition coefficients of disubstituted benzenes: An explanation and its consideration in log P estimation. *Quant. Struct.–Act. Relat.* **2,** 156–163.

Eberhart, R.C. and Dobbins, R.W. (1990). *Neural Network PC Tools. A Practical Guide*. Academic Press, San Diego, p. 414.

Egolf, L.M., Wessel, M.D., and Jurs, P.C. (1994). Prediction of boiling points and critical temperatures of industrially important organic compounds from molecular structure. *J. Chem. Inf. Comput. Sci.* **34,** 947–956.

Finnoff, W., Hergert, F., and Zimmermann, H.G. (1993). Improving model selection by nonconvergent methods. *Neural Networks* **6,** 771–783.

Fujita, T., Iwasa, J., and Hansch, C. (1964). A new substituent constant, π, derived from partition coefficient. *J. Am. Chem. Soc.* **86,** 5175–5180.

Gago, F., Alvarez-Builla, J., and Elguero, J. (1987). Hydrophobicity measurements by HPLC: A new approach to π constants. *J. Liquid Chromat.* **10,** 1031–1047.

Gakh, A.A., Gakh, E.G., Sumpter, B.G., and Noid, D.W. (1994). Neural network-graph theory approach to the prediction of the physical properties of organic compounds. *J. Chem. Inf. Comput. Sci.* **34,** 832–839.

Ghose, A.K. and Crippen, G.M. (1986). Atomic physicochemical parameters for three-dimensional structure-directed quantitative structure–activity relationships I. Partition coefficients as a measure of hydrophobicity. *J. Comput. Chem.* **7,** 565–577.

Ghose, A.K., Pritchett, A., and Crippen, G.M. (1988). Atomic physicochemical parameters for three dimensional structure directed quantitative structure–activity relationships III: Modeling hydrophobic interactions. *J. Comput. Chem.* **9,** 80–90.

Grunenberg, J. and Herges, R. (1995). Prediction of chromatographic retention values (R_M) and partition coefficients (log P_{oct}) using a combination of semiempirical self-consistent reaction field calculations and neural networks. *J. Chem. Inf. Comput. Sci.* **35,** 905–911.

Hansch, C. and Leo, A. (1995). *Exploring QSAR. Fundamentals and Applications in Chemistry and Biology.* American Chemical Society, Washington, DC, p. 557.

Hansch, C., Leo, A., and Hoekman, D. (1995). *Exploring QSAR. Hydrophobic, Electronic, and Steric Constants.* American Chemical Society, Washington, DC, p. 348.

Hine, J. and Mookerjee, P.K. (1975). The intrinsic hydrophilic character of organic compounds. Correlations in terms of structural contributions. *J. Org. Chem.* **40,** 292–298.

Isnard, P. and Lambert, S. (1988). Estimating bioconcentration factors from octanol-water partition coefficient and aqueous solubility. *Chemosphere* **17,** 21–34.

Jodouin, J.-F. (1994). *Les Réseaux de Neurones. Principes et Définitions.* Hermès, Paris, p. 124.

Jordan, S.N., Leach, A.R., and Bradshaw, J. (1995). The application of neural networks in conformational analysis. 1. Prediction of minimum and maximum interatomic distances. *J. Chem. Inf. Comput. Sci.* **35,** 640–650.

Karcher, W. and Devillers, J. (1990). *Practical Applications of Quantitative Structure–Activity Relationships (QSAR) in Environmental Chemistry and Toxicology.* Kluwer Academic Publishers, Dordrecht, p. 475.

Karickhoff, S.W., Brown, D.S., and Scott, T.A. (1979). Sorption of hydrophobic pollutants on natural sediments. *Wat. Res.* **13,** 241–248.

Kenaga, E.E. and Goring, C.A.I. (1980). Relationship between water solubility, soil sorption, octanol–water partitioning, and concentration of chemicals in biota. In, *Aquatic Toxicology* (J.G. Eaton, P.R. Parrish, and A.C. Hendricks, Eds.). ASTM STP 707, pp. 78–115.

Klopman, G. and Wang, S. (1991). A computer automated structure evaluation (CASE) approach to calculation of partition coefficient. *J. Comput. Chem.* **12,** 1025–1032.

Klopman, G., Namboodiri, K., and Schochet, M. (1985). Simple method of computing the partition coefficient. *J. Comput. Chem.* **6,** 28–38.

Klopman, G., Li, J.-Y., Wang, S., and Dimayuga, M. (1994). Computer automated log *P* calculations based on an extended group contribution approach. *J. Chem. Inf. Comput. Sci.* **34,** 752–781.

Kreesuradej, W., Wunsch II, D.C., and Lane, M. (1994). Time delay neural network for small time series data sets. In, *World Congress on Neural Networks*. Town & Country Hotel, San Diego, California, USA, June 5–9, 1994, Lawrence Erlbaum Associates and INNS Press, pp. II-248–II-253.

Kvasnicka, V., Sklenak, S., and Pospichal, J. (1993). Neural network classification of inductive and resonance effects of subtituents. *J. Am. Chem. Soc.* **115,** 1495–1500.

Leo, A.J. (1990). Methods of calculating partition coefficients. In, *Comprehensive Medicinal Chemistry, Vol. 4, Quantitative Drug Design* (C.A. Ramsden, Ed.). Pergamon Press, Oxford, pp. 295–319.

Leo, A., Hansch, C., and Elkins, D. (1971). Partition coefficients and their uses. *Chem. Rev.* **71,** 525–616.

Mackay, D. (1982). Correlation of bioconcentration factors. *Environ. Sci. Technol.* **16,** 274–278.

Mackay, D. (1991). *Multimedia Environmental Models. The Fugacity Approach*. Lewis Publishers, Chelsea, p. 257.

Mackay, D., Shiu, W.Y., and Ma, K.C. (1992). *Illustrated Handbook of Physical–Chemical Properties and Environmental Fate for Organic Chemicals, Vol. 2, Polynuclear Aromatic Hydrocarbons, Polychlorinated Dioxins, and Dibenzofurans.* Lewis Publishers, Chelsea, p. 597.

Moreau, G. and Broto, P. (1980a). The autocorrelation of a topological structure: A new molecular descriptor. *Nouv. J. Chim.* **4,** 359–360.

Moreau, G. and Broto, P. (1980b). Autocorrelation of molecular structures, application to SAR studies. *Nouv. J. Chim.* **4,** 757–764.

Nefati, H., Diawara, B., and Legendre, J.J. (1993). Predicting the impact sensitivity of explosive molecules using neuromimetic networks. *SAR QSAR Environ. Res.* **1,** 131–136.

Nys, G.G. and Rekker, R.F. (1974). The concept of hydrophobic fragmental constants (*f*-values). II. Extension of its applicability to the calculation of lipophilicities of aromatic and heteroaromatic structures. *Eur. J. Med. Chem. – Chim. Ther.* **9,** 361–375.

Panaye, A., Doucet, J.-P., Fan, B.T., Feuilleaubois, E., and Rahali El Azzouzi, S. (1994). Artificial neural network simulation of ^{13}C NMR shifts for methyl substituted cyclohexanes. *Chemom. Intell. Lab. Syst.* **24,** 129–135.

Pao, Y.H. (1989). *Adaptive Pattern Recognition and Neural Networks.* Addison-Wesley Publishing Company, Reading, p. 309.

Rekker, R.F. and de Kort, H.M. (1979). The hydrophobic fragmental constant; an extension to a 1000 data point set. *Eur. J. Med. Chem. – Chim. Ther.* **14,** 479–488.

Rekker, R.F. and Mannhold, R. (1992). *Calculation of Drug Lipophilicity. The Hydrophobic Fragmental Constant Approach*. VCH, Weinheim, p. 112.

Rekker, R.F., ter Laak, A.M., and Mannhold, R. (1993). On the reliability of calculated log P-values: Rekker, Hansch/Leo and Suzuki approach. *Quant. Struct.–Act. Relat.* **12,** 152–157.

Rumelhart, D.E., Hinton, G.E., and Williams, R.J. (1986). Learning representations by back-propagating errors. *Nature* **323,** 533–536.

Shao, J. (1993). Linear model selection by cross-validation. *J. Am. Stat. Assoc.* **88,** 486–494.

Sharma, A.K., Sheikh, S., Pelczer, I., and Levy, G.C. (1994). Classification and clustering: Using neural networks. *J. Chem. Inf. Comput. Sci.* **34,** 1130–1139.

Sigman, M.E. and Rives, S.S. (1994). Prediction of atomic ionization potentials I–III using an artificial neural network. *J. Chem. Inf. Comput. Sci.* **34,** 617–620.

STATQSAR Package (1994). CTIS, Lyon, France.

Sumpter, B.G., Getino, C., and Noid, D.W. (1994). Theory and applications of neural computing in chemical science. *Ann. Rev. Phys. Chem.* **45,** 439–481.

Suzuki, T. and Kudo, Y. (1990). Automatic log *P* estimation based on combined additive modeling methods. *J. Comput. Aid. Mol. Des.* **4,** 155–198.

Svozil, D., Pospichal, J., and Kvasnicka, V. (1995). Neural network prediction of carbon-13 NMR chemical shifts of alkanes. *J. Chem. Inf. Comput. Sci.* **35,** 924–928.

Tsaptsinos, D. and Leigh, J.R. (1993). A step by step approach for the construction of a fermentation process estimator. In, *World Congress on Neural Networks.* Oregon Convention Center, Portland, Oregon, July 11–15, 1993, Lawrence Erlbaum Associates and INNS Press, pp. I-216–I-219.

Tschinke, V. (1995). A neural network study on solvation properties from computed descriptors. Oral presentation in the '*Second International Workshop on Neural Networks and Genetic Algorithms Applied to QSAR and Drug Design*', Lyon, France, June 12–14, 1995.

Veith, G.D., DeFoe, D.L., and Bergstedt, B.V. (1979). Measuring and estimating the bioconcentration factor of chemicals in fish. *J. Fish. Res. Board Can.* **36,** 1040–1048.

Veith, G.D., Macek, K.J., Petrocelli, S.R., and Carroll, J. (1980). An evaluation of using partition coefficients and water solubility to estimate bioconcentration factors for organic chemicals in fish. In, *Aquatic Toxicology* (J.G. Eaton, P.R. Parrish, and A.C. Hendricks, Eds.). ASTM STP 707, pp. 116–129.

Weigend, A.S., Huberman, B.A., and Rumelhart, D.E. (1990). Predicting the future: A connectionist approach. *Int. J. Neural Systems* **1,** 193–209.

Wessel, M.D. and Jurs, P.C. (1995a). Prediction of normal boiling points of hydrocarbons from molecular structure. *J. Chem. Inf. Comput. Sci.* **35,** 68–76.

Wessel, M.D. and Jurs, P.C. (1995b). Prediction of normal boiling points for a diverse set of industrially important organic compounds from molecular structure. *J. Chem. Inf. Comput. Sci.* **35,** 841–850.

West, G.M.J. (1995a). Predicting phosphorus NMR shifts using neural networks. 2. Factors influencing the accuracy of predictions. *J. Chem. Inf. Comput. Sci.* **35,** 21–30.

West, G.M.J. (1995b). Predicting phosphorus nuclear magnetic resonance (NMR) shifts using neural networks. 3. Element-value optimizing (EVO) network architectures. *J. Chem. Inf. Comput. Sci.* **35,** 806–814.

3 Use of a Backpropagation Neural Network and Autocorrelation Descriptors for Predicting the Biodegradation of Organic Chemicals

J. DEVILLERS[1]*, D. DOMINE[1], and R.S. BOETHLING[2]
[1]*CTIS, 21 rue de la Bannière, 69003 Lyon, France*
[2]*Office of Prevention, Pesticides and Toxic Substances, 7406, US Environmental Protection Agency, 401 M St, SW, Washington, DC 20460, USA*

The results of a survey, in which a panel of experts was required to estimate the approximate time for the ultimate biodegradation of a large number of chemicals were used to derive a general quantitative structure–biodegradation relationship (QSBR) model by means of a backpropagation neural network (BNN). Molecules were described from autocorrelation vectors encoding their hydrophobicity. The model was designed from 172 molecules and tested by means of two testing sets of 12 and 57 molecules. The performances of the BNN model compare favorably with those of models obtained using classical linear methods.

KEY WORDS: *structure–biodegradation relationships; autocorrelation method; hydrophobicity; backpropagation neural network.*

INTRODUCTION

There is increasing interest in modeling the environmental fate of chemicals and hence in determining their potential for biodegradation. Indeed, because testing is costly and time-consuming, (quantitative) structure–biodegradability relationship ((Q)SBR) models appear to be potentially valuable tools for

---

* Author to whom all correspondence should be addressed.

In, *Neural Networks in QSAR and Drug Design* (J. Devillers, Ed.)
Academic Press, London, 1996, pp. 65–82.
ISBN 0-12-213815-5

predicting the anaerobic and aerobic biodegradation of organic pollutants (Kuenemann *et al.*, 1990; Mani *et al.*, 1991; Vasseur *et al.*, 1993).

In the environment, or during laboratory tests, the biodegradation of chemicals depends on the duration of exposure, soil type and water content, oxygen and nutrient concentrations, temperature, pH, microbial population present and other variables (Block *et al.*, 1990; Pitter and Chudoba, 1990; Vasseur *et al.*, 1993; Alexander, 1994). Therefore, it is always difficult to collect biodegradation data for a wide range of organic chemicals into a consistent database for modeling purposes. This largely explains the limited size of the training sets that have been used to elaborate QSBR models by means of regression analyses, and the restriction of their predictive power to specific classes of chemicals (for a complete review see Kuenemann *et al.* (1990)). This also helps us to understand why discriminant analysis can represent a satisfying alternative method for modeling biodegradation (Enslein *et al.*, 1984; Niemi *et al.*, 1987; Gombar and Enslein, 1991), since the conversion of quantitative biodegradation data into different classes of activity (e.g., slow and rapid biodegradation) may allow us to overcome the problems of endpoint nonhomogeneity. Recently, it has been suggested that the backpropagation neural network (BNN) is a powerful nonlinear statistical tool for biodegradation modeling (Cambon and Devillers, 1993; Devillers, 1993), due to its ability to learn and generalize from complex, noisy, and partial information (Eberhart and Dobbins, 1990; de Saint Laumer *et al.*, 1991; Bos *et al.*, 1993). Given this, our goal was to confirm this assumption using a very large and heterogeneous set of molecules for which biodegradability had been scored by experts.

BIODEGRADATION DATA

A biodegradation survey was conducted with 17 experts in order to estimate the approximate time required for complete primary and ultimate biodegradation of 200 different molecules in the aquatic environment. The survey chemicals contained a wide diversity of structures (Figure 1 shows some specific structures). The experts were required to estimate biodegradability, not perform experiments, and were instructed to spend only 2–4 minutes on each chemical. Biodegradability was estimated by marking one or more of the following terms: 'Hours', 'Days', 'Weeks', 'Months', 'Longer'. The responses were then used to calculate a score encoding the aerobic biodegradation of the molecules. A score was obtained by associating the integers 1, 2, 3, 4, and 5 with the responses 'Hours', 'Days', 'Weeks', 'Months', and 'Longer', respectively. Thus, for example, the ultimate biodegradation mean score for 1,2-dinitrobenzene was calculated as shown below:

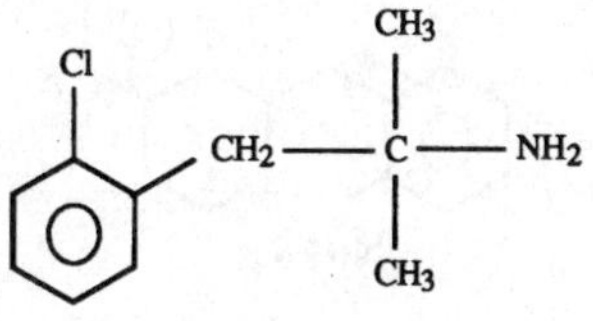

No 8

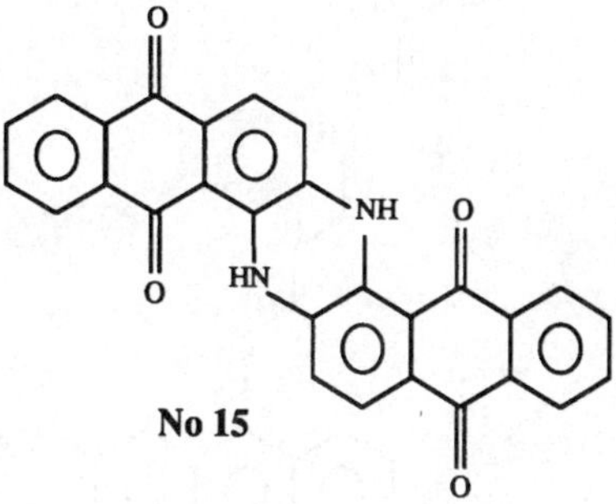

No 15

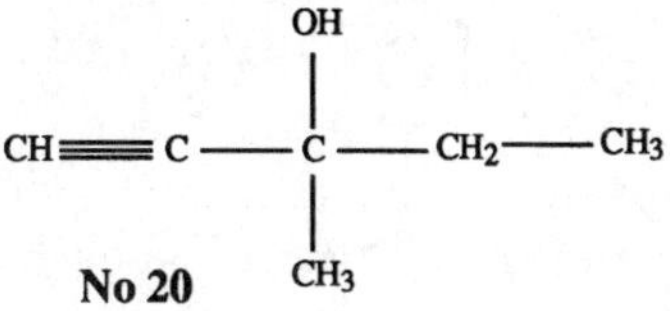

No 20

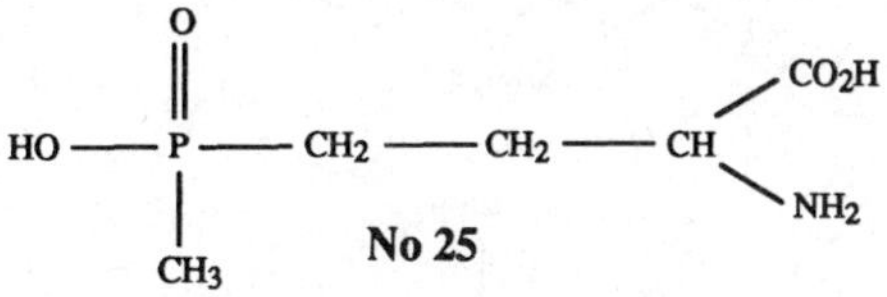

No 25

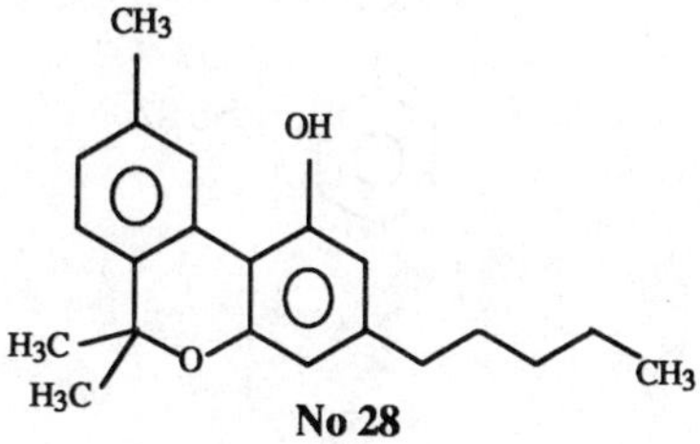

No 28

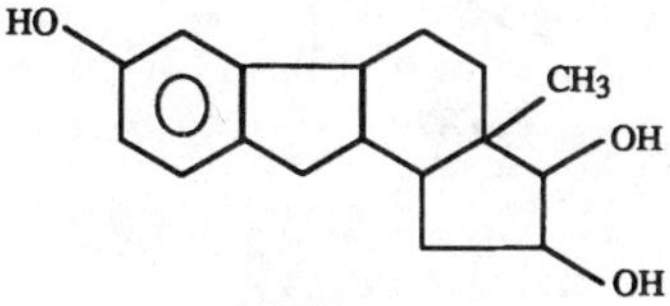

No 29

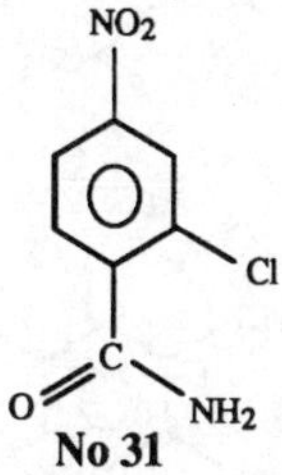

No 31

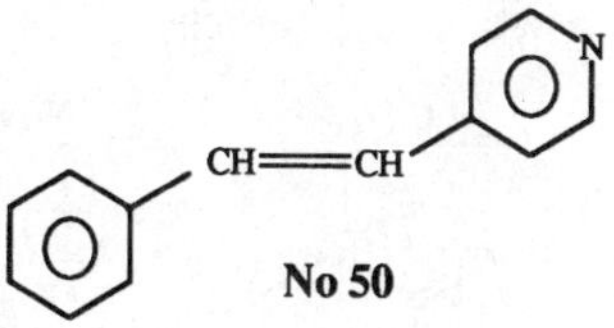

No 50

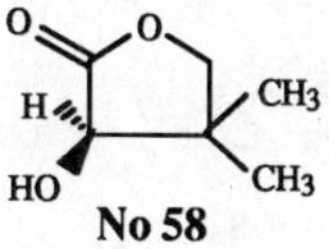

No 58

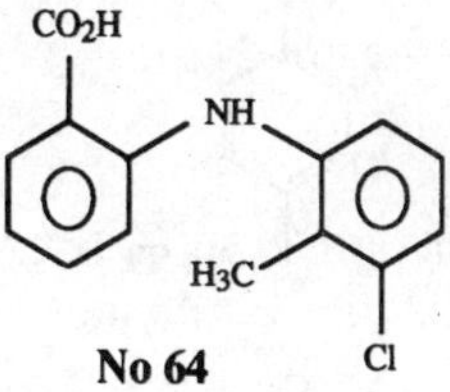

No 64

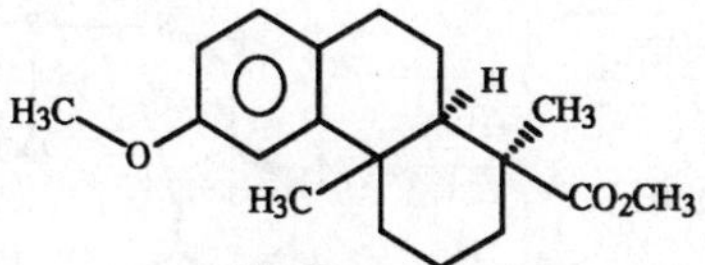

No 68

Figure 1 Representative chemicals from the training and testing sets. See Table I for correspondence between numbers and chemicals (except *; see Table II).

No 76

No 75

No 82

No 87

No 88

No 103

No 89

No 107

No 114

Figure 1 *continued.*

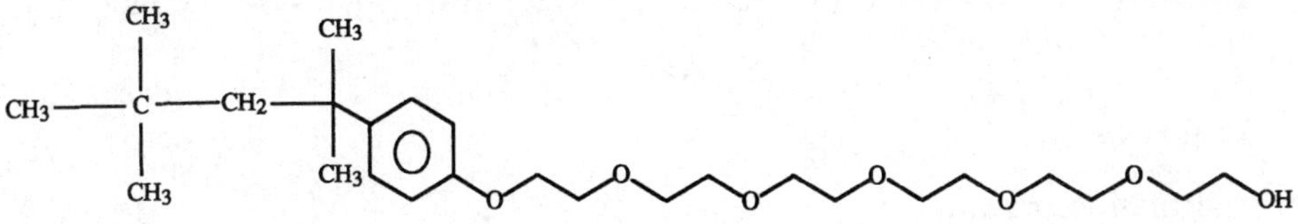

No 123

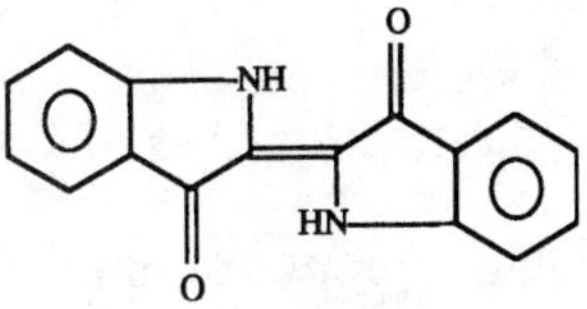

No 126

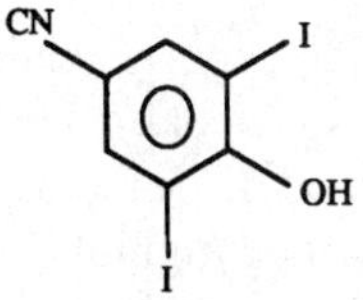

No 132

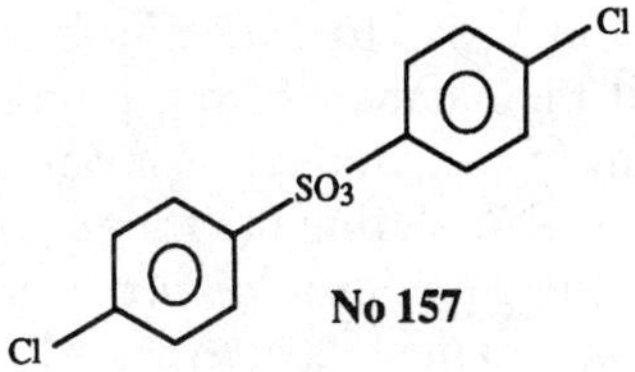

No 157

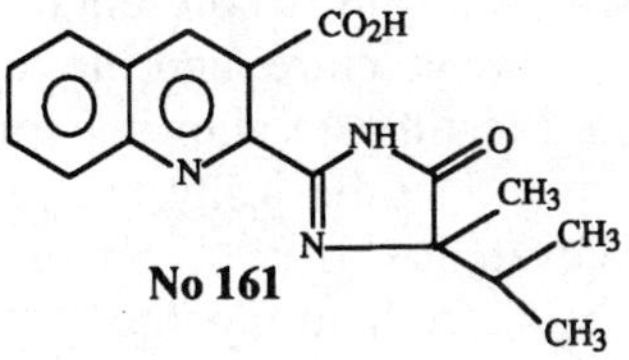

No 161

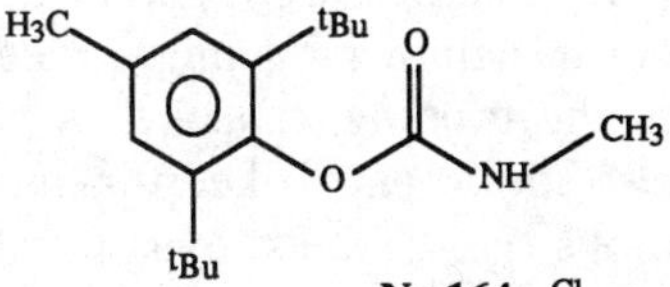

No 164

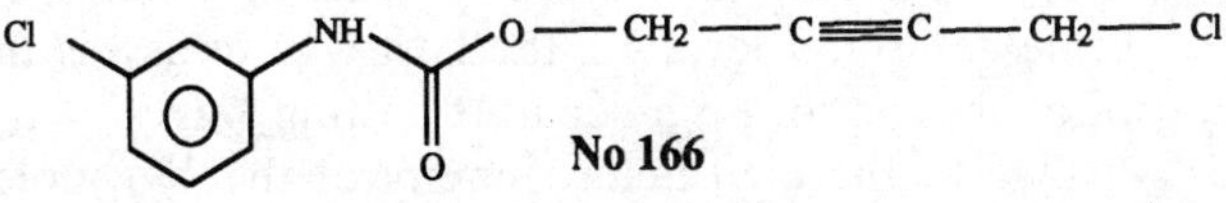

No 166

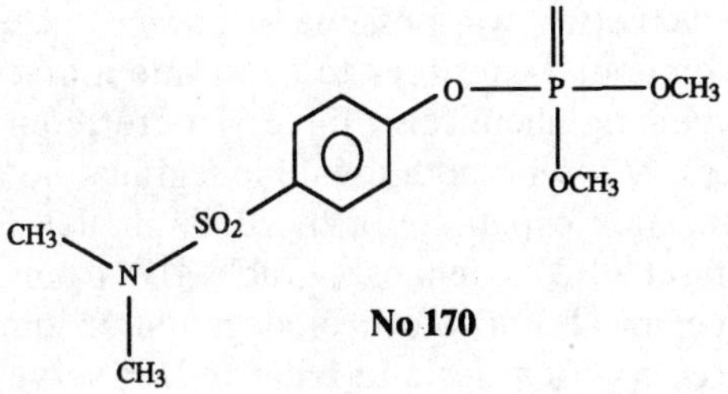

No 170

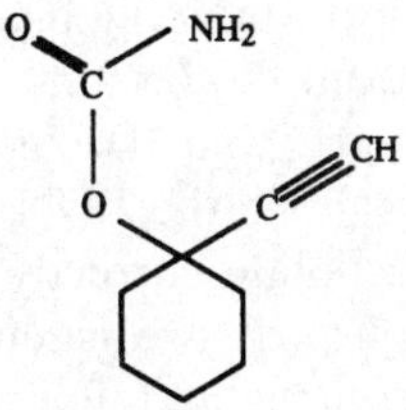

No 10*

Figure 1 *continued.*

Responses (associated integer)	Hours (1)	Days (2)	Weeks (3)	Months (4)	Longer (5)
Number of responses	0	3	8	11	2

Calculation procedure:

$$[(1 \times 0) + (2 \times 3) + (3 \times 8) + (4 \times 11) + (5 \times 2)] \,/\, [0 + 3 + 8 + 11 + 2] = 84/24 = 3.50$$

Note that experts could give multiple responses to express their judgment (e.g., days–weeks). Under these conditions, the total number of responses can exceed 17 for a given chemical.

From the original database of 200 molecules, a set of 184 molecules was selected by removing the molecules presenting problems of coding, such as ionic compounds due to their charge. These molecules were used to constitute a training set (Table I) and a first testing set (Table II). For the constitution of these two sets it was necessary to find a compromise between the need for a large and representative training set in order to avoid problems of BNN architecture, and the necessity to split the database into a training set and a testing set allowing correct evaluation of the predictive power of the BNN model. Basically, the performances of a BNN during the training phase are evaluated from the error value, whereas the generalization performances are assessed during the testing phase from unknown patterns. Although Schüürmann and Müller (1994) recommended leave-n-out procedures for validating BNN models, it is obvious that the leave-one-out (LOO) and leave-n-out techniques are not the most appropriate in neural modeling. Indeed, the removal of patterns can greatly influence the training, especially when the number of individuals is reduced. It is also important to keep in mind that, with a nonlinear method, different solutions (i.e., models) exist for the same dataset. Therefore, the best way to assess the predictive performances of a BNN is to use a separate testing set.

Due to the high heterogeneity of the 184 molecules in the database, the selection of training and testing sets by statistical design did not give satisfactory results. From inspection of the structures of the molecules and their corresponding biodegradation activity, it was possible to extract 12 chemicals from the original set of 184 compounds in order to constitute a first testing set (Table II). As the 172 remaining chemicals (Table I) were required for training the BNN, it was necessary to find additional biodegradation data to estimate correctly the generalization capability of the BNN model. For this purpose, we assembled a dataset of 57 chemicals (Table III) from several sources as follows. First, 19 chemicals for which biodegradation times had also been estimated by experts in an earlier and independent survey, were used (Boethling and Sabljic, 1989). The expert responses had been transformed into a score ranging from 1 to 4. Molecules bearing a score value

Table I *Observed (Obs.) and calculated (Cal.) ultimate biodegradation scores for the training set.*

No.	Name	Obs.	Cal.	Residual
1	1-Iodonaphthalene	2.96	3.27	–0.31
2	2,6-Di-*t*-butylphenol	3.93	4.16	–0.23
3	4-(Phenylthio)-benzoic acid	2.61	3.21	–0.6
4	11-Cyanoundecanoic acid	2.32	2.37	–0.05
5	Chlorpropham	3.04	3.09	–0.05
6	2,5-Pyridinediamine	3.36	2.9	0.46
7	1-Naphthol-3,6-disulfonic acid	3.1	3.1	0
8	Clortermine*	3.44	3.16	0.28
9	Di-*t*-butyl dicarbonate	2.82	2.86	–0.04
10	1,3-Dichloropropanol	2.96	3.09	–0.13
11	Azobenzene	3.57	3.42	0.15
12	Sumithion	3.21	3.18	0.03
13	Isophorone	3.2	3.2	0
14	Pyrazine	3.24	3.22	0.02
15	Vat blue 4*	4	3.93	0.07
16	2-Mercaptobenzothiazole	3.29	3.5	–0.21
17	*n*-Decanal	2.2	2.64	–0.44
18	2-Hydroxy-5-decylbenzaldehyde oxime	3.3	3.35	–0.05
19	4-(1-Butylpentyl)pyridine	3.23	3.14	0.09
20	Meparfynol*	3.24	3.23	0.01
21	4-Butylcyclohexyl 4-butylcyclohexylcarboxylate	3.13	3.19	–0.06
22	Isouron	3.61	3.53	0.08
23	2-(3-Butynyloxy)tetrahydro-2*H*-pyran	3.41	3.61	–0.2
24	*n*-Decylamine	2.4	2.47	–0.07
25	Phosphinothricin*	2.96	2.97	–0.01
26	Methyl-*t*-butyl ether	3.65	3.51	0.14
27	Dimethylformamide	2.59	3.1	–0.51
28	Cannabinol*	3.56	3.57	–0.01
29	16-Epiestriol*	3.43	3.49	–0.06
30	Linolenic acid	2.25	2.43	–0.18
31	Aklomide*	3.58	3.53	0.05
32	3-Cyano-6-methyl-2(1*H*)-pyridinone	3.18	3.44	–0.26
33	2-Amino-5-chloro-4-methyl-benzenesulfonic acid	3.7	3.36	0.34
34	3-Methylene-1,5,7,11-tetraoxa-spiro[5.5]undecane	3.54	3.67	–0.13
35	Di-*n*-decyl ether	3	3.04	–0.04
36	Phenolphthalein	3.56	3.53	0.03
37	6-Nitrocaproic acid	2.55	2.87	–0.32
38	Hexadecyl myristate	2.68	2.68	0
39	5,6,11,12-Tetrahydro-dibenz[*b,f*]azocin-6-one	3.5	3.15	0.35
40	*n*-Hexylbenzene	2.65	2.93	–0.28
41	1,2-Dinitrobenzene	3.5	3.25	0.25
42	3-Acetoxypyridine	2.86	2.75	0.11
43	*p*-Aminophenol	2.83	2.91	–0.08
44	1,4-Dichlorophthalazine	3.85	3.75	0.1
45	(2-Pyrimidylthio)acetic acid	3.38	3	0.38
46	3'-Methyl-4'-chloro-2,2,2-trifluoroacetophenone	4.14	3.96	0.18
47	Acrylonitrile	2.73	3.14	–0.41

Table I *continued.*

No.	Name	Obs.	Cal.	Residual
48	Diisodecyl phthalate	3.3	3.32	–0.02
49	Tris(2,3-dibromopropyl)phosphate	4	4	0
50	Stilbazole*	3.27	3.53	–0.26
51	Abietic acid	3.67	3.88	–0.21
52	2,2,4,4,6,8,8-Heptamethylnonane	4.32	4.43	–0.11
53	Tris(hydroxymethyl)nitromethane	2.91	2.94	–0.03
54	2-Naphthylamine	3.09	3.02	0.07
55	Sulfasimazine	4	3.95	0.05
56	Ethylene glycol diacetate	2.11	2.29	–0.18
57	Pyrazinoic acid	3.18	3.24	–0.06
58	Pantolactone*	2.96	3.01	–0.05
59	1,2-Dichloropropane	3.66	3.39	0.27
60	Alachlor	3.79	3.88	–0.09
61	Maleic hydrazide	2.86	2.85	0.01
62	1,1-Dimethylcyclohexane	3.29	3.38	–0.09
63	Tri-*n*-octylamine	3.17	3.19	–0.02
64	Tolfenamic acid*	3.44	3.65	–0.21
65	Tetraethyl biphenyl-3,3',4,4'-tetracarboxylate	3.52	3.51	0.01
66	Thianthrene	3.82	3.4	0.42
67	*N*-Butyl-*N*-(4-styryl)phenethylamine	3.67	3.71	–0.04
68	Methyl *O*-methylpodocarpate*	3.65	3.71	–0.06
69	Benzanthracene	4.24	4.31	–0.07
70	Fenuron	3.13	3.4	–0.27
71	Acenaphthene	3.8	3.46	0.34
72	3,4-Dimethylaniline	3.26	3.06	0.2
73	2-(2*H*-Benzotriazol-2-yl)-phenol	3.56	3.5	0.06
74	2-Chloro-2-propen-1-ol	3.05	3.16	–0.11
75	Substituted acridine*	4.24	4.15	0.09
76	Berbine*	4.08	3.85	0.23
77	Chlorothalonil	4.43	4.26	0.17
78	Di(methoxypropyl)hydroxylamine	3	2.96	0.04
79	2-Methoxy-4-butylcyclohexanol	3.09	3.26	–0.17
80	10-Chloro-9-anthraldehyde	3.82	3.94	–0.12
81	Picloram	4.4	3.45	0.95
82	Acroteben*	3.44	3.56	–0.12
83	Benefin	4.32	4.06	0.26
84	4-Phenylcyclohexene	3.37	3.43	–0.06
85	Propargyl alcohol	2.77	3.08	–0.31
86	5-Nitro-2-thiophenemethanediol diacetate	3.35	3.32	0.03
87	Pivalaldehyde*	3.38	3.3	0.08
88	Tocol*	3.79	3.82	–0.03
89	Azintamide*	3.69	3.81	–0.12
90	2-Amino-5-thiazolecarboxaldehyde	3.38	3.36	0.02
91	Dichlorvos	3.32	3.32	0
92	2-Phenylethanol	2.48	3.08	–0.6
93	*E*-Caprolactone	2.3	3.09	–0.79
94	Ethyl amyl ketone	2.68	2.99	–0.31
95	1,3-Bis(2-oxazolyl)benzene	3.75	3.74	0.01
96	*N*-Butylpyrrolidine	3.2	3.97	–0.77

Table I *continued.*

No.	Name	Obs.	Cal.	Residual
97	Styrene	2.8	3.11	–0.31
98	*n*-Decane	2.25	2.79	–0.54
99	Atrazine	4	4.19	–0.19
100	2-Acetamido-7-fluorofluorene	3.87	3.66	0.21
101	*N*-(3,4-Dichlorophenyl) propanamide	3.96	3.49	0.47
102	Diphenyl ether	3.21	3.06	0.15
103	Tetradifon*	4.36	4.27	0.09
104	Octacosanol	2.95	2.95	0
105	Cyclophosphamide	3.62	3.39	0.23
106	5-Nitro-2-furaldehyde	3.2	3.17	0.03
107	Dacthal*	4.34	4.13	0.21
108	Dinoseb	3.8	2.95	0.85
109	5-Acetoxymethyl-2-amino-3-cyanopyrazine	3.5	3.38	0.12
110	Hydroxysimazine	3.79	3.75	0.04
111	Dodecylsulfonic acid	2.63	2.65	–0.02
112	Benzoxiquine	3.41	3.27	0.14
113	Dimethyl ethylidenemalonate	2.48	2.96	–0.48
114	Menazon*	3.67	3.7	–0.03
115	Di-2-cyanoethyl ether	3.21	3.15	0.06
116	*o*-Phenylphenol	2.92	2.81	0.11
117	1,2-Butylene oxide	2.88	3.29	–0.41
118	Cycloheximide	3.19	3.19	0
119	*t*-Butylbenzene	3.38	3.45	–0.07
120	Methyl 5,6-dichloro-2-pyrimidine carboxylate	3.92	3.84	0.08
121	3-Nitropentane	3.05	3.39	–0.34
122	4'-(Imidazol-1-yl)acetophenone	3.27	3.16	0.11
123	Fraction of Triton X-100*	3.79	3.82	–0.03
124	1,4-Piperazinebis(ethanesulfonic acid)	3.69	3.87	–0.18
125	*p*-Bromophenethylamine	3.41	3.4	0.01
126	Vat blue 1*	3.92	4.02	–0.1
127	Hexachlorophene	4.23	4.19	0.04
128	1,8,9-Triacetoxyanthracene	3.75	3.79	–0.04
129	Atropic acid	2.8	3.11	–0.31
130	Silvex	4.16	3.82	0.34
131	Trichlorfon	3.73	3.52	0.21
132	Ioxynil*	3.96	4.13	–0.17
133	*N*-Pentyl-4-nitrophthalimide	3.64	3.92	–0.28
134	Pentaerythritol	2.88	2.84	0.04
135	Niclosamide	3.89	3.81	0.08
136	Diethyl dithiolisophthalate	3.12	3.08	0.04
137	Clofibrate	3.54	3.63	–0.09
138	7-Bromoheptanoic acid	2.96	3.11	–0.15
139	2-Chloroethyl vinyl ether	3.17	3.33	–0.16
140	Diallyl phthalate	2.95	3.36	–0.41
141	Cumene	2.96	3.25	–0.29
142	Tripropylene glycol	3.12	3.3	–0.18
143	Cycloate	3.54	3.47	0.07
144	Methyl 3-nitropropionate	2.83	2.98	–0.15
145	*p*-Toluenesulfonic acid	3.04	2.8	0.24

Table I *continued.*

No.	Name	Obs.	Cal.	Residual
146	4,4'-Oxybisbenzeneamine	3.56	3.63	–0.07
147	2-Chloro-*p*-xylene	3.58	3.38	0.2
148	Ethylenediamine-tetra(methylphosphonic acid)	3.71	3.76	–0.05
149	Nornicotine	3.6	3.13	0.47
150	Pentabromoethylbenzene	4.56	4.5	0.06
151	2-Amino-1,3-dibromofluorene	4.23	4.13	0.1
152	2-Nitroresorcinol	3.28	3.09	0.19
153	Vinylidene chloride	3.82	3.35	0.47
154	4-(2-Hydroxyethyl) morpholine	3.3	3.48	–0.18
155	Diphenatrile	3.7	3.34	0.36
156	2,4-Dibromophenol	3.64	4.21	–0.57
157	Ovex*	3.87	3.82	0.05
158	Methyl 2-chloro-α-cyano-hydrocinnamate	3.63	3.57	0.06
159	Methyl oleate	2.55	2.28	0.27
160	Monoacetylbenzidine	3.68	3.76	–0.08
161	Imazaquin*	4.09	3.96	0.13
162	3-(1-Cyanoethyl) benzoic acid	3.24	3.16	0.08
163	Geraniol	2.88	3.21	–0.33
164	Terbutol*	4.24	4.11	0.13
165	Butylcarbitol acetate	3.04	2.68	0.36
166	Barban*	3.9	3.71	0.19
167	2,6-Dichloro-4-nitroaniline	4.22	3.94	0.28
168	1-Iodohexane	3.18	2.93	0.25
169	Propyl 1-piperazinecarboxylate	3.4	3.1	0.3
170	Famphur*	3.54	3.57	–0.03
171	4-Octylaniline	3.1	2.98	0.12
172	Terbacil	3.89	3.88	0.01

* Structure shown in Figure 1.

Table II *Observed (Obs.) and calculated (Cal.) ultimate biodegradation scores for the first testing set.*

No.	Name	Obs.	Cal.	Residual
1	Eicosane	2.68	2.16	0.52
2	Bibenzyl	3.05	3.2	–0.15
3	Pentacene	4.08	3.91	0.17
4	1-Hydroxyethane-1,1-diphosphonic acid	3.14	3.16	–0.02
5	Azulene	3.67	3.21	0.46
6	3-Nitrostyrene	3.42	3.15	0.27
7	1-Methyl-2-pyrrolidinone	2.88	3.36	–0.48
8	2-Benzoylpyridine	3.33	3.36	–0.03
9	Valproic acid	2.71	3.05	–0.34
10	Ethinamate*	3.12	2.93	0.19
11	1,5-Naphthalene disulfonic acid	3.3	2.97	0.33
12	2-Acrylamido-2-methyl-1-propanesulfonic acid	3.43	3.72	–0.29

* Structure shown in Figure 1.

Table III *Observed (Obs.) and calculated (Cal.) ultimate biodegradability for the second testing set.*

No.	Name	Obs.*	Cal.	Code†
1	Phthalic anhydride	BR	3.01	BR
2	Maleic anhydride	BR	3.44	BR
3	2,4-Hexadien-1-ol	BR	3.18	BR
4	*trans*-2-Butene	BR	3.34	BR
5	Trichloroethylene	BS	3.32	BR‡
6	*N*-Ethylbenzamide	BR	3.37	BR
7	*N,N*'-Dimethylpropanamide	BR	3.03	BR
8	*N,N*'-Diphenyl-*p*-phenylenediamine	BS	4.04	BS
9	*n*-Butylbenzene	BR	2.91	BR
10	Hydron yellow	BS	4.14	BS
11	Carbazole vat dye	BS	3.77	BS
12	Epichlorohydrin	BR	3.17	BR
13	Methyl methacrylate	BR	3.23	BR
14	Vinyl acetate	BR	3.48	BR
15	*o*-Benzoylbenzoic acid	BR	3.13	BR
16	2-Methyl-3-hexanone	BR	3.58	BS‡
17	2-Pyrrolidone-5-carboxylic acid	BR	4.25	BS‡
18	Chlorendic anhydride	BS	2.82	BR‡
19	2,4-Dichloro-6-ethoxy-1,3,5-triazine	BS	4.57	BS
20	*p*-Cresol	BR	3.25	BR
21	2,4,5-Trichlorophenol	BS	3.31	BR‡
22	3-Methylcholanthrene	BS	4.9	BS
23	*p,p*'-DDE	BS	3.93	BS
24	Hexachlorobenzene	BS	4.24	BS
25	*n*-Butylamine	BR	3.17	BR
26	4-Methoxyaniline	BR	3.13	BR
27	*o*-Cresol	BR	3.26	BR
28	*m*-Hydroxybenzoic acid	BR	3.19	BR
29	Ethyl 2-hydroxybenzoate	BR	3.18	BR
30	3,4-Dihydroxybenzoic acid	BR	3.08	BR
31	4-Aminopyridine	BS	3.19	BR‡
32	Bromodichloromethane	BS	3.3	BR‡
33	Pentachlorobenzene	BS	3.85	BS
34	3-Methyl-4-chlorophenol	BR	3.09	BR
35	4-Nitrobenzoic acid	BR	3.27	BR
36	Aniline	BR	3.15	BR
37	Pentachlorophenol	BS	3.87	BS
38	2-Methoxyaniline	BS	3.4	BR‡
39	*N,N*-Diethylaniline	BS	4.31	BS
40	*o*-Toluidine	BR	3.3	BR
41	3,4-Dimethylphenol	BR	3.08	BR
42	2,4-Dimethylaniline	BS	3.83	BS
43	*p*-(*tert*-Butyl)benzoic acid	BS	3.7	BS
44	*p*-Hydroxyacetophenone	BR	3.02	BR
45	*p*-Toluic acid	BR	3.2	BR
46	4-Hydroxybenzoic acid	BR	3.22	BR
47	Benzonitrile	BR	3.18	BR
48	*N*-Methylaniline	BS	3.66	BS

Table III *continued.*

No.	Name	Obs.*	Cal.	Code†
49	*n*-Propylbenzene	BR	3.16	BR
50	*p*-Toluidine	BR	3.25	BR
51	*m*-Cresol	BR	3.07	BR
52	Resorcinol	BR	3.1	BR
53	Phenol	BR	3.19	BR
54	Ethyl 4-hydroxybenzoate	BR	3.45	BR
55	*sec*-Butylbenzene	BS	3.25	BR‡
56	2,6-Di-*t*-butyl-4-methylphenol	BR	3.58	BS‡
57	Ethyl 3-hydroxybenzoate	BR	3.46	BR

* BR: Biodegrade rapidly; BS: Biodegrade slowly.
† Code: BR < 3.5; BS ≥ 3.5.
‡ Bad prediction.

lower than 2.5 were considered to biodegrade rapidly (BR) while those having a value above 2.5 were considered to biodegrade slowly (BS). Chemicals with values between 2 and 3 were removed in order to account for the variability in the biodegradation data and facilitate the comparison. Second, 14 chemicals used in the testing set of Cambon and Devillers (1993) and coded BR or BS from the survey of literature results were added. Last, the set was completed by adding 24 chemicals from the BIODEG data bank (Syracuse Research Corporation, 1992) whose biodegradation behavior had been classified as BR or BS from the review of literature results.

MOLECULAR DESCRIPTORS

Structural fragments represent valuable descriptors for deriving SBR and QSBR (Degner *et al.*, 1991; Cambon and Devillers, 1993; Devillers, 1993; Boethling *et al.*, 1994). They appear to be particularly efficient when using a BNN (Cambon and Devillers, 1993; Devillers, 1993; Domine *et al.*, 1993). Due to the heterogeneity of our set of molecules (Table I), it was impossible to find a reduced number of structural descriptors possessing a high discrimination power and available with a sufficient frequency. As we recently showed the usefulness of the autocorrelation method for modeling large heterogeneous sets of molecules by means of a BNN (Devillers *et al.*, 1995), this method was used in the present study. The calculation procedure of an autocorrelation vector is illustrated in Figure 2. Numerous physicochemical properties have been investigated to generate autocorrelation vectors (for an extensive review, see Broto and Devillers (1990)). By using the fragmental constants of Rekker and Mannhold (1992) and by strictly following their methodological approach (Nys and Rekker, 1974; Rekker and de Kort, 1979; Rekker *et al.*, 1993), it was possible to derive 66 atomic and group

Step 1. Definition of the molecular graph (B) from the formula (A).

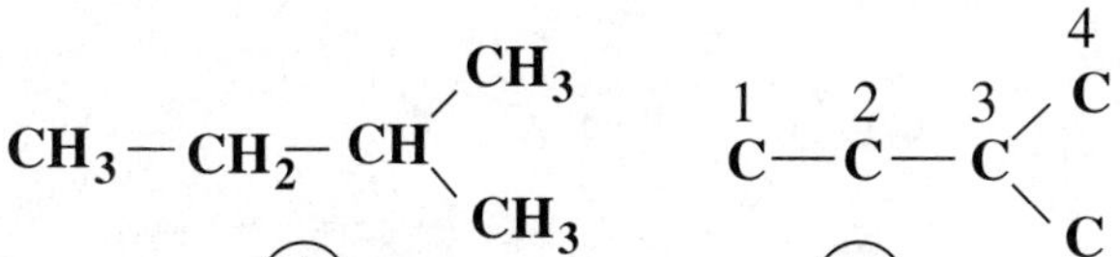

$$CH_3-CH_2-CH(CH_3)_2 \quad \text{(A)}$$

$$\overset{1}{C}-\overset{2}{C}-\overset{3}{C}(\overset{4}{C})(\underset{5}{C}) \quad \text{(B)}$$

Step 2. Calculation of the shortest interatomic distances d(i,j) (expressed as number of bonds) between each couple of atoms i and j.

d(1, 1) = 0	d(1, 2) = 1	d(1, 3) = 2	d(1, 4) = 3	d(1, 5) = 3
	d(2, 2) = 0	d(2, 3) = 1	d(2, 4) = 2	d(2, 5) = 2
		d(3, 3) = 0	d(3, 4) = 1	d(3, 5) = 1
			d(4, 4) = 0	d(4, 5) = 2
				d(5, 5) = 0

Step 3. Selection of an atomic property G and calculation of the autocorrelation vector AV = $(C_0, C_1, C_2, ..., C_n)$ corresponding to a given property and for which the components C of order 0, 1, 2, ..., n are calculated by means of the following equation:

$$C_n = \Sigma\, g(i) \times g(j)$$

where g(i) and g(j) are the contributions attributed to atoms i and j.

If connectivity (i.e.; number of neighbors of each atom) is chosen as atomic property then we obtain:

$$g(1) = 1,\ g(2) = 2,\ g(3) = 3,\ g(4) = 1, \text{ and } g(5) = 1$$

Step 4. Calculation of the components Ti of the vector T encoding the connectivity.

$T_0 = (1 \times 1) + (2 \times 2) + (3 \times 3) + (1 \times 1) + (1 \times 1) = 16$
$T_1 = (1 \times 2) + (2 \times 3) + (3 \times 1) + (3 \times 1) = 14$
$T_2 = (1 \times 3) + (2 \times 1) + (2 \times 1) + (1 \times 1) = 8$
$T_3 = (1 \times 1) + (1 \times 1) = 2$

For isopentane, higher order components equal zero.

$$T = (16, 14, 8, 2)$$

Figure 2 Principle of the autocorrelation method.

contributions encoding log P of molecules. An example of coding is shown in Figure 3 for 1,2-dinitrobenzene (no. 41, Table I). The different contributions were used to compute the first 14 components of the autocorrelation vector H encoding the hydrophobicity of chemicals. Calculations were performed by means of an algorithm using a modified autocorrelation function which calculates the first component of the autocorrelation vector

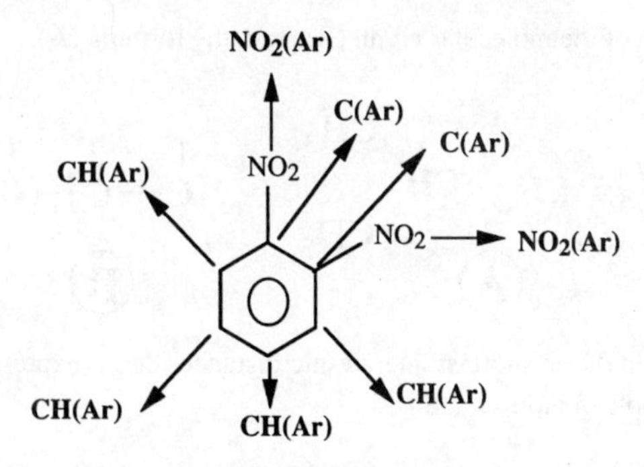

Figure 3 Autocorrelation description of 1,2-dinitrobenzene. Coding of this molecule requires the use of the three following atomic or group contributions: CH(Ar), C(Ar), and NO_2(Ar) where Ar stands for aromatic.

by simply summing the positive and negative contributions attributed to the atoms and functional groups constituting a molecule (Devillers *et al.*, 1992). In addition, the calculation procedure takes into account the signs of the different contributions in order to increase the physicochemical significance of the descriptors.

STATISTICS

A normed principal components analysis (zero mean and unit variance) was used to scale the input data represented by the first 14 components of the autocorrelation vector encoding the hydrophobicity of chemicals. The chemicals of the testing sets were treated as supplementary individuals. For the training set (Table I) and the first testing set (Table II), the biodegradation data were scaled using the following equation:

$$x'_i = [a(x_i - x_{min})/(x_{max} - x_{min})] + b \qquad (1)$$

In Eq. (1), x'_i is the scaled value, x_i the original value, x_{min} and x_{max} are respectively the minimum and maximum values in each column of the data matrix. Due to the fact that the biodegradation scores are necessarily comprised between 1 and 5, these values were taken as x_{min} and x_{max}, respectively (even if the actual minimum and maximum values were different). The constants a and b were taken as 0.9 and 0.05, respectively.

A three-layer BNN including biases was used. Numerous trials were performed to determine the optimal set of parameters for the network (i.e., number of principal components, number of hidden neurons, learning rate (η), momentum term (α), number of learning cycles). For this purpose, the STATQSAR (1994) package was used. The program automatically tests several times all the possible combinations of the values of the above parameters.

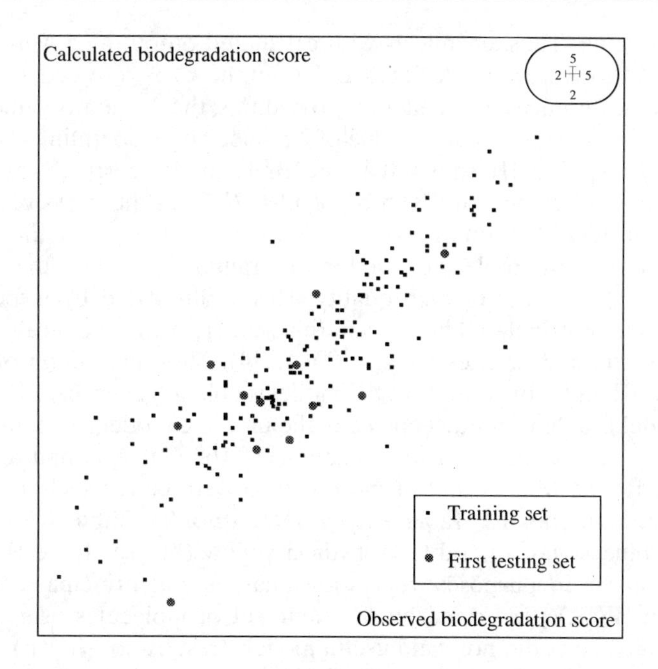

Figure 4 Observed versus calculated biodegradation scores for the training and first testing sets.

MODELING RESULTS

The best results are obtained using the nine first principal components (accounting for 96.9 per cent of the total variance) as inputs and nine neurons in the hidden layer (9/9/1 BNN, $\eta = 0.2$, $\alpha = 0.9$, and <2000 cycles). As regards the training set (Table I), it is noteworthy that the correlation between observed and calculated outputs is satisfactory ($r^2 = 0.76$) if we consider the complexity of the studied activity. Inspection of the residuals (Table I) confirms the high quality of the BNN model. Indeed, only two molecules (i.e., picloram (no. 81, Table I), dinoseb (no. 108, Table I)) have residuals greater than 0.8 whereas 94.8 per cent of the chemicals have residuals lower than 0.5 (absolute value). The good results obtained for the testing sets (Tables II and III) reveal the high predictive power of the BNN model. Thus, for the first testing set (Table II), all absolute residual values except one (0.52) are lower than 0.5. For the second testing set (Table III), as a first step it was necessary for purposes of comparison to transform the numerical outputs calculated by the BNN model into two classes encoding slow (BS) and rapid (BR) biodegradability. Due to the recommendations given in Boethling *et al.* (1994) and the calculation procedure used to compute the scores (see biodegradation data section), 3.5 was selected as the cutoff value. Thus, chemicals with calculated output ≥3.5 were classified as BS, and those with a calculated BNN output <3.5 were coded BR. Note that due to differences in calculation procedures, the 3.5 cutoff value adopted here, corresponds to the cutoff value of 2.5 selected in Boethling *et al.* (1994). Inspection of Table III shows that the BNN model correctly predicts the biodegradation behavior for 82.5 per cent (47/57) of the molecules.

If we consider the complexity of the studied activity and the structural heterogeneity of the molecules used in the training set, it becomes clear that the proposed model is of high quality. This is illustrated by inspecting the observed versus calculated biodegradation scores for the chemicals belonging to the training and first testing sets (Figure 4). Thus, inspection of Figure 4 and Tables II and III reveals that the model is able to generalize. The selected BNN model is able to predict correctly the ultimate biodegradation of simple organic molecules such as 1-iodohexane (no. 168, Table I) and aniline (no. 36, Table III) as well as that of more complex structures such as pesticides (e.g., sumithion (no. 12, Table I), *p,p'*-DDE (no. 23, Table III)) and dyes (e.g., vat blue 4 (no. 15, Table I), hydron yellow (no. 10, Table III)).

For comparison purposes, regression analysis and principal components regression (PCR) performed on the same set of molecules using the autocorrelation vectors did not yield useful models (results not shown). However, it has been shown (Boethling *et al.*, 1994) that a regression model using the same database but developed from 36 different structural descriptors and molecular weight produced significant results ($n = 200$, $r^2 = 0.72$). Note that, according to Boethling *et al.* (1994), the molecular weight of the chemicals was included as an independent variable in order to make predictions possible for all organic chemicals even if they do not contain any of the 36 structural fragments. On the basis of the results obtained in the two studies, it appears that the BNN model compares favorably with that obtained from regression analysis. We have shown that our BNN model is able to generalize, while the predictive power of the regression model proposed by Boethling *et al.* (1994) has not been formally tested. A criticism of their model is that it contains several descriptors that have negligible frequencies (e.g., 0.5 per cent) in the dataset and this could considerably influence its predictive power (Devillers *et al.*, 1989).

Recently, we showed that the usefulness of autocorrelation vectors encoding the molar refractivity (MR) of molecules for modeling the toxicity of chemicals in the Microtox™ test (Devillers *et al.*, 1995). MR, which can be viewed as both an electronic and steric molecular descriptor, could be a powerful parameter for modeling biodegradation. Therefore, the aim of our future research will be to test this autocorrelation descriptor on the same dataset. We similarly intend to apply this methodology to develop model(s) for the scores encoding primary biodegradation of the 184 molecules.

Acknowledgment

The work presented in this paper was supported by a contract from the French Ministry of the Environment.

REFERENCES

Alexander, M. (1994). *Biodegradation and Bioremediation.* Academic Press, San Diego, p. 302.

Block, J.C., Férard, J.F., Flambeau, J.P., Kuenemann, P., Thouand, G., and Vasseur, P. (1990). Aspects microbiologiques de la dégradation: Une revue. In, *Evaluation de la Dégradation des Substances Organiques dans l'Environnement* (Séfa, Ed.). Paris, pp. 25–73.

Boethling, R.S., Howard, P.H., Meylan, W., Stiteler, W., Beauman, J., and Tirado, N. (1994). Group contribution method for predicting probability and rate of aerobic biodegradation. *Environ. Sci. Technol.* **28,** 459–465.

Boethling, R.S. and Sabljic, A. (1989). Screening-level model for aerobic biodegradability based on a survey of expert knowledge. *Environ. Sci. Technol.* **23,** 672–679.

Bos, M., Bos, A., and van der Linden, W.E. (1993). Data processing by neural networks in quantitative chemical analysis. *Analyst* **118,** 323–328.

Broto, P. and Devillers, J. (1990). Autocorrelation of properties distributed on molecular graphs. In, *Practical Applications of Quantitative Structure–Activity Relationships (QSAR) in Environmental Chemistry and Toxicology* (W. Karcher and J. Devillers, Eds.). Kluwer Academic Publishers, Dordrecht, pp. 105–127.

Cambon, B. and Devillers, J. (1993). New trends in structure–biodegradability relationships. *Quant. Struct.–Act. Relat.* **12,** 49–56.

Degner, P., Nendza, M., and Klein, W. (1991). Predictive QSAR models for estimating biodegradation of aromatic compounds. In, *QSAR in Environmental Toxicology–IV* (J.L.M. Hermens and A. Opperhuizen, Eds.). Elsevier, Amsterdam, pp. 253–259.

de Saint Laumer, J.Y., Chastrette, M., and Devillers, J. (1991). Multilayer neural networks applied to structure–activity relationships. In, *Applied Multivariate Analysis in SAR and Environmental Studies* (J. Devillers and W. Karcher, Eds.). Kluwer Academic Publishers, Dordrecht, pp. 479–521.

Devillers, J. (1993). Neural modelling of the biodegradability of benzene derivatives. *SAR QSAR Environ. Res.* **1,** 161–167.

Devillers, J., Bintein, S., Domine, D., and Karcher, W. (1995). A general QSAR model for predicting the toxicity of organic chemicals to luminescent bacteria (Microtox® test). *SAR QSAR Environ. Res.* **4,** 29–38.

Devillers, J., Domine, D., and Chastrette, M. (1992). A new method of computing the octanol/water partition coefficient. In, *Proceedings of QSAR92.* The University of Minnesota Duluth, Duluth, p. 12.

Devillers, J., Zakarya, D., Chastrette, M., and Doré, J.C. (1989). The stochastic regression analysis as a tool in ecotoxicological QSAR studies. *Biomed. Environ. Sci.* **2,** 385–393.

Domine, D., Devillers, J., Chastrette, M., and Karcher, W. (1993). Estimating pesticide field half-lives from a backpropagation neural network. *SAR QSAR Environ. Res.* **1,** 211–219.

Eberhart, R.C. and Dobbins, R.W. (1990). *Neural Network PC Tools. A Practical Guide.* Academic Press, San Diego, p. 414.

Enslein, K., Tomb, M.E., and Lander, T.R. (1984). Structure–activity models of biological oxygen demand. In, *QSAR in Environmental Toxicology* (K.L.E. Kaiser, Ed.). D. Reidel Publishing Company, Dordrecht, pp. 89–109.

Gombar, V.K. and Enslein, K. (1991). A structure–biodegradability relationship model by discriminant analysis. In, *Applied Multivariate Analysis in SAR and Environmental Studies* (J. Devillers and W. Karcher, Eds.). Kluwer Academic Publishers, Dordrecht, pp. 377–414.

Kuenemann, P., Vasseur, P., and Devillers, J. (1990). Structure-biodegradability relationships. In, *Practical Applications of Quantitative Structure–Activity Relationships (QSAR) in Environmental Chemistry and Toxicology* (W. Karcher and J. Devillers, Eds.). Kluwer Academic Publishers, Dordrecht, pp. 343–370.

Mani, S.V., Connell, D.W., and Braddock, R.D. (1991). Structure activity relationships for the prediction of biodegradability of environmental pollutants. *Critic. Rev. Environ. Control* **21,** 217–236.

Niemi, G.J., Veith, G.D., Regal, R.R., and Vaishnav, D.D. (1987). Structural features associated with degradable and persistent chemicals. *Environ. Toxicol. Chem.* **6,** 515–527.

Nys, G.G. and Rekker, R.F. (1974). The concept of hydrophobic fragmental constants (*f*-values). II. Extension of its applicability to the calculation of lipophilicities of aromatic and heteroaromatic structures. *Eur. J. Med. Chem. – Chim. Ther.* **9,** 361–375.

Pitter, P. and Chudoba, J. (1990). *Biodegradability of Organic Substances in the Aquatic Environment.* CRC Press, Boca Raton, p. 306.

Rekker, R.F. and de Kort, H.M. (1979). The hydrophobic fragmental constant; an extension to a 1000 data point set. *Eur. J. Med. Chem. – Chim. Ther.* **14,** 479–488.

Rekker, R.F. and Mannhold, R. (1992). *Calculation of Drug Lipophilicity. The Hydrophobic Fragmental Constant Approach.* VCH, Weinheim, p. 112.

Rekker, R.F., ter Laak, A.M., and Mannhold, R. (1993). On the reliability of calculated log P-values: Rekker, Hansch/Leo and Suzuki approach. *Quant. Struct.–Act. Relat.* **12,** 152–157.

Schüürmann, G. and Müller, E. (1994). Back-propagation neural networks – recognition vs. prediction capability. *Environ. Toxicol. Chem.* **13,** 743–747.

STATQSAR Package (1994). CTIS, Lyon, France.

Vasseur, P., Kuenemann, P., and Devillers, J. (1993). Quantitative structure–biodegradability relationships for predictive purposes. In, *Chemical Exposure Predictions* (D. Calamari, Ed.). Lewis Publishers, Boca Raton, pp. 47–61.

4 Structure–Bell-Pepper Odor Relationships for Pyrazines and Pyridines Using Neural Networks

M. CHASTRETTE* and C. EL AIDI
Laboratoire de Chimie Organique Physique et Synthétique, URA 463, Université Claude Bernard-Lyon I, 69622 Villeurbanne CEDEX, France

Structure–green bell-pepper odor relationships were established by means of neural networks (NN) on a set of 104 pyrazines and 11 pyridines. Each molecule was coded by describing its four substituents located on a common ring. Substituents were coded using Charton's steric hindrance descriptor (one substituent also had an electronegativity descriptor). Odor was coded with a binary variable. The neural network trained on 84 pyrazines gave an excellent classification rate (97.6%) that was slightly better than that obtained by discriminant analysis (92.9%). To test its reliability, training was made on nine learning subsets (LS) of 74–75 pyrazines, with a mean correct-classification rate of 97.6%. For the nine corresponding test subsets (TS) of 10 or 9 molecules, a correct-prediction rate of 97.6% was obtained for the 84 compounds. A further test was made using 20 new pyrazines bearing substituents for which Charton's upsilon had to be estimated. A correct-prediction rate of 75% (85% using an electronegativity descriptor) was obtained. All pyridines were correctly predicted using the same network. Finally, the contribution of each descriptor was evaluated. The results show structure–green bell-pepper odor relationships for currently known pyrazines and validate a previously proposed interaction model involving interactions with the receptor through hydrogen bonding.

KEY WORDS: *SAR; bell-pepper odor; pyrazines; pyridines; neural networks.*

---

* Author to whom all correspondence should be addressed.

In, *Neural Networks in QSAR and Drug Design* (J. Devillers, Ed.)
Academic Press, London, 1996, pp. 83–96.
ISBN 0-12-213815-5

INTRODUCTION

Pyrazines, a class of heterocyclic nitrogen-containing compounds, have been reported as important flavor ingredients (Bramwell *et al.*, 1969; Maga and Sizer, 1973; Nursteen and Sheen, 1974; Murray and Whitfield, 1975; Takken *et al.*, 1975; Parliment *et al.*, 1977; Fors and Olofsson, 1986; Mihara and Masuda, 1988) and arouse great interest because of their high odor potency and characteristic aroma. Correlations have been established between their olfactory thresholds and their physico-chemical parameters, such as lipophilicity, aqueous solubility or polarity (Mozell and Jagodowicz, 1973, 1974; Fors and Olofsson, 1985; Laffort and Patte, 1987; Mihara and Masuda, 1987). Differences, both in quality and intensity, have been reported between the odors of enantiomers (Mihara *et al.*, 1990).

Some of these pyrazine derivatives exhibit a characteristic bell-pepper-like odor associated with extremely low olfactory thresholds. The relationships between the bell-pepper character or odor potency, and chemical structure have been explored by many authors. According to Beets (1978), an odor generally results from the activation of several types of olfactory receptor sites. However, the bell-pepper note, found in compounds selectively complexing odorant-binding proteins (Pelosi *et al.*, 1981, 1982; Bignetti *et al.*, 1988), seems to belong to the more specific olfactory notes where only one or a few receptor sites are preferentially involved, thus leading to easier structure–odor relationships.

Pittet and Hruza (1974) have suggested an osmophore pattern comprising an unsaturated nitrogen-containing heterocycle with an alkyl group in position 2 (preferably an isobutyl group) and a methoxy group in position 3 (Figure 1). Substitutions in other positions (especially in position 6) are unfavorable to the bell-pepper odor.

Buttery *et al.* (1976) proposed a more general structural pattern, based on the study of a number of very potent alkylthiazoles described with a bell-pepper-like odor. More recently, Masuda and Mihara (1988) studied the effects of the length of the alkyl chain on the odor of 3-substituted

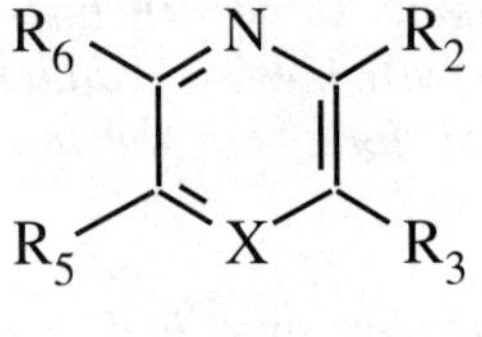

R_6 N R_2

R_5 X R_3

X = CH or N

R_i = H, alkyl, alkoxy, carbonyl ...

Figure 1 Underlying skeleton and substituents for the 115 substituted pyrazines or pyridines.

2-alkylpyrazines. They proposed a model of interaction between the pyrazines derivatives and the receptor sites consisting of three hydrogen bonds and a hydrophobic interaction corresponding to an *n*-pentyl group.

Rognon and Chastrette (1994) studied a large set of heterocylic compounds and proposed an interaction model based on the HBD theory which considers mainly hydrogen bonding and dispersion forces (Chastrette and Zakarya, 1988, 1992; Chastrette, 1990; Chastrette *et al.*, 1990, 1992b). This model accounts for both the quality and intensity of the odor.

In the present work, a nonlinear approach using a neural network (NN) was used to explore the relationships between chemical structure and green bell-pepper odor on a set of 115 pyrazines or pyridines containing bell-pepper compounds with known olfactory thresholds. Several applications of neural networks in structure–activity relationships have been developed (Aoyama and Ichikawa, 1991; Zupan and Gasteiger, 1991; So and Richards, 1992; Song and Yu, 1993; Song *et al.*, 1993; Villemin *et al.*, 1993). Chastrette and de Saint Laumer (1991) and Chastrette *et al.* (1992a) have modelled the structure–musk odor relationship of 79 nitrobenzene compounds using a multilayer backpropagation neural network. Chastrette *et al.* (1994) have also used the same method, with a three-layer neural network, to study a series of tetralins and indans. Each molecule was coded with steric hindrance and electronegativity descriptors of substituents on a common skeleton.

To be able to compare our results with those obtained with a linear approach, we also carried out a discriminant analysis (Lefevre, 1976; Dagnelie, 1984) on the same set. Thereafter, we sought to measure the contribution of each descriptor to the classification.

MATERIALS AND METHODS

Materials

The set comprised 115 compounds (Table I), 47 of which were described as green bell-pepper. The structure and olfactory properties of these compounds have already been described (Rognon and Chastrette, 1994, and references cited therein).

The odor was coded using a binary variable: absence or presence, without indication of intensity, although olfactory thresholds were used to estimate the importance of classification errors.

Description of the structure

The molecules were described from a common underlying skeleton comprising a pyrazine ring with four substituents in positions 2, 3, 5, and 6 (Figure 1).

Each substituent was described using one steric hindrance descriptor (with the exception of the substituent in position 3 also described by an

Table I *Substitution patterns and odor coding of pyrazines and pyridines.*

No.	R2	R3	R5	R6	Odor	Ref
1	H	H	H	H	0.2	a,d,s
2	H	OMe	H	H	0.2	a,s
3	Me	OMe	H	H	0.2	d,j,s
4	H	SMe	H	H	0.2	a
5	H	$N(Me)_2$	Me	H	0.2	b
6	Me	H	H	H	0.2	d,q,s
7	Me	Me	H	H	0.2	c,j,s
8	Me	H	Me	H	0.2	d,k,s
9	Me	H	H	Me	0.2	d,k,s
10	Me	Me	Me	H	0.2	b,d,q
11	Me	Me	Me	Me	0.2	d,e,f
12	Me	OMe	H	Me	0.2	b,f,g,v
13	Me	H	OMe	H	0.2	d,h,k
14	Me	H	H	OMe	0.2	i,k
15	Me	OEt	H	H	0.2	a,d,j
16	Me	H	OEt	H	0.2	k
17	Me	H	H	OEt	0.2	d,k,l
18	Me	OBu	H	H	0.2	m
19	Me	SMe	H	H	0.2	a,h,j
20	Me	H	SMe	H	0.2	h,k
21	Me	H	H	SMe	0.2	k,n
22	Me	SEt	H	H	0.2	a
23	Me	COMe	H	H	0.2	k,o
24	Me	H	COMe	H	0.2	p
25	Me	H	H	Ac	0.2	k,p,q
26	Me	NHMe	H	H	0.2	b
27	Me	$N(Me)_2$	H	H	0.2	b
28	Et	H	H	H	0.2	a,q,s
29	Et	Me	H	H	0.2	o,q,s
30	Et	H	Me	H	0.2	k,o,r,s
31	Et	H	H	Me	0.2	k,r
32	Et	Me	Me	H	0.2	e,q
33	Et	Me	H	Me	0.2	d,e,f
34	Et	H	Et	H	0.2	s
35	Et	H	H	Et	0.2	s
36	Et	OMe	H	H	0.2	a,d,m,s
37	Et	OEt	H	H	0.2	a,t
38	Et	SMe	H	H	0.2	a
39	Et	SEt	H	H	0.2	a
40	Pr	H	H	H	0.2	a
41	Pr	Me	H	H	0.8	k
42	Pr	OMe	H	H	0.8	a,d,f,m,s
43	Pr	OBu	H	H	0.2	m
44	Pr	SMe	H	H	0.8	k
45	iPr	Me	H	H	0.8	k,u
46	iPr	OMe	H	H	0.8	d,h,m
47	iPr	OMe	Me	H	0.8	b
48	iPr	OMe	H	Me	0.8	b
49	iPr	OMe	Me	OMe	0.8	b

Table I *continued.*

No.	R2	R3	R5	R6	Odor	Ref
50	iPr	OMe	OMe	Me	0.8	b
51	iPr	OMe	iPr	OMe	0.8	b
52	iPr	SMe	H	H	0.8	k
53	Bu	OMe	H	H	0.8	a,m
54	Bu	SEt	H	H	0.8	a
55	iBu	H	H	H	0.2	e,q,s
56	iBu	Me	H	H	0.8	b,d,s
57	iBu	OMe	H	H	0.8	b,d,f,g,h,s
58	iBu	OMe	Me	H	0.8	f,s,t
59	iBu	OMe	H	Me	0.8	f,s,t
60	iBu	OMe	Me	Me	0.8	f,s,t
61	iBu	OEt	H	H	0.8	m
62	iBu	SMe	H	H	0.8	b,m
63	iBu	$N(Me)_2$	H	H	0.2	b
64	Am	OMe	H	H	0.8	a
65	Am	OEt	H	H	0.8	a
66	Am	SMe	H	H	0.8	a
67	Am	SEt	H	H	0.8	a
68	$(CH_2)_2$–iPr	OMe	H	H	0.8	m
69	CH_2–sBu	OMe	H	H	0.8	k
70	CHMePr	OMe	H	H	0.8	m
71	$Vi(CH_2)_3$–	OMe	H	H	0.8	k
72	Hex	H	H	H	0.2	a
73	Hex	OMe	H	H	0.8	a,d,f,s
74	$(CH_2)_3$–iPr	OMe	H	H	0.8	k
75	Hept	OMe	H	H	0.8	a
76	Oct	OMe	H	H	0.8	a
77	Oct	OEt	H	H	0.8	a
78	Oct	SMe	H	H	0.8	a
79	Oct	SEt	H	H	0.8	a
80	Ac	H	H	H	0.2	b,d
81	Ac	OMe	H	H	0.2	b
82	Ac	OMe	Me	H	0.2	b
83	Ac	OMe	H	Me	0.2	b
84	Ac	OMe	Me	OMe	0.2	b
85	Me	OPh	H	H	0.2	a
86	Me	SPh	H	H	0.2	a
87	Pr	SPh	H	H	0.8	a
88	sBu	OMe	H	H	0.8	m
89	Am	OPh	H	H	0.2	a
90	Am	SPh	H	H	0.2	a
91	$(CH_2)_2CH{=}CHMe$	OMe	H	H	0.8	k
92	$CH_2CHMePr$	OMe	H	H	0.8	k
93	Oct	SPh	H	H	0.2	a
94	$(CH_2)_6$–iPr	OMe	H	H	0.8	a,m
95	$CH_2CHMeHex$	OMe	H	H	0.8	a,m
96	Dec	H	H	H	0.2	a
97	Dec	OMe	H	H	0.8	a
98	Dec	OEt	H	H	0.8	a

Table I *continued.*

No.	R2	R3	R5	R6	Odor	Ref
99	Dec	OPh	H	H	0.2	a
100	Dec	SMe	H	H	0.2	a
101	Dec	SEt	H	H	0.2	a
102	Dec	SPh	H	H	0.2	a
103	$C(Me)_2OH$	OMe	H	Me	0.8	b
104	$C(Me)_2OH$	OMe	Me	H	0.8	b
105*	H	Me	H	H	0.2	q
106*	Et	H	H	H	0.2	q
107*	Pr	H	H	H	0.2	q
108*	iPr	H	H	H	0.2	q
109*	iBu	H	H	H	0.2	q
110*	iBu	OMe	H	H	0.8	q
111*	iBu	OH	H	H	0.8	q
112*	OMe	iBu	H	H	0.2	t
113*	OMe	H	H	H	0.2	q
114*	OEt	H	H	H	0.2	q
115*	Ac	H	H	H	0.2	q

* Compounds with a pyridine skeleton.
a: Masuda and Mihara (1988). b: Takken *et al.* (1975). c: Mihara and Masuda (1987). d: Vernin (1979). e: Koehler *et al.* (1971). f: Maga and Sizer (1973). g: Seifert *et al.* (1970). h: Calabretta (1978). i: Nakel and Haynes (1972). j: Masuda and Mihara (1986). k: Mihara and Masuda (1988). l: Kung and Epstein (1974). m: Parliment and Epstein (1973). n: Kolor and Rizzo (1971). o: Coleman and Ho (1980). p: Roberts (1968). q: Pittet and Hruza (1974). r: Goldman *et al.* (1967). s: Teranishi *et al.* (1974). t: Seifert *et al.* (1972). u: Flament and Stoll (1967). v: Nursteen and Sheen (1974).

electronegativity descriptor). By convention, we decided that all substituents containing a heteroatom should occupy position 3. Even using this convention on position numbering, it was still possible to obtain two different descriptions for 13 molecules and three for one (i.e., for 84 pyrazines, 84 or 99 descriptions using respectively one or more input vector per compound). These descriptions can be related to different orientations of the molecule relative to the receptor. If only one of these orientations corresponds to a bell-pepper pattern, the receptor can recognize it, and therefore the particular molecule should be considered as possessing at least one bell-pepper pattern.

The substituent steric hindrance was described with effective upsilons υ_{ef} or, where impracticable, with the minimum values υ_{min} of Charton's upsilon (Charton, 1983). The electronic effects of substituent R_2 were described with Boyd and Boyd (1992) group electronegativity values and, where impracticable, with other group electronegativity scales (Bratsch, 1985; Xie *et al.*, 1995). The electronegativity of the substituents was adjusted by subtracting the hydrogen electronegativity value.

Table II *Charton's effective upsilon values representing the steric hindrance of different substituents.*

Substituent	Upsilon	Substituent	Upsilon	Substituent	Upsilon
H	0	Dec	0.68*	–NHMe	0.39
Me	0.52	MePrCH–	1.05	$–N(Me)_2$	0.43
Et	0.56	$MePrCHCH_2–$	0.71*	–SMe	0.64
Pr	0.68	$iPr(CH_2)_2–$	0.68	–SEt	0.94
iPr	0.76	$iPr(CH_2)_3–$	0.68	–SPh	1.8*
Bu	0.68	$iPr(CH_2)_6–$	0.68*	$–COCH_3$	0.5†
iBu	0.98	iPrCHMe	1.29	$–C(Me)_2OH$	0.76
sBu	1.02	$sBu–CH_2$	1.00	OH	0.32
Am	0.68	tBu–CHMe	2.11*	OMe	0.36
Hex	0.73	$CH_2CHMeHex$	0.99*	OEt	0.48
Hept	0.73	$(CH_2)_2CH=CHMe$	0.75*	OBu	0.58
Oct	0.68	$Vi(CH_2)_3–$	0.75	OPh	0.73*

* Our estimate.
† Minimum upsilon value given by Charton (1983).

Constitution of the sets

Charton's upsilon values were known for all the substituents of the 84 pyrazines defining set A. However, they were not available for seven substituents found in 20 other compounds and had to be estimated from values found for similar groups (Table II). These pyrazines constituted a test set designated B. Thus, the upsilon values of substituents $–(CH_2)_2CH=CHMe$ in molecule no. 91, $–CH_2CHMePr$ in molecule no. 92, $–(CH_2)_6$ iPr in molecule no. 94, $–CH_2CHMeHex$ in molecule no. 95 and $–C_{10}H_{21}$ in molecules nos. 96–102, were estimated at 0.75, 0.71, 0.68, 0.99 and 0.68, respectively. The υ_{ef} value of the OPh, SPh substituents in molecules 85–89, 90 and 93 were estimated at 0.73 and 1.80, respectively, from a comparison with similar groups. Their electronegativity values were estimated at 3.57 and 2.52, respectively.

Neural networks

A three-layer backpropagation NN was used to study the sets described above. The connection between the three layers was complete (Figure 2).

The number of neurons in the input layer was equal to the number of molecular descriptors whereas the output layer had only one neuron. The number of neurons in the hidden layer was determined by trial and error. The molecules/connections ratio (So and Richards, 1992) was kept between 2.3 and 3.4.

The activation sigmoid function was of the form: $f(x) = a/(1 + \exp(-b(x - c)))$, where a is the upper limit of the output values (0, a), b the non-linearity

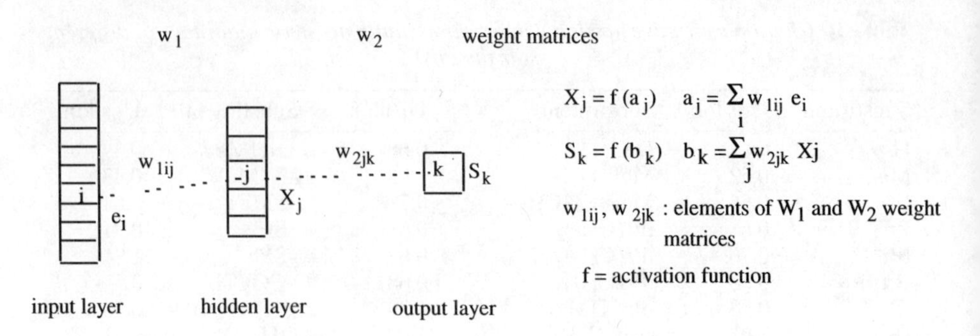

Figure 2 Structure of the multilayer network.

parameter and c the threshold value. The olfactory activity represented by the output of the network was coded 0.8 (or alternatively 0.9 and 1) if the molecule was bell-pepper and 0.2 (or alternatively 0.1 and 0) if the molecule was not bell-pepper. A bell-pepper compound was considered as correctly classified (or predicted) if the value of the output neuron corresponding to its activity was higher than a threshold value ($c = 0.5$).

In the learning (classification) phase, the error resulting from the comparison between the output computed by the network and the desired output was reduced by adjusting the weights inside the neural network through an iterative process of gradient backpropagation (Figure 2). First, we started by classifying 84 molecules (set A). The results reported were obtained after 5000 cycles.

Then for the prediction phase, we constituted nine learning subsets LS (three with 74 molecules and six with 75), and nine corresponding test subsets TS (three with 10 molecules and six with 9) were also constituted. The compounds of the TS sets were those withdrawn from set A for the purpose of constituting the LS sets. The network was successively trained on each of the LS, then used under the best configuration to predict the TS. Each molecule of set A was therefore predicted once.

Finally, a set B comprising 20 pyrazines, for which the values of the steric hindrance of some substituents were estimated, was used as test set. A set C of 11 substituted pyridines with known Charton's upsilon values was also used as a second test set.

RESULTS AND DISCUSSION

Classification

The best configuration obtained had 4 to 5 neurons on both the input and hidden layers, and only 1 on the output layer (i.e., 4-4-1 or 5-5-1 structures),

Table III *Results obtained in the classification and prediction phases for different sets.*

Set	A	LS	TS	B	C
Correct classification (4 desc)	97.6%*	97.6%*			
Correct classification (5 desc)	100%	97.6%*			
Correct prediction (4 desc)			97.6%*	75%†	100%
Correct prediction (5 desc)			97.6%*	85%‡	–

* Molecules nos. 43 and 63 were not correctly classified or predicted.
† Molecules nos. 48, 71, 85, 97, and 98 were not correctly predicted.
‡ Molecules nos. 87, 100, and 101 were not correctly predicted.

with the following activation function parameter values: a = 1, b = 0.3, and c = 0.5.

Set A

The classification phase yielded 97.6% (82/84) of correctly classified molecules (Table III). All compounds, with the exception of two non bell-pepper molecules (nos. 43 and 63) were correctly classified. Compound no. 63 with its *N,N*-dimethyl amino substituent has already been discussed in a previous work (Rognon and Chastrette, 1994). The same results were obtained when 84 or 99 descriptions were used for the 84 molecules of set A. The classification rate rose from 97.6% to 100% when a neural network with 5 neurons on the input layer (5-5-1) was used to ensure a better description by considering the electronegativity of substituent R_3.

The discriminant analysis of the 84 compounds of set A yielded a result of 92.9% (78/84) of correctly classified molecules. Other linear methods such as principal component analysis (PCA) or correspondence factor analysis (CFA) (Lefevre, 1976; Dagnelie, 1984) gave rather poor results.

Learning Subsets (LS)

The number of incorrectly classified molecules varied between one and two per subset of 74 or 75 pyrazines, that is a correct classification rate of 97.3% to 98.6% for all the nine LS subsets.

Prediction

Test Sets (TS)

The neural network with a 4-4-1 structure yielded a result of two incorrectly predicted molecules for the nine TS, that is 97.6% (82/84) of correctly predicted molecules (Table III). The 5-5-1 neural network yielded exactly the same result.

Sets B and C

Fifteen molecules of set B (75%) were correctly predicted using the 4-4-1 network (Table III). The 5-5-1 network gave a better result as 17 molecules

Table IV *Contribution of the different steric and electronic descriptors to the classification of pyrazines in set A.*

Descriptor *	% of misclassified molecules (4–4–1)	(5–5–1)
S_2	25	23
S_3	6	2.4
S_5	2.4	0
S_6	2.4	0
E_3	–	2.4

* S_2, S_3, S_5, S_6 correspond to the steric descriptors at the positions given in subscript; E_3 corresponds to the electronic descriptor of the subsituent R_3.

(85%) of B were correctly predicted. All the 11 molecules in set C were correctly predicted using the 4-4-1 network (Table III).

Contribution of the different descriptors

The contribution of each descriptor to the classification was estimated using the strategy outlined by Chastrette *et al.* (1994). All input values corresponding to the descriptor under evaluation were taken as zero. The results of the four (or five) computations made by successively removing one of the four steric hindrance descriptors (and electronic descriptor) show that the relative importance of the different descriptors varied in the order: S2 > S3 = E3 (Table IV).

CONCLUSION

We have shown in the present study that it is possible to model the structure–green bell-pepper odor relationship for a set of 115 molecules with substitutions on the four available positions on a pyrazine or pyridine ring.

Three-layer backpropagation NNs (4-4-1 or 5-5-1) were used to obtain a correct-classification rate of 97.6% (as against the 92.9% obtained by discriminant analysis) and a correct-prediction rate of 97.6% with subsets obtained through a leave-*n*-out procedure. With two sets of independent molecules the correct-prediction rate was 75 to 85% for 20 substituted pyrazines and 100% for 11 substituted pyridines. These results show that a very simple description of the substituents, such as the one presented here, which uses classical steric and electronic descriptors could provide enough information for the NN to learn structure–odor rules.

From the two compounds which were misclassified by the NN, one was already singled out in a classical molecular mechanics analysis and its behavior was explained. Furthermore, a quick analysis of the influence of substituents in pertinent positions confirms previous conclusions (Rognon and Chastrette, 1994). It suggests that the use of neural networks can be extended to the study of other sets of odorant compounds.

Acknowledgement

We thank the Fondation Roudnitska for the financial support to C. E. A.

REFERENCES

Aoyama, T. and Ichikawa, H. (1991). Neural network applied to pharmaceutical problems. V. Obtaining the correlation indexes between drugs activity and structural parameters using a neural network. *Chem. Pharm. Bull.* **39,** 372–378.

Beets, M.G.J. (1978). *Structure–Activity Relationships in Human Chemoreception.* Applied Science Publishers, London.

Bignetti, E., Cattaneo, P., Cavaggioni, A., Damiani, G., and Tirindelli, R. (1988). The pyrazine-binding protein and olfaction. *Comp. Biochem. Physiol.* **90 B,** 1–5.

Boyd, R.J. and Boyd, S.L. (1992). Group electonegativities from the bond critical point model. *J. Am. Chem. Soc.* **114,** 1652–1655.

Bramwell, A.F., Burrel, J.W.K., and Riezebos, G. (1969). Characterization of pyrazines in galbanum oil. *Tetrahedron Lett.* **37**, 3215–3216.

Bratsch, S.G. (1985). A group electronegativity method with Pauling units. *J. Chem. Educ.* **62,** 101–103.

Buttery, R.G., Guadagni, D.E., and Ludin, R.E. (1976). Some 4,5-dialkylthiazoles with potent bell-pepper-like aromas. *J. Agric. Food Chem.* **24,** 1–3.

Calabretta, P.J. (1978). Synthesis of some substituted pyrazines and their olfactive properties. *Perfum. Flavor* **3,** 33–34.

Charton, M. (1983). The upsilon steric parameter – Definition and determination. *Top. Curr. Chem.* **114,** 57–91.

Chastrette, M. (1990). Stimulus properties and binding to receptors. In, *NATO ASI Series, Chemosensory Information Processing*, Volume 39 (D.H. Schild, Ed.). NATO, pp. 97–107.

Chastrette, M. and de Saint Laumer, J.Y. (1991). Structure–odor relationships using neural networks. *Eur. J. Med. Chem.* **26,** 829–833.

Chastrette, M., de Saint Laumer, J.Y., and Peyraud, J.F. (1992a). Adapting the structure of a neural network to extract chemical information. Applications to structure–odour relationships. *SAR QSAR Environ. Res.* **1,** 221–231.

Chastrette, M., Rognon, C., Sauvegrain, P., and Amouroux, R. (1992b). On the role of chirality in structure–odor relationships. *Chem. Senses* **17,** 555–572.

Chastrette, M. and Zakarya, D. (1988). Sur le rôle de la liaison hydrogène dans l'interaction entre les récepteurs olfactifs et les molécules à odeur de musc. *C. R. Acad. Sci. Paris* **307(II),** 1185–1188.

Chastrette, M. and Zakarya, D. (1992). Molecular structure and smell. In, *The Human Sense of Smell* (D.G. Laing, R.L. Doty, and W. Breipohl, Eds.). Springer-Verlag, Heidelberg, pp. 77–92.

Chastrette, M., Zakarya, D., and Peyraud, J.F. (1994). Structure–musk odor relationships for tetralins and indans using neural networks. On the contribution of descriptors to the classification. *Eur. J. Med. Chem.* **29,** 343–348.

Chastrette, M., Zakarya, D., and Pierre, C. (1990). Structure–odor relationships in sandalwood: Search for an interaction model based on the concept of santalophore superpattern. *Eur. J. Med. Chem.* **24,** 433–440.

Coleman, E.C. and Ho, C.T. (1980). Chemistry of baked potato flavor. 1. Pyrazines and thiazoles identified in the volatile flavor of baked potato. *J. Agric. Food Chem.* **28,** 66–68.

Dagnelie, P. (1984). *Analyse Statistique à Plusieurs Variables*. Presses agronomiques de Gembloux, Gembloux, p. 167.

Flament, I. and Stoll, M. (1967). Pyrazines. I. Synthesis of 3-alkyl-2-methylpyrazine by condensation of ethylenediamine with 2,3-dioxoalkanes. *Helv. Chim. Acta* **50,** 1754–1758.

Fors, S.M. and Olofsson, B.K. (1985). Alkylpyrazines, volatiles formed in the Maillard reaction. I. Determination of odor detection thresholds and odor intensity functions by dynamic olfactometry. *Chem. Senses* **10,** 287–296.

Fors, S.M. and Olofsson, B.K. (1986). Alkylpyrazines, volatiles formed in the Maillard reaction. II. Sensory properties of five alkylpyrazines. *Chem. Senses* **11,** 65–67.

Goldman, I.M., Seibl, J., Flament, I., Gautshi, F., Winter, M., Whillhalm, B., and Stoll, M. (1967). Recherches sur les arômes. Sur l'arôme du café, II. Pyrazines et pyridines. *Helv. Chim. Acta* **50,** 694–705.

Koehler, P.E., Mason, M.E., and Odell, G.V. (1971). Odor threshold levels of pyrazine compounds and assessment of their role in the flavor of roasted foods. *J. Food Sci.* **36,** 816–818.

Kolor, M.G. and Rizzo, D.J. (1971). Electron-impact induced skeletal rearrangement of 2-methoxy-3-methylpyrazine and 2-(methylthio)-3-methylpyrazine. *Org. Mass. Spectrom.* **5,** 959–966.

Kung, J.F. and Epstein, M.F. (1974). U.S. Patent 3 803 331.

Laffort, P. and Patte, F. (1987). Solubility factors established by gas–liquid chromatography. A balance sheet. *J. Chromatogr.* **406,** 51–74.

Lefevre, J. (1976). *Introduction aux Analyses Statistiques Multidimensionnelles.* Masson, Paris, p. 137.

Maga, J.A. and Sizer, C.E. (1973). Pyrazines in foods. A review. *J. Agric. Food Chem.* **21,** 22–30.

Masuda, H. and Mihara, S. (1986). Synthesis of alkoxy-, (alkylthio-)-, Phenoxy-, and (Phenylthio) pyrazines and their olfactive properties. *J. Agric. Food Chem.* **34,** 377–381.

Masuda, H. and Mihara, S. (1988). Olfactive properties of alkylpyrazines and 3-substituted-2-alkylpyrazines. *J. Agric. Food Chem.* **36,** 584–587.

Mihara, S. and Masuda, H. (1987). Correlation between molecular structures and retention indexes of pyrazines. *J. Chromatogr.* **402,** 309–317.

Mihara, S. and Masuda, H. (1988). Structure–odor relationships for disubstituted pyrazines. *J. Agric. Food Chem.* **36,** 1242–1247.

Mihara, S., Masuda, H., Nishimura, O., and Tabeba, H. (1990). Determination of the enantiomeric composition of 2-methoxy-3-(1'-methylpropyl) pyrazine from galbanum oil using achiral and chiral lanthanide shift reagents. *J. Agric. Food Chem.* **38,** 465–467.

Mozell, M.M. and Jagodowicz, M. (1973). Chromatographic separation of odorants by the nose. Retention times measured across *in vivo* olfactory mucosa. *Science* **181,** 1247–1249.

Mozell, M.M. and Jagodowicz, M. (1974). Mechanisms underlying the analysis of odorant quality at the level of the olfactory mucosa. I. Spatiotemporal sorption pattern. *Ann. N.Y. Acad. Sci.* **237,** 76–90.

Murray, K.E. and Whitfield, F.B. (1975). Occurrence of 3-alkyl-2 methoxypyrazines in raw vegetables. *J. Sci. Food Agric.* **26,** 973–986.

Nakel, G.M. and Haynes, L.V. (1972). Structural identification of the methoxy-methylpyrazine isomers. *J. Agric. Food Chem.* **20,** 682–684.

Nursten, H.E. and Sheen, M.R.J. (1974). Volatile flavor components of cooked potato. *J. Sci. Food Agric.* **25,** 643–663.

Parliment, T.H. and Epstein, M.F. (1973). Organoleptic properties of some alkyl-substituted alkoxy- and alkylthiopyrazines. *J. Agric. Food Chem.* **21,** 714–716.

Parliment, T.H., Kolor, M.G., and Maing, Y.J. (1977). Identification of the major volatile components of cooked beets. *J. Food Sci.* **42,** 1592–1593.

Pelosi, P., Baldaccini, N.E., and Pisanelli, A.M. (1982). Identification of a specific olfactory receptor for 2-isobutyl-3-methoxypyrazine. *Biochem. J.* **201,** 245–248.

Pelosi, P., Pisanelli, A.M., Baldaccini, N.E., and Gagliardo, A. (1981). Binding of [^{3}H]-2-isobutyl-3-methoxypyrazine to cow olfactory mucosa. *Chem. Senses* **6,** 77–85.

Pittet, A.O. and Hruza, D.E. (1974). Comparative studies of flavor properties of thiazoles derivatives. *J. Agric. Food Chem.* **22,** 264–269.

Roberts, D.L. (1968). U.S. Patent 3 402 051.

Rognon, C. and Chastrette, M. (1994). Structure–odor relationships: A highly predictive tridimensional interaction model for the bell-pepper note. *Eur. J. Med. Chem.* **29,** 595–609.

Seifert, R.M., Buttery, R.G., Guadagni, D.G., Black, D.R., and Harris, J.G. (1970).

Synthesis of some 2-methoxy-3-alkylpyrazines with strong bell pepper-like odors. *J. Agric. Food Chem.* **8,** 246–249.

Seifert, R.M., Buttery, R.G., Guadagni, D.G., Black, D.R., and Harris, J.G. (1972). Synthesis and odor properties of some additional compounds related to 2-isobutyl-3-methoxypyrazine. *J. Agric. Food Chem.* **20,** 135–137.

So, S. and Richards, W.G. (1992). Applications of neural networks: Quantitative structure–activity relationships of the derivatives of 2,4-diamino-5-(substituted-benzyl)pyrimidines as DHFR inhibitors. *J. Med. Chem.* **35,** 3201–3207.

Song, X.H., Chen, Z., and Yu, R.Q. (1993). Artificial neural networks applied to odor classification for chemical compounds. *Comput. Chem.* **3,** 303–308.

Song, X.H. and Yu, R.Q. (1993). Artificial neural network applied to the quantitative structure–activity relationship study of dihydropteridine reductase inhibitors. *Chemom. Intell. Lab. Syst.* **19,** 101–109.

Takken, H.J., van der Lind, L.M., Boelens, M., and van Dort, J.M. (1975). Olfactive properties of a number of polysubstituted pyrazines. *J. Agric. Food Chem.* **23,** 638–642.

Teranishi, R., Buttery, R.G., and Guadagni, D.G. (1974). Odor quality and chemical structure in fruit and vegetable flavors. *Ann. N.Y. Acad. Sci.* **237,** 209–216

Vernin, G. (1979). Heterocycles in food aromas. I. Structure and organoleptic properties. *Parfums, Cosmet., Aromes* **27,** 77–86.

Villemin, D., Cherqaoui, D., and Cense, J.M. (1993). Neural network studies: Quantitative structure–activity relationship of mutagenic aromatic nitro compounds. *J. Chim. Phys.* **90,** 1505–1519.

Xie, Q., Sun, H., Xie, G., and Zhou, J. (1995). An iterative method for calculation of group electronegativities. *J. Chem. Inf. Comput. Sci.* **35,** 106–109.

Zupan, J. and Gasteiger, J. (1991). Neural networks: A new method for solving chemical problems or just a passing phase? *Anal. Chim. Acta* **248,** 1–30.

5 A Neural Structure–Odor Threshold Model for Chemicals of Environmental and Industrial Concern

J. DEVILLERS*, C. GUILLON, and D. DOMINE
CTIS, 21 rue de la Bannière, 69003 Lyon, France

The human sense of smell provides useful information about the chemicals in the environment. Its functions (e.g., pleasure, repulsion) are various and the phenomena governing the odor detection are multiple and complex. As it is impossible, in terms of both time and cost to measure all odor threshold values, there is a need for tools allowing prediction of these data. It is well recognized that besides physiological, anatomical, and psychological parameters, odor detection generally depends on the structure and physicochemical properties of the chemicals. Under these conditions, the aim of this study was to derive a model allowing one to predict air odor thresholds from the structure of the chemicals. For this purpose, a set of 173 chemicals, for which the air odor threshold values were available, was described by means of 18 structural fragments known to have an influence on the odor thresholds of chemicals. Due to the complexity and the multidimensional character of the studied activity, backpropagation neural networks (BNNs) with one or three output neurons were used to estimate the air odor threshold of organic molecules. Different scaling transformations were assayed. Best results were obtained with a new formula specifically designed for frequency data tables and a BNN containing three output neurons. The obtained model was tested on a heterogeneous set of 57 molecules. The BNN results outperformed those obtained with a discriminant analysis.

KEY WORDS: *odor threshold; backpropagation neural network; scaling transformation; discriminant analysis.*

* Author to whom all correspondence should be addressed.

In, *Neural Networks in QSAR and Drug Design* (J. Devillers, Ed.)
Academic Press, London, 1996, pp. 97–117.
ISBN 0-12-213815-5

INTRODUCTION

Chemical signals are indispensable for the survival of numerous invertebrates which use chemoreceptors to find their way, to hunt for and inspect food, to detect enemies and predators, and to find members of the opposite sex. Numerous agricultural and crop protection techniques are now based on the rational use of these chemical signals (Doré and Renou, 1985; Renou *et al.*, 1988a,b; Devillers and Domine, 1995; Domine and Devillers, 1995; Domine *et al.*, 1995; Petroski and Vaz, 1995). It is obvious that chemoreception also plays a key role in the life of vertebrates. Thus, for example, it has been shown that an abortion can appear in a mouse in contact with the odor of a non-familiar male. This syndrome, called the Bruce effect, is directly related to the functioning of the vomeronasal organ (Rajendren and Dominic, 1984).

Human behaviors are also affected by sensory information obtained from the interaction of molecules with olfactory and taste receptors. However, the importance of fragrance and flavor substances in humans has evolved to become quantitatively and qualitatively different from that in other mammals. This is firstly because humans depend to a greater extent on acoustic and optical signals for orientation. This is also due to the fact that olfaction differs fundamentally from vision and hearing. Indeed, while the latter senses can be easily described by means of physical parameters, the description and characterization of an odor is always difficult since we have to integrate ethnological components, take into account socio-cultural events and also pathologic situations. Thus, for example, it has been shown that human breast-fed neonates are able to rely on their olfactory ability for the detection and discrimination of the odor complex originating from their mother's breast areola (Schaal, 1988). Olfactory dysfunctions have been noted in Parkinson's disease (Ward *et al.*, 1983; Doty *et al.*, 1988), the Korsakoff syndrome (Potter and Butters, 1980) and Alzheimer's disease (Roberts, 1986; Warner *et al.*, 1986; Doty *et al.*, 1987; Serby, 1987).

Last, we can note that the use of fragrance and flavor substances in perfumery, as well as for the flavoring of foods and beverages, is primarily directed toward invoking pleasant and emotional sensations.

As regards human behaviors affected by sensory information, it is also interesting to stress that the human sense of smell can also be used to detect pollutants. Indeed, the human nose is highly sensitive to certain repulsive-smelling chemicals produced in trace amounts by numerous pathogenic and putrefying microorganisms which can contaminate the different compartments of the biosphere (Mouchet, 1978; Bouscaren, 1984). The human nose can serve as a first-line warning system to detect hazardous chemicals. Indeed, actual concentrations of organic chemicals in air can be sampled and analyzed by various techniques. However, the necessary equipment is often expensive, cumbersome and slow, and requires professional skills to operate and interpret. In contrast, the human nose is perfectly placed to sample the inspired air, monitors rapidly and continuously and may even exceed the sensitivity of the best instrument (Maës, 1990). However, it requires calibration. This calibration is very difficult because basically, the ability of the members of a population to detect and estimate the odor threshold value of a given chemical is strongly influenced by the innate variability of different persons' olfactory powers, their prior experience with that odor, the degree of attention they accord to the matter, and so on (Hoshika *et al.*, 1993). As a consequence, these important data are difficult to manipulate and model. The scope of this work is to show that neural networks can be useful tools to overcome this methodological problem.

MATERIALS AND METHODS

Literature search for odor threshold data

Odor threshold data are available for a large number of chemicals (Bouscaren, 1984). However, they are widely scattered in the literature and there is little conformity in the choice of units for expressing the results. Thus, for example, if we consider the 29 reported thresholds for *n*-butyl alcohol in Table I, we can note that they are gathered from the works of a large number of authors, who used many different systems of concentration units (Amoore and Hautala, 1983). Furthermore, no two of these 29 thresholds were measured by precisely the same experimental method. The lack of standardization, taken in conjunction with the inconsistent purity of the chemical samples and the variability of human sensitivity, is responsible for the rather wide range of threshold concentrations found for this chemical (Table I). This summarizes the situation that can be observed for all the organic chemicals. It is problematic since the first requirement when we want to elaborate a (quantitative) structure–activity relationship ((Q)SAR) model is to constitute a valuable training set with representative chemicals for which the biological activity has been measured under standardized conditions. In addition, when we want to use a backpropagation neural network (BNN) for deriving a (Q)SAR model, it is also required to constitute different testing sets (Devillers *et al.*, 1995). Therefore, secure odor threshold data for modeling purposes is always a difficult task. To overcome this problem, a classical data transformation widely used in air pollution modeling can be adopted (Amoore and Hautala, 1983). First of all, after data compilation, extreme values must be discarded when they widely diverge from the other threshold data found for the same chemical. In a second step, the original threshold data, in a variety of concentration units, must be converted into common units of grams per liter. Any water dilution thresholds have to be further converted to the equivalent air dilution threshold, through multiplication by the air–water partition coefficient. Last, the relationship between

Table I *Literature odor thresholds for* n-butyl alcohol *and calculation of the air odor threshold for this chemical (transformed from Amoore and Hautala, 1983).*

Medium	Dilution threshold original data	g l^{-1} (air)	First reference*
Air	1 µg l^{-1}	1.00×10^{-6}	Passy, 1892
Air	0.565×10^{-8} mol l^{-1}	4.18×10^{-7}	Backman, 1917
Air	0.000223 mg l^{-1}	2.23×10^{-7}	Jung, 1936
Air	$Act_{25} = 6 \times 10^{-6}$	1.61×10^{-7}	Gavaudan, 1948
Air	$Act_{37} = 7 \times 10^{-4}$	4.09×10^{-5}	Mullins, 1955
Water	0.005% (v/v)	1.45×10^{-5}	Moncrieff, 1957
Air	15 ppm (v/v)	4.56×10^{-5}	Scherberger, 1958
Water	1 mg l^{-1}	3.60×10^{-7}	Nazarenko, 1962
Water	1.00 ppm (w/v)	3.60×10^{-7}	Rosen, 1962
Water	2.5 ppm (v/v)	7.24×10^{-7}	Baker, 1963
Air	$Act_{25} = 5 \times 10^{-5}$	1.40×10^{-6}	Gavaudan, 1966
Air	33 mg m^{-3}	3.30×10^{-5}	May, 1966
Water	0.50 ppm (v/v)	1.45×10^{-7}	Flath, 1967
Air	1.10×10^{13} mol cc^{-1}	1.34×10^{-6}	Dravnieks, 1968
Air	1.2 mg m^{-3}	1.20×10^{-6}	Khachaturyan, 1969
Air	0.013 mg l^{-1}	1.30×10^{-5}	Corbitt, 1971
Air	$- \log_{10}$ M l^{-1} = 7.91	9.12×10^{-7}	Laffort, 1973
Air	0.30 ppm (v/v)	9.11×10^{-7}	Hellman, 1974
Air	3.16 ppm (v/v)	9.60×10^{-6}	Moskowitz, 1974
Air	62 ppm (v/v)	1.88×10^{-4}	Moskowitz, 1974
Water	2.0 mg kg^{-1}	7.20×10^{-7}	de Grunt, 1975
Water	3.6×10^{-4} M l^{-1}	9.61×10^{-6}	Hertz, 1975
Water	2.77 ppm (w/v)	9.97×10^{-7}	Lillard, 1975
Air	0.0231 mm Hg	9.23×10^{-5}	Piggott, 1975
Air	0.390 ppm (v/v)	1.18×10^{-6}	Dravnieks, 1976
Air	2.8×10^{-1} ppm (v/v)	8.50×10^{-7}	Williams, 1977
Water	6.5×10^{-3} g l^{-1}	2.34×10^{-6}	Amoore, 1978
Air	3.5 ppm (v/v)	1.06×10^{-5}	Laing, 1978
Air	$\log_2$ ppb = 10.42	4.15×10^{-6}	Punter, 1980

Geometric mean, air odor threshold = 2.54×10^{-6} g l^{-1} (N = 29)
= 2.54 mg m^{-3}
= 0.84 ppm (v/v)
Air odor threshold (v/v) = $2.54 \times 24.4/74.1 = 0.84$

*Cited in Amoore and Hautala (1983).

odor-intensity sensation and odorant being exponential, in order to preserve the normal distributions of olfactory-threshold measurements, all concentrations of odorants must be calculated in a logarithmic scale. The above methodology was applied by Amoore and Hautala (1983) in order to produce a compilation of odor thresholds for more than 200 chemicals. The calculation procedure is exemplified for *n*-butyl alcohol in Table I. Note that the mean air odor threshold in grams per liter is converted to mg m^{-3} and then to ppm by volume in order to have the possibility to compare the molecules. From the valuable compilation of Amoore and Hautala (1983), a training set of

173 organic chemicals was selected (Table II). Due to the general degree of noise of the odor threshold data, it was decided to convert them into the three following classes of activity:

- class 1: op ≤ -1.00
- class 2: $-1.00 <$ op ≤ 1.08
- class 3: $1.08 <$ op

where op equals – log (threshold in $\mu l\ l^{-1}$).

Thus, chemicals belonging to class 3 were the most odorous. In order to test the performances of the models obtained, a compilation of Devos *et al.* (1990) constituted of compatible data was used to design a testing set of 57 molecules (Table III).

Molecular descriptors

Numerous studies have underlined the importance of particular functional groups and structural fragments in the fragrant activity of the organic molecules (Bouscaren, 1984). Thus, for example, the influence of double bonds and some functional groups on the odor recognition threshold of some molecules is clearly underlined in Table IV.

From an analysis of the literature, 18 structural parameters were selected for describing the sets of molecules (Tables II and III). They are listed in Table V.

Statistical analysis

A classical three-layer BNN was used (Rumelhart *et al.*, 1986; Eberhart and Dobbins, 1990; Cambon and Devillers, 1993). The input layer contained 18 neurons corresponding to the 18 structural descriptors (Table V). The output layer was constituted of three neurons encoding the classes of activity (100, 010, and 001). A three-layer BNN with one single output neuron was also tested. In this case, the three classes of activity were coded 0, 0.5, and 1, respectively. Last, a hidden layer constituted of an adjustable number of neurons was used.

Two different scaling procedures were tested:

– a classical min/max transformation (Eq. (1)),

$$x'_i = [a(x_i - x_{min})/(x_{max} - x_{min})] + b \tag{1}$$

In Eq. (1), x'_i is the scaled value, x_i the original value, x_{min} and x_{max} are respectively the minimum and maximum values in each column of the data matrix, and a and b are constants allowing one to fix the limits of the interval for the scaled values. In the present study the values of a and b equaled 0.9 and 0.05, respectively, so that the scaled data ranged between 0.05 and 0.95.

Table II *Training set.*

No.	Name	Class of activity
1	acetaldehyde	3
2	acetic acid	2
3	acetic anhydride	2
4	acetone	1
5	acetonitrile	1
6	acetylene	1
7	acrolein	2
8	acrylic acid	2
9	acrylonitrile	1
10	allyl alcohol	2
11	allyl chloride	2
12	*n*-amyl acetate	3
13	*sec*-amyl acetate	3
14	aniline	2
15	benzene	1
16	benzyl chloride	3
17	biphenyl	3
18	bromoform	2
19	1,3-butadiene	2
20	butane	1
21	2-butoxyethanol	2
22	*n*-butyl acetate	2
23	*n*-butyl acrylate	3
24	*n*-butyl alcohol	2
25	*sec*-butyl alcohol	2
26	*tert*-butyl alcohol	1
27	*n*-butylamine	2
28	*n*-butyl lactate	2
29	*n*-butyl mercaptan	3
30	*p*-*tert*-butyltoluene	2
31	camphor	2
32	α-chloroacetophenone	3
33	chlorobenzene	2
34	chlorobromomethane	1
35	chloroform	1
36	chloropicrin	2
37	β-chloroprene	1
38	*o*-chlorotoluene	2
39	*m*-cresol	3
40	*trans*-crotonaldehyde	2
41	cumene	2
42	cyclohexane	1
43	cyclohexanol	2
44	cyclohexanone	2
45	cyclohexene	2
46	cyclohexylamine	2
47	cyclopentadiene	2
48	diacetone alcohol	2
49	*o*-dichlorobenzene	2

Table II *continued.*

No.	Name	Class of activity
50	*p*-dichlorobenzene	2
51	*trans*-1,2-dichloroethylene	1
52	β,β'-dichloroethyl ether	3
53	dicyclopentadiene	3
54	diethanolamine	2
55	diethylamine	2
56	diethylaminoethanol	3
57	diisobutyl ketone	2
58	diisopropylamine	2
59	*N,N*-dimethylacetamide	1
60	dimethylamine	2
61	*N,N*-dimethylaniline	3
62	*N,N*-dimethylformamide	2
63	1,1-dimethylhydrazine	2
64	1,4-dioxane	1
65	epichlorhydrin	2
66	ethanolamine	2
67	2-ethoxyethanol	2
68	2-ethoxyethyl acetate	3
69	ethyl acetate	2
70	ethyl acrylate	3
71	ethyl alcohol	1
72	ethylamine	2
73	ethyl *n*-amyl ketone	2
74	ethyl benzene	2
75	ethyl bromide	2
76	ethyl chloride	2
77	ethylene	1
78	ethylenediamine	2
79	ethylene dichloride	1
80	ethylene oxide	1
81	ethylenimine	2
82	ethyl ether	2
83	ethyl formate	1
84	ethylidene norbornene	3
85	ethyl mercaptan	3
86	*N*-ethylmorpholine	2
87	formaldehyde	2
88	formic acid	1
89	furfural	3
90	furfuryl alcohol	2
91	halothane	1
92	heptane	1
93	hexachlorocyclopentadiene	3
94	hexachloroethane	2
95	hexane	1
96	hexylene glycol	1
97	indene	3
98	isoamyl acetate	3

Table II *continued.*

No.	Name	Class of activity
99	isoamyl alcohol	3
100	isobutyl acetate	2
101	isobutyl alcohol	2
102	isophorone	2
103	isopropyl acetate	2
104	isopropyl alcohol	1
105	isopropyl ether	3
106	maleic anhydride	2
107	mesityl oxide	2
108	2-methoxyethanol	2
109	methyl acetate	2
110	methyl acrylonitrile	2
111	methyl alcohol	1
112	methylamine	2
113	methyl *n*-amyl ketone	2
114	*N*-methylaniline	2
115	methyl *n*-butyl ketone	3
116	methyl chloroform	1
117	methyl 2-cyanoacrylate	2
118	methylcyclohexane	1
119	*cis*-3-methylcyclohexanol	1
120	methylene chloride	1
121	methyl ethyl ketone	2
122	methyl formate	1
123	methyl hydrazine	2
124	methyl isoamyl ketone	3
125	methyl isobutyl carbinol	3
126	methyl isobutyl ketone	2
127	methyl isocyanate	2
128	methyl isopropyl ketone	2
129	methyl mercaptan	3
130	methyl methacrylate	2
131	morpholine	3
132	naphthalene	2
133	nitrobenzene	3
134	nitroethane	2
135	nitromethane	2
136	1-nitropropane	1
137	2-nitropropane	1
138	*m*-nitrotoluene	3
139	nonane	1
140	octane	1
141	pentane	1
142	perchloroethylene	1
143	phenol	3
144	phenyl ether	3
145	phenyl mercaptan	3
146	phthalic anhydride	3
147	propionic acid	2

Table II *continued.*

No.	Name	Class of activity
148	*n*-propyl acetate	2
149	*n*-propyl alcohol	2
150	propylene	1
151	propylene dichloride	2
152	propylene glycol 1-methyl ether	1
153	propylene oxide	1
154	*n*-propyl nitrate	1
155	pyridine	2
156	quinone	2
157	styrene	2
158	1,1,2,2-tetrachloroethane	2
159	tetrahydrofuran	2
160	toluene	2
161	toluene-2,4-diisocyanate	2
162	*o*-toluidine	2
163	1,2,4-trichlorobenzene	2
164	trichloroethylene	1
165	trichlorofluoromethane	2
166	1,1,2-trichloro-1,2,2-trifluoroethane	1
167	triethylamine	2
168	1,3,5-trimethylbenzene	2
169	*n*-valeraldehyde	3
170	vinyl chloride	1
171	vinylidene chloride	1
172	*m*-xylene	2
173	2,4-xylidine	3

Table III *Testing set.*

No.	Name	Class of activity
1	acetophenone	2
2	4-allyl-2-methoxyphenol	3
3	benzyl acetate	2
4	benzylthiotoluene	3
5	bromobenzene	2
6	bromoethane	2
7	2-butanethiol	3
8	1-butene	2
9	2-butene	2
10	2-butene-1-thiol	3
11	butoxybutane	3
12	1-*tert*-butyl-3,5-dimethyl-2,4,6-trinitrobenzene	3
13	butylthiobutane	3
14	carbonyl chloride	2
15	4-chloroaniline	2
16	cyclopentanol	1

Table III *continued.*

No.	Name	Class of activity
17	cyclopentanone	2
18	decanal	3
19	dibutylamine	2
20	2,4-diisocyanato-1-methylbenzene	2
21	1,2-dimethylbenzene	2
22	1,4-dimethylbenzene	2
23	3,3-dimethylbutan-2-one	2
24	2,4-dimethylpentane	1
25	2,3-dimethylphenol	3
26	1,3-dioxolane	1
27	1-dodecanethiol	3
28	1,2-ethanediamine	2
29	3-ethoxy-4-hydroxybenzaldehyde	3
30	diethyl succinate	3
31	*N*-ethylaniline	2
32	ethyldithioethane	3
33	ethyl 2-hydroxypropanoate	2
34	ethylthioethane	3
35	2-heptanone	2
36	4-hydroxy-3-methoxybenzaldehyde	3
37	2-isopropyl-5-methylphenol	3
38	5-isopropyl-2-methylphenol	3
39	diisopropyl sulfide	3
40	limonene	2
41	3-methylaniline	2
42	4-methylaniline	2
43	3-methylbutane-1-thiol	3
44	2-methylbutane-2-thiol	3
45	2-methylpropane	1
46	phenyl acetate	2
47	α-pinene	2
48	propane	1
49	benzaldehyde	3
50	2-chlorophenol	3
51	4-chlorophenol	3
52	1-decanol	3
53	2,4-dimethylphenol	3
54	2,5-dimethylphenol	3
55	2,6-dimethylphenol	3
56	3,4-dimethylphenol	3
57	3,5-dimethylphenol	3

Table IV *Influence of double bonds and functional groups on the odor recognition threshold of organic chemicals. The data were retrieved from Verschueren (1983).*

Molecule	100% Recognition threshold* (ppm)
ethane	1500
ethanol	6000
acetaldehyde	0.3
ethylmercaptan	0.002
ethene	800
propane	11000
propionaldehyde	0.08
propylmercaptan	0.0007
propene	80
acrolein	20
allylmercaptan	0.00005
butane	5000
isobutane	2
butyraldehyde	0.04
butylmercaptan	0.0008
1-butene	0.07
isobutene	0.6
crotonaldehyde	0.2
crotylmercaptan	0.000055
pentane	900
pentanol	10
2-pentanone	8
1-pentene	0.002
benzene	300
phenol	20
phenylmercaptan	0.0002

*Concentration at which 100% of the odor panel defined the odor as being representative of the odorant being studied (Verschueren, 1983).

– a new equation specially designed for frequency data (Eq. (2)),

$$x'_i = 1/\{1 + \exp\{[-4/(x_{max} - x_{min})] \times [x_i - ((x_{max} - x_{min})/2)]\}\} \quad (2)$$

A trial and error procedure was used for selecting the number of neurons on the hidden layer, the learning rate (η), the momentum term (α), and the number of learning cycles. For this, the STATQSAR package (STATQSAR, 1994) was employed.

In order to evaluate the performances of the BNN models with the two scaling procedures, different parameters were calculated. Indeed, for a given class of activity associated to an output neuron, four possible alternatives exist. These are true positive (TP), false positive (FP), true negative (TN), and false negative (FN). A TP response for an output neuron corresponds to the case in which the neuron is expected to give a positive response (i.e.,

Table V *List of 18 structural descriptors used to code the molecules.*

Total number of non-hydrogen atoms
Number of C
Number of double or triple bonds between 2 C
Number of S
Number of –N< and –N=
Number of C≡N
Number of NO_2 and NO_3
Number of OH
Number of COOH
Number of COOR
Number of C–O–C
Number of C=O
Number of F
Number of Cl
Number of Br
Cyclic/non cyclic (1/0)
Number of quaternary atoms
Number of tertiary atoms

expected output equals 1) and it actually gives the correct response (i.e., calculated output superior to the selected cutoff value). An FP case occurs when a neuron gives a positive calculated output while it is expected to give a negative one (i.e., expected output equals 0). A TN corresponds to the case where the target is negative and the calculated output is in agreement (i.e., inferior to the selected cutoff value). Last, in the FN case, the calculated output is negative while the expected output is positive.

From these four alternatives, the following metrics were calculated (Eberhart and Dobbins, 1990; Pattichis *et al.*, 1995).

$$\text{False positive rate (in \%): } \%FP = 100 \times FPs/(FPs + TNs) \quad (3)$$

$$\text{False negative rate (in \%): } \%FN = 100 \times FNs/(FNs + TPs) \quad (4)$$

$$\text{Sensitivity (in \%): } \%TP = 100 \times TPs/(FNs + TPs) \quad (5)$$

$$\text{Specificity (in \%): } \%TN = 100 \times TNs/(FPs + TNs) \quad (6)$$

For the three-output neuron BNNs, the above parameters were directly calculated for each output neuron and a total was computed. It is obvious that in some instances the calculated outputs can be undetermined. Indeed, in our case study, to count the total number of correct classifications, a TN had to produce a calculated output below 0.4 while a TP had to be above 0.6. When undetermined cases occurred, such as for example an output of 0.55, 0.48, 0.1 for an expected output 0, 1, 0, the output of neuron 1 is FP because it is above 0.4, that of neuron 2 is FN because it is below 0.6 and the last is TN because it is below 0.4. Note that we also tested the influence of other cutoff values (e.g., 0.5).

For the one-output neuron BNNs, the notion of TP, FP, TN, and FN was slightly changed since it was not possible, due to the three states (i.e., 0, 0.5, and 1) of the output neuron, to oppose positive and negative states. Thus, by means of equations similar to Eqs. (3)–(6), the parameters %FP*, %FN*, %TP*, and %TN* were calculated for each class of activity. The total for all the three classes was also computed. These parameters were not based on each neuron output as for the three-output neuron BNNs, but on each class of activity. Thus, for a given class of activity, an FP* output occurs when the calculated output corresponds to this given class while the expected output is different. An FN* is recorded when the calculated output is different from this expected. For example, if a chemical is actually in class 3 (i.e., expected output equals 1) and the BNN response corresponds to class 2 (i.e., calculated output between 0.33 and 0.66) then it is an FN* for class 3 and it is an FP* for class 2. For the determination of the classes of activities, the cutoff values were 0.33 and 0.66. Thus, a chemical was assigned to class 1 if its calculated output was inferior to 0.33, class 2 if its output was between 0.33 and 0.66, and class 3 if its output was larger than 0.66.

RESULTS AND DISCUSSION

Three output neurons

For each scaling procedure, the search for the best model with three output neurons was based on 1350 runs, for which the number of hidden neurons varied between 6 and 8. Due to the size of the training set, most of the runs were performed with 6 or 7 hidden neurons. Indeed, with 8 neurons in the hidden layer, the total number of connections in the BNNs (i.e., 179) was larger than the number of chemicals in the training set. When using Eq. (1), two different configurations provided interesting results. The former consisted of an 18/7/3 BNN ($\eta = 0.8$, $\alpha = 0.8$, 1154 cycles) and the latter was also represented by an 18/7/3 BNN ($\eta = 0.4$, $\alpha = 0.8$, 4946 cycles). The latter presented results of lesser quality during the training phase but yielded better performances during the testing phase. With Eq. (2), the best results were obtained with an 18/7/3 BNN ($\eta = 0.2$, $\alpha = 0.8$, 4985 cycles). Inspection of Table VI indicates that if the percentages of correct classifications are considered, the best results are obtained with Eq. (2) for both the training and testing sets. When using 0.4 and 0.6 as cutoff values, it is noteworthy that Eq. (2) yields fewer undetermined results (Table VI). If a cutoff value of 0.5 is chosen (results not shown), the results are unchanged except for the second model obtained with Eq. (1) for which we have only 9 undetermined chemicals in the training set versus 15 with 0.4 and 0.6 as cutoff values (Table VI) and 1 versus 2 in the testing set. The percentage of correct classifications for the training set reaches 88.4% in this case and that for the testing set is

Table VI *Summary of the results obtained with three output neurons.*

Scaling	% of correct classifications	Number of undetermined	Total			
			%FP	%FN	%TP	%TN
Eq. (1)						
Training set	90.2	7	3	10	90	97
Testing set	86.0	1	6	14	86	94
Training set	86.7	15	3	13	87	97
Testing set	91.2	2	4	7	93	96
Eq. (2)						
Training set	94.8	3	2	5	95	98
Testing set	93.0	0	4	7	93	96

Table VII *Detailed analysis of the responses of each output neuron for the results obtained with three output neurons.*

Scaling	Neuron 1				Neuron 2				Neuron 3			
	%FP	%FN	%TP	%TN	%FP	%FN	%TP	%TN	%FP	%FN	%TP	%TN
Eq. (1)												
Training set	2	9	91	98	11	4	96	89	0	25	75	100
Testing set	2	40	60	98	6	17	83	94	14	7	93	86
Training set	1	13	87	99	12	9	91	88	0	25	75	100
Testing set	0	60	40	100	6	4	96	94	11	0	100	89
Eq. (2)												
Training set	2	7	93	98	4	4	96	96	1	6	94	99
Testing set	0	40	60	100	6	4	96	94	7	3	97	93

unchanged. Analysis of the results by means of the four different calculated parameters (Table VI) shows that all the models present the same trends in the training and testing sets. Thus, the %TN is greater than the %TP and the %FN greater than the %FP. This seems to denote a tendency of the different models to give negative outputs preferentially (i.e., calculated output inferior to the cutoff value). Table VII allows one to depict the classification behavior of each model neuron by neuron. It underlines differences between results obtained for the training set and the testing set, but also between each output neuron. This has to be linked to the population of each class of activity for the training and testing sets. Indeed, we have to keep in mind for example, that there are only five chemicals belonging to class 1 in the testing set. Therefore, the obtained percentages must be analyzed with care. From Table VII, it also appears that the scaling procedures influence the performances and the classification behaviors of the models. Thus, for example, for neuron 3 in the training set for the models using Eq. (1), the %FN is very high. This denotes a tendency for these models to classify the most odorous chemicals incorrectly. They will tend to be assigned to class 1 or 2 or be undetermined. In contrast, chemicals belonging to class 1 or 2 will not be assigned to class 3 by the models since the %FP equals 0 and %TN equals 100. However, this tendency is not confirmed in the testing set. Last, from a general point of view, Eq. (2) generally provides lower %FN and %FP values for all neurons indicating its better overall performances. With Eq. (2), 94.8% of good classifications are reached (i.e., 164/173). Indeed, the only nine chemicals badly classified by the BNN model are acetaldehyde, benzene, benzyl chloride, *N,N*-dimethylacetamide, ethyl chloride, *cis*-3-methylcyclohexanol, naphthalene, propylene dichloride, and trichlorofluoromethane (chemicals no. 1, 15, 16, 59, 76, 119, 132, 151, and 165 in Table II). As regards the testing set, 93.0% of good predictions (i.e., 53/57) are obtained. The four molecules badly predicted by the BNN model are cyclopentanol, 2,4-dimethylpentane, limonene, and benzaldehyde (chemicals no. 16, 24, 40, and 49 in Table III). With Eq. (1), the incorrectly classified chemicals are generally the same. Thus, in the training set, for example, chemicals no. 1, 15, 16, 59, 76, and 119 are always misclassified or give undetermined outputs. Chemical no. 165 is incorrectly classified by the model using Eq. (2) and one of the models using Eq. (1). All of the above incorrect classifications can be explained by inspecting the other chemicals of the training set. Thus, for example, chemical no. 1 is the sole aldehyde containing fewer than 5 carbon atoms, which belongs to the third class of activity. Only chemicals no. 132 and 151 are misclassified when using Eq. (2) and well classified when using Eq. (1). In contrast, many other chemicals are misclassified by at least one of the models using Eq. (1) and correctly predicted with Eq. (2) (i.e., chemicals no. 4, 21, 25, 42, 43, 45, 47, 52, 70, 84, 90, 97, 99, 101, 104, 115, 118, 124, 131, 159, and 169 in Table II). For molecules no. 132 and 151, inspection of the chemical structures included in the training set allows one to find a rationale for their incorrect classification by the BNN, but it is not always possible for the chemicals only classified badly by the BNNs using Eq. (1).

As regards the testing set, there are only four chemicals whose class of activity is not correctly predicted with the BNN using Eq. (2) while the models using Eq. (1) do not allow one to predict the activity of eight and five chemicals, respectively. It is noteworthy that two molecules are never assigned the correct class of activity by the models (i.e., chemicals no. 16 and 24 in Table III). Molecule no. 40 is incorrectly predicted by the model using Eq. (2) and one of the models using Eq. (1). Compound no. 49 is the sole one for which only the BNN model using Eq. (2) does not yield the correct class of activity while both models derived from Eq. (1) give a correct simulation. Molecules no. 1, 11, 18, 19, 23, 45, and 47 have been found to be incorrectly predicted by at least one of the models using Eq. (1) while the BNN model derived after scaling by Eq. (2) correctly predicted their activity. For all the above molecules of the testing set, rationales for the incorrect predictions are less evident. Therefore, on the basis of the percentages of correct predictions in the testing set and also on the performances of the BNN during the training phase, it appears obvious that the BNN model derived with Eq. (2) compares favorably with those derived with Eq. (1).

One output neuron

In a BNN model, the use of a single output neuron instead of more outputs is always promising since it allows one to reduce the number of connections within the network. For comparison purposes, the search for the best model with one output neuron was also based on 1350 runs for which the number of hidden neurons varied between 6 and 8. When using Eq. (1), an 18/8/1 BNN ($\eta = 0.6$, $\alpha = 0.9$, 2373 cycles) was found to give the best results. With Eq. (2), two configurations gave interesting results. The former consisted of an 18/7/1 BNN ($\eta = 0.6$, $\alpha = 0.9$, 2431 cycles) and the latter, for which the training was of lesser quality but the testing provided better results, was obtained with an 18/6/1 BNN ($\eta = 0.5$, $\alpha = 0.9$, 4866 cycles). Tables VIII and IX report the parameters that were calculated to assess the results. If the percentages of correct classifications are considered (Table VIII), it is noteworthy that the results obtained with the linear min/max scaling procedure (Eq. (1)) compare favorably with those obtained when using Eq. (2), especially for the training set. However, inspection of the other parameters reported in Table VIII shows that all the models present the same tendencies. Thus, the %TN* is greater than the %TP* and the %FN* greater than the %FP*. Table IX, which details the classification behavior of each model class by class, underlines that the models give different results in the training and testing sets. Thus, for the training set, all the models produce low values of %FP* and %FN* and therefore high %TN* and %TP* values except in class 3 for the second model using Eq. (2). In contrast, as regards the testing set, the reported parameters indicate different trends in the prediction

Table VIII *Summary of the results obtained with one output neuron.*

Scaling	% of correct classifications	Total			
		%FP*	%FN*	%TP*	%TN*
Eq. (1)					
Training set	95.4	2	5	95	98
Testing set	87.7	6	12	88	94
Eq. (2)					
Training set	94.8	3	5	95	97
Testing set	86.0	7	14	86	93
Training set	93.6	3	6	94	97
Testing set	87.7	6	12	88	94

Table IX *Detailed analysis for each class of activity of the results obtained with one output neuron.*

Scaling	Class 1				Class 2				Class 3			
	%FP*	%FN*	%TP*	%TN*	%FP*	%FN*	%TP*	%TN*	%FP*	%FN*	%TP*	%TN*
Eq. (1)												
Training set	2	0	100	98	2	7	93	98	3	6	94	97
Testing set	4	0	100	96	3	26	74	97	14	3	97	86
Eq. (2)												
Training set	2	4	96	98	6	4	96	94	1	8	92	99
Testing set	8	20	80	92	3	22	78	97	11	7	93	89
Training set	3	2	98	97	7	5	95	93	1	14	86	99
Testing set	4	40	60	96	9	13	87	91	7	7	93	93

behavior obtained for each class. In addition, the scaling procedures influence the results. Indeed, for example, high values of %FN* are recorded for class 1 with only the models using Eq. (2) while for class 2, all the models give large %FN* values. For class 3, in the testing set, the %FP* value for Eq. (1) is relatively important.

Due to the larger percentages obtained in the training and testing sets (Table VIII), it seems that Eq. (1) gives the best results. A more detailed inspection of the chemicals incorrectly predicted by the different models, and the search for a rationale for these compounds do not generally yield definitive conclusions. We can therefore consider that the model using the linear min/max equation (Eq. (1)) should be chosen for its superiority in the total percentages of correct classifications. This model does not allow one to find the correct class of activity for molecules no. 1, 7, 11, 33, 38, 40, 97, and 121 in the training set (Table II) and molecules no. 3, 14, 19, 33, 40, 47, and 49 in the testing set (Table III).

Comparative analysis

From the above results, but also from the mean percentages of correct classifications obtained during the different runs, the study of the influence of different cutoff values, and the analysis of the effects of the few molecules of the testing set presenting the same coding as some molecules of the training set (results not shown), it appears that Eq. (1) gives better results with the BNNs having only one output neuron and Eq. (2) provides simulations of higher quality when three output neurons are used. The comparison of the best model obtained with one output neuron and that obtained with three output neurons shows that they present slightly different trends. Thus, the former is better during the training phase but the latter yields more correct predictions during the testing phase. If the sum of the incorrectly classified chemicals in the training and testing sets is made, this favors the model obtained with three output neurons (i.e., only 13 vs. 15 for the BNN with one output neuron). As stressed earlier, it is a bit more difficult to find rationales for the incorrect predictions produced by the BNN model having one output neuron and using Eq. (1). Therefore, from all the above, it appears that the model to be preferred is that derived with Eq. (2) as scaling procedure and comprising three output neurons.

Despite the fact that the distribution of some molecular descriptors was not statistically optimal to perform a discriminant factor analysis (DFA) (Tomassone *et al.*, 1988), this statistical approach was assayed on the data matrices (Tables II and III). The obtained results (not reproduced here) were widely inferior to those obtained with the BNN. In a recent study, we showed that a BNN compared favorably with a DFA for the estimation of pesticide field half-lives (Domine *et al.*, 1993). Similar conclusions were made by Weinstein and coworkers (Weinstein *et al.*, 1992) in a study on drugs' mechanisms of action against cancer.

CONCLUDING REMARKS

In the United States, more than 50% of telephone complaints coming from the residential areas located near industries concern bad odors and stench. The situation is the same in Europe. In order to use this valuable source of information, different methodologies have been tested (Clarenburg, 1987). In all cases, they provide data which are difficult to interpret and model.

Our study reveals that a BNN can be a useful tool for modeling this activity and that results obtained with three output neurons are better than those obtained with one output neuron. This agrees with the statement of Eberhart and Dobbins (1990, page 43): '*In general, output neurodes cost you relatively little computationally, so be generous with them*'. More generally, our study confirms the usefulness of the BNNs for modeling complex biological activities. They also support the idea that a BNN outperforms DA and DFA for allocating individuals into classes.

Acknowledgments

Part of this work was initiated through a collaboration with Pr M. Chastrette (Laboratoire de Chimie Organique Physique et Synthétique, Université Lyon-I, France). Some of the results reproduced here were presented at the 209th ACS National Meeting (Abstract COMP-134, April 2–6, 1995, Anaheim, USA). We wish to thank Pr J. Gasteiger (Computer-Chemie-Centrum, Universität Erlangen-Nürnberg, Germany) for comforting us through critical comments to pursue our study on the comparison of the results obtained with single and multiple output neuron BNNs. We are also grateful to Dr D. Manallack (Chiroscience Limited, Cambridge, England) for his valuable comments and suggestions on our work. His ideas and results on the problem of the comparison of the results obtained with one- or three-output BNNs were very helpful.

REFERENCES

Amoore, J.E. and Hautala, E. (1983). Odor as an aid to chemical safety: Odor thresholds compared with threshold limit values and volatilities for 214 industrial chemicals in air and water dilution. *J. Appl. Toxicol.* **3,** 272–290.

Bouscaren, R. (1984). Les produits odorants. Leurs origines. *T.S.M.–L'Eau* **June**, 313–320.

Cambon, B. and Devillers, J. (1993). New trends in structure–biodegradability relationships. *Quant. Struct.-Act. Relat.* **12,** 49–56.

Clarenburg, L.A. (1987). Odour: A mathematical approach to perception and nuisance. In, *Environmental Annoyance: Characterization, Measurement, and Control* (H.S. Koelega, Ed.). Elsevier Science Publishers B.V., Amsterdam, pp. 75–92.

Devillers, J., Bintein, S., Domine, D., and Karcher, W. (1995). A general QSAR model for predicting the toxicity of organic chemicals to luminescent bacteria (Microtox® test). *SAR QSAR Environ. Res.* **4,** 29–38.

Devillers, J. and Domine, D. (1995). Deriving structure–chemoreception relationships from the combined use of linear and nonlinear multivariate analyses. In, *QSAR and Molecular Modelling: Concepts, Computational Tools and Biological Applications* (F. Sanz, J. Giraldo, and F. Manaut, Eds.). J.R. Prous Science Publishers, Barcelona, pp. 57–60.

Devos, M., Patte, F., Rouault, J., Laffort, P., and van Gemert, L.J. (1990). *Standardized Human Olfactory Thresholds*. IRL Press, Oxford, p. 165.

Domine, D. and Devillers, J. (1995). Nonlinear multivariate SAR of Lepidoptera pheromones. *SAR QSAR Environ. Res.* **4,** 51–58.

Domine, D., Devillers, J., Chastrette, M., and Doré, J.C. (1995). Combined use of linear and nonlinear multivariate analyses in structure–activity relationship studies: Application to chemoreception. In, *Computer-Aided Molecular Design. Applications in Agrochemicals, Materials, and Pharmaceuticals* (C.H. Reynolds, M.K. Holloway, and H.K. Cox, Eds.). ACS Symposium Series 589, American Chemical Society, Washington, DC, pp. 267–280.

Domine, D., Devillers, J., Chastrette, M., and Karcher, W. (1993). Estimating pesticide field half-lives from a backpropagation neural network. *SAR QSAR Environ. Res.* **1**, 211–219.

Doré, J.C. and Renou, M. (1985). Analyse multivariée des relations entre 9 molécules phéromonales et 11 espèces de lépidoptères Noctuidae. *Acta Œcol. Œcol. Applic.* **6,** 269–284.

Doty, R.L., Deems, D.A., and Stellar, S. (1988). Olfactory dysfunction in parkinsonism: A general deficit unrelated to neurologic signs, disease stage, or disease duration. *Neurology* **38,** 1237–1244.

Doty, R.L., Reyes, P.F., and Gregor, T. (1987). Presence of both odor identification and detection deficits in Alzheimer's disease. *Brain Res. Bull.* **18,** 597–600.

Eberhart, R.C. and Dobbins, R.W. (1990). *Neural Network PC Tools. A Practical Guide.* Academic Press, San Diego, p. 414.

Hoshika, Y., Imamura, T., Muto, G., van Gemert, L.J., Don, J.A., and Walpot, J.I. (1993). International comparison of odor threshold values of several odorants in Japan and in The Netherlands. *Environ. Res.* **61,** 78–83.

Maës, M. (1990). Odeurs nuisibles: Proust, olfactométrie et désodorisation. *Eau Indus. Nuis.* **141,** 43–47.

Mouchet, P. (1978). Recherches bibliographiques sur les goûts et les odeurs d'origine biologique dans les eaux potables. Identification des organismes et de leurs métabolites: Remèdes possibles. *T.S.M.-L'Eau* **March,** 145–153.

Pattichis, C.S., Charalambous, C., and Middleton, L.T. (1995). Efficient training of neural network models in classification of electromyographic data. *Med. Biol. Eng. Comput.* **33**, 499–503.

Petroski, R.J. and Vaz, R. (1995). Insect aggregation pheromone response synergized by 'host-type' volatiles. Molecular modeling evidence for close proximity binding of pheromone and coattractant in *Carpophilushemipterus* (*L.*) (*Coleoptera: Nitidulidae*). In, *Computer-Aided Molecular Design. Applications in Agrochemicals, Materials, and Pharmaceuticals* (C.H. Reynolds, M.K. Holloway, and H.K. Cox, Eds.). ACS Symposium Series 589, American Chemical Society, Washington, DC, pp. 197–210.

Potter, H. and Butters, N. (1980). An assessment of olfactory deficits in patients with damage to prefrontal cortex. *Neuropsychol.* **18,** 621–628.

Rajendren, G. and Dominic, C.J. (1984). Role of the vomeronasal organ in the male-induced implantation failure (the Bruce effect) in mice. *Arch. Biol. (Bruxelles)* **95,** 1–9.

Renou, M., Lalanne-Cassou, B., Doré, J.C., and Milat, M.L. (1988a). Electro-antennographic analysis of sex pheromone specificity in neotropical Catocalinae (Lepidoptera: Noctuidae): A multivariate approach. *J. Insect Physiol.* **34,** 481–488.

Renou, M., Lalanne-Cassou, B., Michelot, D., Gordon, G., and Doré, J.C. (1988b). Multivariate analysis of the correlation between Noctuidae subfamilies and the chemical structure of their sex pheromones or male attractants. *J. Chem. Ecol.* **14,** 1187–1215.

Roberts, E. (1986). Alzheimer's disease may begin in the nose and may be caused by aluminosilicates. *Neurobiol. Aging* **7,** 561–567.

Rumelhart, D.E., Hinton, G.E., and Williams, R.J. (1986). Learning representations by back-propagating errors. *Nature* **323,** 533–536.

Schaal, B. (1988). Olfaction in infants and children: Developmental and functional perspectives. *Chem. Senses* **13,** 145–190.

Serby, M. (1987). Olfactory deficits in Alzheimer's disease. *J. Neural. Transm.* **24**, 69–77.

STATQSAR Package (1994). CTIS, Lyon, France.

Tomassone, R., Danzart, M., Daudin, J.J., and Masson, J.P. (1988). *Discrimination et Classement.* Masson, Paris, p. 172.

Verschueren, K. (1983). *Handbook of Environmental Data on Organic Chemicals.* Van Nostrand Reinhold Company, New York, p. 1310.

Ward, C.D., Hess, W.A., and Calne, D.B. (1983). Olfactory impairment in Parkinson's disease. *Neurology* **33,** 943–946.

Warner, M.D., Peabody, C.A., Flattery, J.J., and Tinklenberg, J.R. (1986). Olfactory deficits and Alzheimer's disease. *Biol. Psychiatry* **21,** 116–118.

Weinstein, J.N., Kohn, K.W., Grever, M.R., Viswanadhan, V.N., Rubinstein, L.V., Monks, A.P., Scudiero, D.A., Welch, L., Koutsoukos, A.D., Chiausa, A.J., and Paull, K.D. (1992). Neural computing in cancer drug development: Predicting mechanism of action. *Science* **258,** 447–451.

6 Adaptive Resonance Theory Based Neural Networks Explored for Pattern Recognition Analysis of QSAR data

D. WIENKE[1]*, D. DOMINE[2], L. BUYDENS[1], and J. DEVILLERS[2]

[1]*Catholic University of Nijmegen, Laboratory for Analytical Chemistry, Toernooiveld 1, 6525 ED Nijmegen, The Netherlands*

[2]*CTIS, 21 rue de la Bannière, 69003 Lyon, France*

The family of adaptive resonance theory (ART) based artificial neural networks concerns distinct theoretical neural models for unsupervised and supervised pattern recognition. After a brief presentation of the neurophysiological basis of ART and the different ART paradigms, the algorithms of ART-2a and FuzzyART are presented, and the heuristic potency of these methods is illustrated from two case studies. The former dealing with the headspace analysis of roses clearly shows in a didactic manner how to set the different parameters and how the results can be interpreted. The second case study is a real-life application showing the potency of these techniques for optimal test series selection. The results are compared with those obtained from hierarchical cluster analysis and visual mapping methods. The advantages and drawbacks of each method are discussed. Features of the ART methodology such as rapid training speed, self-organization behavior, and interpretability of the network weights are also discussed.

KEY WORDS: *adaptive resonance theory; ART networks; hierarchical cluster analysis; nonlinear mapping; headspace analysis of roses; selection of test series.*

* Author to whom all correspondence should be addressed.

In, *Neural Networks in QSAR and Drug Design* (J. Devillers, Ed.)
Academic Press, London, 1996, pp. 119–156.
ISBN 0-12-213815-5

INTRODUCTION

Historically, chemometrics was dominated by linear methods such as principal components analysis, correspondence factor analysis, discriminant analysis, and hierarchical cluster analysis (Wold, 1972, 1974; Albano *et al.*, 1978; Varmuza, 1978; Lewi, 1982; Sharaf *et al.*, 1986; Massart *et al.*, 1988). However, nonlinear multivariate techniques have been increasingly infiltrating the field (Kowalski and Bender, 1972, 1973; Sjöström and Kowalski, 1979; Forina *et al.*, 1983; Zitko, 1986; Rauret *et al.*, 1988; Vriend *et al.*, 1988; Nicholson and Wilson, 1989; Fang *et al.*, 1990; Henrion *et al.*, 1990; Rose *et al.*, 1990; Domine *et al.*, 1994c; Treiger *et al.*, 1994; Budzinski *et al.*, 1995). Among them, recently, neural networks have become privileged statistical tools for modeling complex biological activities or properties of chemicals (Borman, 1989; Thomson and Meyer, 1989; Zupan and Gasteiger, 1991, 1993; Bos *et al.*, 1992, 1993; Tusar *et al.*, 1992; Bruchmann *et al.*, 1993; Melssen *et al.*, 1993; Simon *et al.*, 1993; Smits *et al.*, 1994a,b; Sumpter *et al.*, 1994; Wessel and Jurs, 1994, 1995; Wienke and Hopke, 1994a,b; Wienke *et al.*, 1994a, 1995; Zupan *et al.*, 1994). The same evolution has been observed in quantitative structure–activity relationships (QSAR) and drug design. Thus, the original Hansch approach consisted of the design of multiple linear regression equations linking the logarithm of the biological activities of chemicals to their physicochemical descriptors encoding lipophilic, steric, and electronic effects (Hansch and Fujita, 1964). Even if, since the early 1970s, multivariate methods have been increasingly used in structure–activity relationships (SAR) and QSAR (see reviews in Devillers and Karcher (1991) and van de Waterbeemd (1994, 1995)), the consecration of these approaches has been represented by the introduction of the partial least squares (PLS) method in the field (Geladi and Tosato, 1990). Only recently, the heuristic potency of the nonlinear methods in bioactive compound research has been clearly demonstrated first by the use of the nonlinear mapping method for visualizing and interpreting numerous QSAR data (Court *et al.*, 1988; Livingstone *et al.*, 1988; Hudson *et al.*, 1989; Livingstone, 1989; Domine *et al.*, 1993a, 1994b,c, 1995; Chastrette *et al.*, 1994; Devillers *et al.*, 1994; Devillers, 1995; Devillers and Domine, 1995; Domine and Devillers, 1995) and second by the application of different neural network paradigms for solving numerous SAR and QSAR problems (Aoyama *et al.*, 1990; Andrea and Kalayeh, 1991; Aoyama and Ichikawa, 1991a,b,c, 1992; de Saint Laumer *et al.*, 1991; Liu *et al.*, 1992a,b; Livingstone and Salt, 1992; Peterson, 1992, 1995; Cambon and Devillers, 1993; Chastrette *et al.*, 1993; Devillers, 1993; Devillers and Cambon, 1993; Domine *et al.*, 1993b, 1996; Feuilleaubois *et al.*, 1993; Livingstone and Manallack, 1993; Wiese and Schaper, 1993; Bienfait, 1994; Manallack and Livingstone, 1994; Devillers *et al.*, 1996).

Despite this flowering of new statistical tools, chemometricians are still seeking methods presenting the following properties:

- increased speed of classification;
- improved power of discrimination and prediction, in particular for nonlinear data separation problems;
- ease of interpretation in chemical terms (no black box desired!);
- robustness against outliers, extra- and interpolations;
- simplicity to recalibrate and to configure for other applications;
- ability to fulfill on-line and real-time classification tasks in production processes and in automated analytical set-ups.

In this context, the usefulness of adaptive resonance theory (ART) based neural networks in chemometrics has recently been stressed by Wienke and coworkers (Wienke, 1994; Wienke and Kateman, 1994; Wienke *et al.*, 1994b, 1996a,b; Xie *et al.*, 1994; Wienke and Buydens, 1995, 1996). The aim of this paper is to explore the interest of the ART algorithms in QSAR. Thus, after a brief presentation of the neuro-physiological basis of ART and the different ART paradigms, the ART-2a and FuzzyART algorithms are presented. The usefulness of these methods is illustrated in QSAR from two case studies.

NEURO-PHYSIOLOGICAL BASIS OF ART

By means of his adaptive resonance theory (ART), Grossberg (1976a,b) started to model selected aspects of real-time pattern recognition in the brain. This theory rather provided neural models that were not explicitly designed for multivariate data analysis. ART can explain how a pattern vector is classified by a trained biological neural network, when this pattern vector does not fit into any existing learned category. ART states for this typical extrapolation case, that a brain extends its knowledge by an initialization of up-to-now unused regions of 'fresh' neurons. The new region is dynamically linked to the active region of the brain. In the brain, this ability avoids collapses when confronted, for example, with an unexpected visual novelty. This novelty is, rather, dynamically detected as deviating from all recorded categories of knowledge and it is immediately (in real-time) stored by the brain as a new knowledge. Thus, the ART system is divided into two subsystems: an attentional subsystem, which processes familiar stimuli, and an orienting subsystem that detects unfamiliar input patterns, and resets the attentional subsystem when it detects such a pattern. The essence of ART is therefore a comparison of input patterns with the existing network structure to check whether they fit into ('resonate with') it. In case of resonance, an 'adaptation' of the existing weights is performed otherwise the structure is adapted by creation of new weight vector(s). Hence, the network 'learns' by modification or creation of a set of coefficients (weights). Thus, ART networks are designed to resolve the stability–plasticity dilemma. They are stable enough to preserve past learning, but nevertheless remain adaptable enough to incorporate new information whenever it might appear.

TAXONOMY AND STATE-OF-THE-ART

The research group of Grossberg and Carpenter developed several ART based classifiers (Figure 1). The methods mainly differ in approaching pattern similarity by either Euclidean angle based or fuzzy theory based distance concepts. Selected ART classifiers (ART-1, ARTMAP) are restricted to binary (0–1) coded input descriptors while ART-2, ART-3 and FuzzyART can stably learn to categorize either binary or analog input patterns. ART-3 can additionally carry out parallel search, or hypothesis testing, of distributed recognition codes in a multilevel network hierarchy. The FuzzyARTMAP can rapidly learn stable categorical mappings between analog or binary input and output vectors. Historically, the group of unsupervised ART methods ART-1 (Grossberg, 1976a,b, 1982), ART-2 (Carpenter and Grossberg, 1987), ART-2a (Carpenter *et al.*, 1991b), ART-3 (Carpenter and Grossberg, 1990), FuzzyART (Carpenter *et al.*, 1991c) was developed before the supervised ART techniques called ARTMAP (Carpenter *et al.*, 1991a) and FuzzyARTMAP (Carpenter *et al.*, 1992). Many of the important papers on the topic have been compiled in a book by Carpenter and Grossberg (1991) dedicated to self-organizing neural networks.

The present study will concentrate on the application of ART-2a and FuzzyART in QSAR studies. However, outside the field of QSAR, numerous applications of ART models were reported, such as automated identification

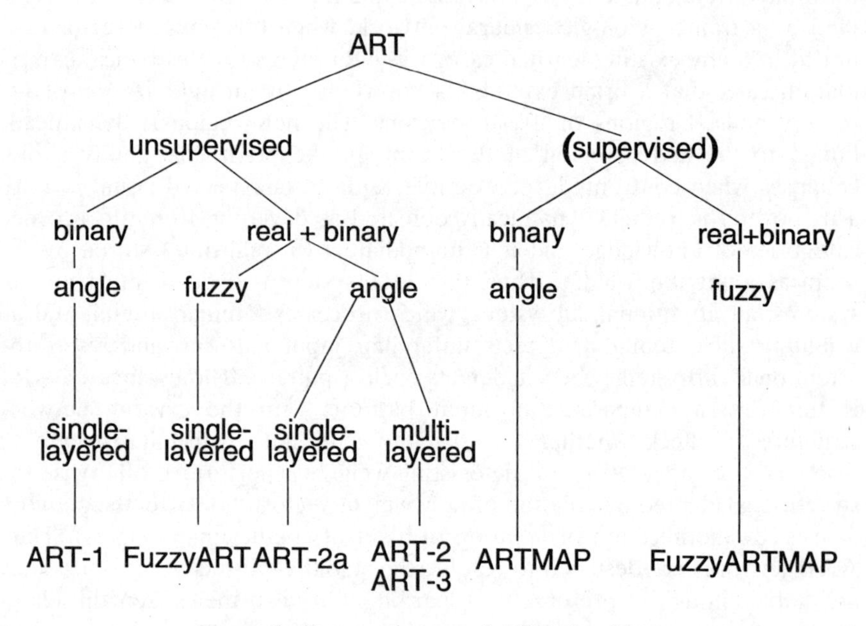

Figure 1 Taxonomic tree of ART algorithms.

of hand-written characters (Gan and Lua, 1992), comparisons with the k-means clustering method (Burke, 1991), and combination of ART with parallel operating optoelectronic multichannel detector arrays for remote control tasks (Caudell, 1992). Optical, optoelectronic and electronic hardware implementations of ART-1 and ART-2 were realized (Kane and Paquin, 1993; Wunsch *et al.*, 1993a,b; Ho *et al.*, 1994). An application of ART-2 to pattern recognition with image data was also published (Resch and Szabo, 1994). Further, it was found (Benjamin *et al.*, 1995) that multilayer feed-forward backpropagation (MLF-BP) neural networks have a better discrimination power if compared with ART-2. The applicability of FuzzyARTMAP to medical data analysis (Hohenstein, 1994) provided the result that FuzzyARTMAP discriminates as well as the MLF-BP neural network. A theoretical overview of all ART methods from the point of view of information science, can be found in a report of Willems (1994).

During the last two years, chemists have started to explore the usefulness of ART neural networks. Lin and Wang (1993) modeled process-analytical time series data by an autoregressive function. The obtained characteristic pattern vectors of autoregressive parameters were successfully classified using ART-2. Whiteley and Davis (1993, 1994) proposed the use of a trained ART-2 network for the monitoring and control of chemical reactors. They showed how the ART network correctly predicted 'normal' and 'not normal' situations of the reactor during the running chemical reaction. Wienke and Kateman (1994) used ART-1 for classification and interpretation of UV/VIS- and IR-spectra. These authors demonstrated the quantitative chemical interpretability of the ART weights in terms of spectral absorbances after a suitable step of decoding. Wienke *et al.* (1994b) and Xie *et al.* (1994) proposed to use ART-2a in environmental monitoring of ambient air. In these two particular studies, it has been demonstrated that the network weights can be interpreted in terms of their environmental-chemical meaning. Also it has been shown how a trained ART network managed an extrapolation of the training subspace by linking additional neurons to its structure. A comparison of ART-2a versus SIMCA (Geladi and Tosato, 1990) and MLF-BP neural network has been given by Wienke *et al.* (1996b). They applied ART-2a in this comparative study to rapid sorting of post-consumer plastics by remote NIR spectroscopy. By a large crossvalidation study with distinct training and testing sets it was quantitatively shown, that the results obtained with ART-2a compared favorably with those produced by the SIMCA method, but better results were obtained with MLF-BP neural network. However, advantages of ART-2a against MLF-BP network were higher training speed, built-in detector against outliers, and chemical interpretability of network weights. A comparison of FuzzyARTMAP against MLF-BP networks and the PLS method for rapid sorting of post-consumer plastic waste was reported by Wienke *et al.* (1996a). After careful data preprocessing by scaling and features selection it was found, that FuzzyARTMAP was able to train faster and to discriminate better than an MLF-BP neural network in this particular application to near infrared spectra. Additionally, the trained ART network could be interpreted in spectroscopic terms. An overview of ART neural networks for chemists has been recently published (Wienke and Buydens, 1995, 1996).

THEORY OF ART-2a AND FUZZYART

As with all ART paradigms, the mathematics behind ART-2a and FuzzyART are complicated but the fundamental ideas are not. Under these conditions, we only outline below the main ideas and algorithmic information. Readers who are more mathematically inclined will find an abundance of theory in the different references of Carpenter and Grossberg cited in this article. ART networks allow one to derive clusters from a data matrix $\mathbf{X}$ of n individuals described by m descriptors. Each individual $\mathbf{x}_i$ is therefore considered as a directed vector in the m-dimensional space of variables. Some of the individuals can be closer to each other in this space forming groups if the values of their m variables are similar. The aim of ART networks such as ART-2a and FuzzyART is to find such groups (clusters) of similar individuals, whereby the number c of groups in the data set $\mathbf{X}$ is not known *a priori*. They incorporate the basic features of all ART systems, notably, pattern matching between bottom-up and top-down learned prototype vectors. There are bottom-up and top-down connections between each input and output neurons (layers F1 and F2, respectively) which comprise the 'adaptive filters'. Both ART-2a and FuzzyART networks are trained according to the general scheme shown in Figure 2. First, an initialization of the so-called long-term memory (LTM), which is a matrix of weight coefficients, is performed. For this purpose, all the weights are assigned the same value of $1/\sqrt{m}$. The number of potential categories is arbitrary and initially each category is said to be 'uncommitted'. Then, a single individual $\mathbf{x}_i$ from the data matrix $\mathbf{X}$ is randomly selected and preprocessed. ART-2a and FuzzyART require that each individual is normalized between 0 and 1. It is important to know that ART-2a internally scales each individual to unit length (i.e., the norm of each vector (individual) $\mathbf{x}_i$ equals one). In the case of FuzzyART, complement coding is used as an additional preprocessing step. This preprocessing procedure is presented later in a special section. The clustering process starts at the very beginning (when no active weight vector exists) by simply copying the preprocessed individual to the initialized future LTM. In the easiest case, this will be a weight vector $\mathbf{w}_0$ of length m representing a raw estimate of a first (new) cluster (Figure 3, step 1). The weights in ART thus represent vectors assigning the directions for particular clusters in the m-dimensional space of variables (Figure 3). After this initialization of a new (or first) cluster, another individual $\mathbf{x}_i$ is randomly selected and sequentially (or in parallel) compared with all the already existing $k = 1$ to c clusters $\mathbf{W}$ (dimension $c \times m$). The clusters, represented by a set of weight vectors, correspond to 'committed' or 'active' ART neurons. Grossberg's neuro-physiological ART approach states that, for such sequential comparison, a thought or virtual image $\mathbf{x}_i^*$ of $\mathbf{x}_i$ is generated internally within the network. Indeed, F1 activates neuron(s) in F2 and F2 gives a feedback to F1 *via* the top-down connections which creates the virtual image $\mathbf{x}_i^*$. This virtual image can fit to $\mathbf{x}_i$ (resonance case) or deviate (novelty detection). Both situations will be discussed below. Simultaneously, $\mathbf{x}_i^*$ also forms a kind of short-term memory (STM) for its own original $\mathbf{x}_i$. After comparison of $\mathbf{x}_i$ with all c clusters (weight vectors), a winner among them can be found, providing a virtual image $\mathbf{x}_i^{*,\mathrm{winner}}$ having the highest calculated similarity $\max(\rho^{\mathrm{calc}}) = \rho^{\mathrm{max,calc}}$ (lowest dissimilarity) to $\mathbf{x}_i$. In the neural network terminology, the search for the winning neuron followed by its training is called competitive learning. This notion is also known, for example, in Kohonen networks (Kohonen, 1989, 1990). In the ART-2a algorithm, vector similarity is expressed as:

$$\rho^{\mathrm{calc}} = \mathbf{r}_i \cdot \mathbf{w}_k \tag{1}$$

where $\mathbf{r}_i$ represents a preprocessed input descriptor $\mathbf{x}_i$. Because $\mathbf{r}_i$ and $\mathbf{w}_k$ have unit length, Eq. (1) comes down to the cosine of the angle between both vectors. In case of FuzzyART, we have:

$$\rho^{\mathrm{calc}} = |\mathbf{r}_i \wedge \mathbf{w}_k| / (\alpha + |\mathbf{r}_i|) \tag{2}$$

whereby

$$|\mathbf{r}_i \wedge \mathbf{w}_k| = \mathrm{length}(\min(\mathbf{r}_i, \mathbf{w}_k)) \tag{3}$$

where $\wedge$ is the fuzzy AND operator and α is a constant in the interval [0, 1].

In contrast to classical cluster analysis or to a Kohonen neural network, highest similarity in ART between $\mathbf{x}_i$ and cluster $\mathbf{w}_k^{\mathrm{winner}}$ is not the only satisfying criterion for their future fusion. In a second, crucial, step, it is checked whether the virtual image $\mathbf{x}_i^*$ generated within the network, is close enough to the original actual input $\mathbf{x}_i$. In other words, the network asks itself (like we do *via* our brain): 'have I ever seen this input or is it a novelty?' Technically, this is achieved by comparing $\rho^{\mathrm{calc,winner}}$ between $\mathbf{x}_i$ and $\mathbf{w}_k^{\mathrm{winner}}$ with a so called 'vigilance' parameter ρ^{max}. Graphically expressed, ρ^{max} forms the spatial limit of each cluster. In case of ART-2a, ρ^{max} is a fixed constant chosen by the user in the interval [0, 1]. Thus, for ART-2a if Eq. (4) is satisfied, $\mathbf{x}_i^{*,\mathrm{winner}}$ is similar enough to $\mathbf{x}_i$ (classically expressed $\mathbf{x}_i$ falls inside the borders of the winning neuron (cluster) $\mathbf{w}_k^{\mathrm{winner}}$).

$$\rho^{\mathrm{calc,winner}} \geq \rho^{\mathrm{max}} \tag{4}$$

A ρ^{max} close to 1 predefines a higher critical level for vector similarity which will be more difficult to satisfy. In such a case, a small deviation between

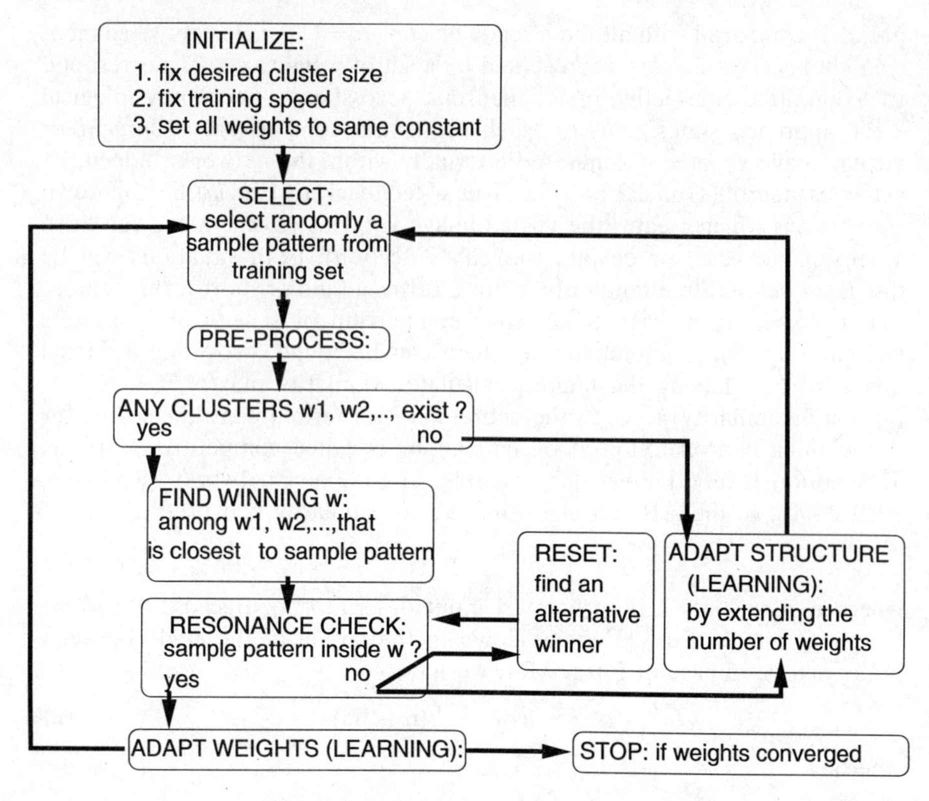

Figure 2 Program flowchart of basic calculation steps during training of a FuzzyART (or ART-2a) neural network.

$\mathbf{w}_k^{\text{winner}}$ and $\mathbf{x}_i$ is already enough for additional linking of a new neuron $\mathbf{w}_{c+1}$ to the network structure. A large $\rho^{\max}$ will thus provide many new clusters having a small diameter. In the reverse case where $\rho^{\max}$ is chosen close to 0 (low similarity vector), only a few but large clusters are generated.

For FuzzyART, $\rho^{\max}$ is not directly compared with $\rho^{\text{calc,winner}}$ but the additional following criterion is used:

$$\rho^{\text{calc}} \geq |\mathbf{r}_i \wedge \mathbf{w}_k|/(\alpha + |\mathbf{w}_k|) \tag{5}$$

If Eq. (4) (or Eq. (5) in the case of FuzzyART) is fulfilled, one says that the network comes into 'resonance' with this type of input or it already 'knows' it. In the reverse case, when Eq. (4) (respectively Eq. (5)) is not fulfilled, the network discovers a 'novelty'. In comparison to all other neurons, $\mathbf{x}_i$* has been closest to $\mathbf{w}_k^{\text{winner}}$, but it lies outside the spatial cluster limits of the winning neuron. After the resonance check (Eq. (4) or Eq. (5) for ART-2a or FuzzyART, respectively), the third crucial step which follows

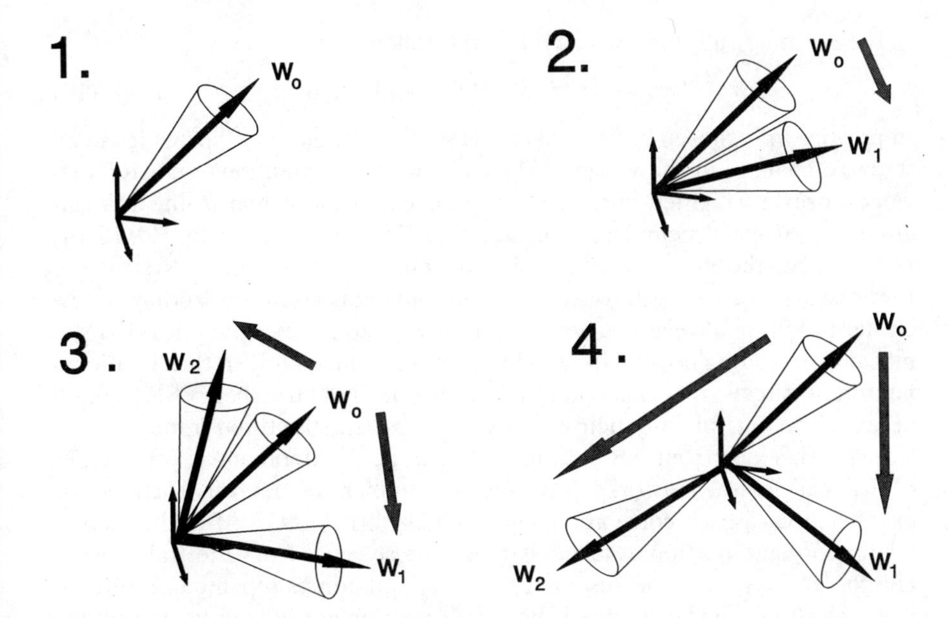

Figure 3 Cluster initialization, cluster formation, competitive learning and mutual cluster contrast enhancement during the training of an ART neural network. For caption, see text.

is called 'adaptation'. If Eq. (4) (respectively Eq. (5)) is satisfied, the weights of the network are modified so that the spatial position of the winning cluster becomes closer to the presented $\mathbf{x}_i$. The learning rate (η) determines the magnitude of the modification towards the presented $\mathbf{x}_i$. For ART-2a, the formula for weight adaptation (learning rule), containing η, is the following:

$$\mathbf{w}_k^{\text{new,winner}} = \mathbf{t}_k/\|\mathbf{t}_k\| \tag{6}$$

with

$$\mathbf{t}_k = \mathbf{u}_k + (1 - \eta)\ \mathbf{w}_k^{\text{old,winner}} \tag{7}$$

and

$$\mathbf{u}_k = \eta\ (\mathbf{v}_k/\|\mathbf{v}_k\|) \tag{8}$$

as well as

$$\begin{aligned} \mathbf{v}_k &= \mathbf{r}_i \quad \text{if } \mathbf{w}_k^{\text{old,winner}} > \theta \\ \mathbf{v}_k &= 0 \text{ otherwise} \end{aligned} \tag{9}$$

θ is a threshold that smooths out low values, providing in this way a nonlinear transfer of the weight vector.

The learning rule for FuzzyART is the following:

$$\mathbf{w}^{\text{new,winner}} = \eta(\mathbf{r}_i \wedge \mathbf{w}^{\text{old,winner}}) + (1 - \eta)\mathbf{w}^{\text{old,winner}} \tag{10}$$

In practice, η is chosen between 0 and 1. An η value close to 0 provides small stepwise changes in the weights. Thus, *via* 'weight adaptation', an ART network stores a weighted part of the present input individual in the LTM as any other artificial neural network does. If Eq. (4) (respectively Eq. (5)) is not fulfilled, the network adapts first, in place of its weights, its structure towards the discovered novelty. 'Structure adaptation' means adding a new cluster (additional weight vector or neuron $\mathbf{w}_{c+1}$) to the existing ones. The significantly deviating novel individual $\mathbf{x}_i$ is stored immediately in this additional neuron by direct copy. This is another original idea of Grossberg: ART neural networks do not only use their weights but also their structure for information storage. This idea from ART stimulated also the development of other types of artificial neural networks with flexible numbers of neurons such as, for example, the cascade correlation algorithm (Jankrift, 1993). After this step of adaptation, another input vector is randomly selected from the training set, and the entire process of 'resonance' and 'adaptation' is repeated (Figure 2), whereby the former content of the STM is continuously overwritten by the new generated virtual input vectors. This is similar to the behavior of the STM in the brain. When the random sampling of the training data matrix $\mathbf{X}$ has been performed n times, one has performed one epoch. Simultaneously, the contrast between clusters in the LTM increases ('generalization', Figure 3, steps 2 to 4). The process converges within a few epochs with the formation of c clusters. The chosen constants of learning rate η and vigilance parameter ρ determine how many clusters c are formed (Figures 2 and 3). In other words, by a suitable choice of η and ρ, the number c of clusters to derive from the multivariate data cloud $\mathbf{X}$ can be monitored. In this way, a variation of η and ρ can serve for active exploration of a data set $\mathbf{X}$ by resolving it into distinct numbers of clusters presenting distinct sizes. As with classical pattern recognition methods, for ART, the more compact and well separated the clusters are, the less subjective the user's choice of η and ρ is and the more stable the results are. In this situation, ART will always find the true number c of hidden clusters. In contrast, the more the data scatter or the more they form a continuous hyper-surface of equidistant points, the higher the influence of η and ρ on the clustering result is. The extreme case of a data cloud $\mathbf{X}$ concerning n uniform and equidistant points (thus without any cluster structure) will give arbitrarily different results from run to run.

DATA PREPROCESSING BY COMPLEMENT CODING

Complement coding is a data preprocessing procedure used for the FuzzyART algorithm (or FuzzyARTMAP). Complement coding means that the number of variables of each $\mathbf{x}_i$ is doubled by appending its complement individual $1 - \mathbf{x}_i$ before offering it to the FuzzyART network. It has to be ensured that the elements of individuals $\mathbf{x}_i$ are scaled between 0 and 1. For this purpose, any scaling can be used (e.g., normalization to length one). The following example illustrates the complement coding of an already scaled vector. Assume an arbitrarily chosen six-dimensional individual

$$\mathbf{x}_i = [0.8, 0.2, 0.6, 0.3, 0.5, 0.0]$$

After complement coding the new twelve-element individual $\mathbf{r}_i$ is:

$$[\mathbf{x}_i, 1 - \mathbf{x}_i] = \mathbf{r}_i = [0.8, 0.2, 0.6, 0.3, 0.5, 0.0, 0.2, 0.8, 0.4, 0.7, 0.5, 1]$$

For the supervised method, FuzzyARTMAP, the desired output $\mathbf{y}_i$ is also complement coded by its complement $1 - \mathbf{y}_i$. After complement coding, the first half-part of the vector encodes features that are critically 'present', while the second part encodes those that are 'absent'. It leads to a symmetric theory in which the MIN and MAX operators of the fuzzy set theory play complementary roles (Carpenter *et al.*, 1991c). In addition, this prevents category proliferation (Carpenter *et al.*, 1991c). After this simple preprocessing step, each complement coded input $\mathbf{r}_i$ is normalized but the amplitude of each feature is not lost. Thus, the length of the vectors (individuals) does not play any role during the step of discrimination in the fuzzy distance measure (Eq. (2)). Complement coding can be considered as a technique that does not simply double the number of features from m to $2m$, but it rather substitutes a single crisp feature by a 'feature range' with an upper and a lower limit. This explains why dynamically growing hyper-rectangular shaped boxes are obtained (Figure 4B) instead of hyper-circular ones (Figure 4A) to model a data cloud $\mathbf{X}$. The boxes grow during training from a small point vector to a large hyper-rectangle, because, *via* Eq. (3) and the learning rule (Eq. (10)), simultaneously both limits of the range are manipulated. These outer limits of such a hyper-rectangle are called in the present work 'upper fuzzy bounds' and 'lower fuzzy bounds'. After network training, these ranges can be decoded again (and rescaled) providing for each neural weight and the m features, an upper and a lower cluster limit in terms of the original variables, while ART-1, ART-2, ART-2a, ART-3 and ARTMAP only yield single crisp values.

QUANTIFICATION OR QUALIFICATION?

From the above description of the theory of ART-2a and FuzzyART, it appears obvious that both types of networks are more applicable for pattern recognition (qualification) than for function fitting (quantification). This is because each ART cluster (or neuron) models a limited local area in the feature space (Figure 4). In this way, complicated and irregularly shaped

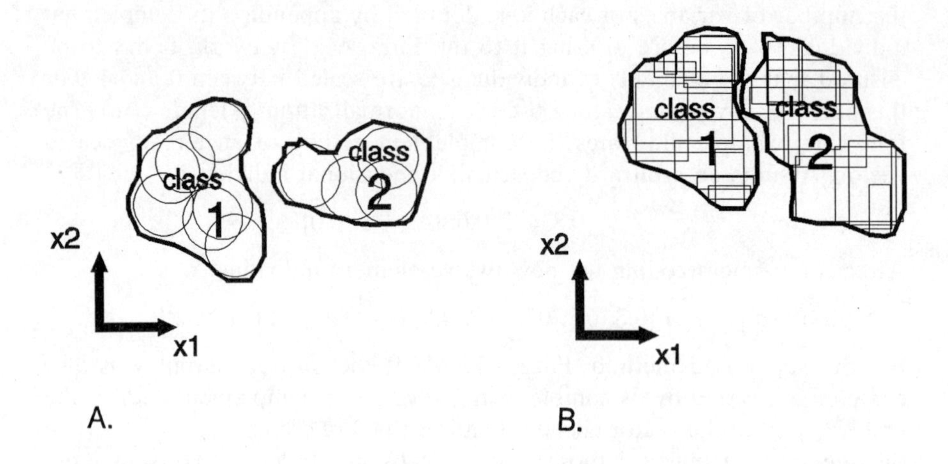

Figure 4 ART-2a and FuzzyART are locally working modeling techniques. ART-2a approximates a data cloud by radial shaped clusters (hyper-circles) based on its similarity of Euclidean angle (A). In contrast, FuzzyART approximates a data cloud by hyper-rectangles (B) based on the data preprocessing using complement coding and based on the intersection operator as similarity concept originating from fuzzy set theory.

clouds of points can be approximated in principle, by a sequence of overlapping ART hyper-clusters (radial or rectangular, as shown in Figure 4). In theoretical studies (Carpenter *et al.*, 1991c, 1992; Hohenstein, 1994; Willems, 1994), it has been shown that ART networks can model clouds of points that are located within each other or that are formed from winded spirals. However, the more a data set forms a continuous function in the variable space, the more new single weight vectors are required. Function fitting is thus always possible with ART, but the cost one has to pay for this, is a rapid proliferation of an enormous number of new single clusters.

CASE STUDY I: CLASSIFICATION OF ROSE VARIETIES FROM THEIR HEADSPACE ANALYSIS

To illustrate the ART-2a and FuzzyART paradigms, a small data set dealing with headspace analysis of 27 varieties of roses was retrieved from a publication of Flament and coworkers (Flament *et al.*, 1992). The aim of this application is mainly to show the functioning of the above paradigms, provide guidelines for the interpretation of the results and compare them with those obtained by a classical hierarchical cluster analysis (HCA). In their paper,

Table I *List of the 27 studied varieties of roses. Trade marks are capitalized.*

No.	Rose variety
1	MONICA
2	SONIA MEILLAND
3	CHARLES DE GAULLE
4	*Rosa rugosa rubra*
5	*Rosa muscosa purpurea*
6	MARGARET MERRIL
7	*Rose à parfum de l'Hay*
8	*Rosa gallica*
9	WESTERLAND
10	*Rosa damascena*
11	CHATELAINE DE LULLIER
12	BARONNE E. DE ROTHSCHILD
13	SUTTERS GOLD
14	CONCORDE 92
15	COCKTAIL 80
16	SUSAN ANN
17	HIDALGO
18	YONINA
19	NUAGE PARFUMÉ
20	ELISABETH OF GLAMIS
21	JARDIN DE BAGATELLE
22	PRELUDE
23	SYLVIA
24	YOUKI SAN
25	GINA LOLLOBRIGIDA
26	PAPA MEILLAND
27	WHITE SUCCESS

Flament and coworkers (Flament *et al.*, 1992) reported the results of a head-space analysis for 27 varieties of roses (Table I) performed on living flowers by means of gas-chromatography for the separation of the constituents and mass spectrometry for their identification. Rather than listing all the compounds, they preferred a classification of the identified constituents into six groups according to their chemical functionality and reported the results (% of the total amount of volatile constituents) for the four main categories (i.e., hydrocarbons (R–H), alcohols (R–OH), esters (–COOR), phenols and aromatic ethers (Ar–O–)). To analyze the results, but also to simplify their representation, Flament and coworkers (Flament *et al.*, 1992) used a graphical illustration of a flower with six petals (based on the six groups they studied), where the petals were proportional in size to the percentages of the total amount of volatile constituents. By a simple visual inspection of the flowers, they proposed a classification of odoriferous roses consisting of five types (i.e., hydrocarbon, alcohol without orcinol dimethylether, alcohol containing orcinol dimethylether, ester, aromatic ether). Under these conditions, due to its simplicity and also the information provided by Flament *et al.* (1992) regarding the classification and composition of the roses, this example is ideal to show how ART-2a and FuzzyART work and compare the results to the classification proposed by Flament *et al.* (1992) as well as that obtained from HCA. Obviously, because we only have the percentages for the four main groups, our classification can only be based on these groups, and the distinction between the types, alcohol without orcinol dimethylether and alcohol containing orcinol dimethylether, cannot be uncovered by our technique since this information is not encoded in the variables used.

In order to analyze the data matrix of 27 varieties of roses described by the percentages of four chemical functionalities, in a first step, each line (i.e., rose variety) was scaled to length one (i.e., Euclidean norm = 1) (Table II). We used both ART-2a and FuzzyART with a learning rate of $\eta = 0.01$ and

Table II *Relative percentage profiles (individuals* $\mathbf{x}_i$*) for four volatile compound types (i.e., R–H, R–OH, –COOR, Ar–O–) found in 27 distinct cultivars of roses by headspace analysis. The original data (percentages of each component), taken from Flament* et al. *(1992), were scaled by normalizing each row to length one (i.e., Euclidean norm = 1).*

No.	R–H	R–OH	–COOR	Ar–O–
1	0.966	0.0309	0.0995	0.2368
2	0.8993	0.4116	0.1405	0.0449
3	0.8839	0.0738	0.4604	0.0369
4	0	0.9989	0.0469	0.0074
5	0.1333	0.9788	0.1555	0.0056
6	0.1097	0.9661	0.0989	0.2118
7	0.0458	0.9247	0.378	0.0016
8	0.1972	0.8438	0.4992	0
9	0.0018	0.8406	0.539	0.0533
10	0.0068	0.6399	0.7683	0.0135
11	0.004	0.7556	0.4622	0.4642
12	0.0198	0	0.9998	0.0055
13	0.031	0.1022	0.9767	0.1863
14	0.0321	0.0549	0.9805	0.1862
15	0.0137	0.1052	0.9691	0.2228
16	0.0062	0.3777	0.9242	0.0557
17	0.0383	0.0367	0.8892	0.4545
18	0.0173	0.0895	0.8758	0.474
19	0.0802	0.3762	0.87	0.3084
20	0.1326	0.3811	0.9079	0.1139
21	0.5365	0.0047	0.8047	0.2541
22	0.1569	0.6278	0.6643	0.3741
23	0.1698	0.0044	0.0102	0.9854
24	0.0878	0	0.0844	0.9926
25	0.0927	0.1975	0.3194	0.9222
26	0.0245	0.5514	0.2836	0.7842
27	0.3618	0.0949	0.5132	0.7724

various values of the vigilance parameter (ρ^{max}) in order to derive clusters of roses presenting similar percentages of volatile constituents. As with the other classical neural network paradigms, it is necessary to run the network several times since the random presentation of the individuals implies that from one run to another, slightly varying results can be obtained. With ART paradigms, this particularly appears when it is difficult to allocate definitively an individual to a particular cluster since it is on the borderline of two or more clusters. In that case, the individual will appear sometimes in one of the clusters and sometimes in the other(s). Thus, in this case study dealing with clustering of rose varieties from the analysis of their volatile components, 10 trials were performed for each tested value of the vigilance parameter. It may also be noted that the numbering of the clusters will vary from one run to the other (e.g., cluster 1 can become cluster 3) due to the random presentation of the individuals. However, the interpretation of the results remains obviously the same. Tables III and IV list the results obtained with ART-2a and FuzzyART, respectively. Inspection of these tables shows that the number of clusters formed can be monitored by adjusting the vigilance parameter at different levels ρ^{max}. Indeed, when ρ^{max} increases, the size of the clusters is reduced and some of the clusters are split. It can also be noted that in this particular case study, the results obtained with ART-2a (Table III) were more stable than those obtained with FuzzyART (Table IV). Indeed, Tables III and IV indicate that the possibility of assignment to different clusters for separate runs is less likely with ART-2a than with FuzzyART. It only occurs once with ART-2a for the rose variety no. 11 when $\rho^{max} = 0.6$.

A more detailed inspection of the results obtained with ART-2a (Table III) shows that when the value of ρ^{max} ranges between 0.1 and 0.3, all the roses are grouped into one cluster. For ρ^{max} between 0.4 and 0.5, three clusters can be distinguished. At this step, the rose varieties 1 to 3 are grouped in cluster 3. These are characterized by high percentages of hydrocarbons in their volatile constituents (Table I). Rose varieties 4 to 22 are grouped in cluster 1 and rose varieties 23 to 27 in cluster 2. Cluster 1 mixes the varieties principally characterized by high percentages of alcohols or esters, while cluster 2 groups the varieties containing a large percentage of aromatic ethers. Except for the grouping of alcohol and ester type roses, it is noteworthy that this classification is in accordance with that proposed by Flament *et al.* (1992). When ρ^{max} equals 0.6, the same repartition is observed but the previous cluster 1 is split into two clusters (i.e., clusters 1 and 2). Cluster 1 groups the rose varieties 10 to 22 and cluster 2, the rose varieties 4 to 9. The former is principally characterized by high percentages of esters and the latter by high percentages of alcohols. As stressed earlier, the rose variety 11 has been found to fall into cluster 1 or 2 during different runs. This is due to the fact that the percentages of alcohols and esters are both rather high. Compared with the classification proposed by Flament *et al.* (1992) (see Table III) the

Table III *ART-2a results obtained with different vigilance parameter values ρ^{max} (learning rate $\eta = 0.01$). The numbers in columns 2–7 are the indices of the clusters to which the rose varieties have been assigned. Indices between brackets indicate that the considered variety has been assigned to different clusters for separate runs. The last two columns give the classification obtained by HCA when cutting the dendrogram at such a level that four clusters are obtained.*

Rose no.	ρ^{max}= *0.1–0.3*	ρ^{max}= *0.4–0.5*	ρ^{max}= *0.6*	Flament*	HCA *Av.*†	HCA Ward‡
1	1	3	4	4	4	4
2	1	3	4	4	4	4
3	1	3	4	4	4	4
4	1	1	2	2	2	2
5	1	1	2	2	2	2
6	1	1	2	2	2	2
7	1	1	2	2	2	2
8	1	1	2	2	2	2
9	1	1	2	2	2	2
10	1	1	1	2	1	1
11	1	1	(1,2)	2	1	1
12	1	1	1	1	1	1
13	1	1	1	1	1	1
14	1	1	1	1	1	1
15	1	1	1	1	1	1
16	1	1	1	1	1	1
17	1	1	1	1	1	1
18	1	1	1	1	1	1
19	1	1	1	1	1	1
20	1	1	1	1	1	1
21	1	1	1	1	1	1
22	1	1	1	1	1	1
23	1	2	3	3	3	3
24	1	2	3	3	3	3
25	1	2	3	3	3	3
26	1	2	3	3	3	3
27	1	2	3	3	3	3

*After Flament *et al.* (1992); †Average linkage; ‡Ward's algorithm.

only difference therefore concerns the rose variety 10 which is classified as an ester type with ART-2a. This is not surprising since its percentage in esters is superior to that in alcohols. When ρ^{max} increases, the four clusters corresponding to $\rho^{max} = 0.6$ are split and rearrangements can be observed (results not shown). A finer analysis of the profiles is therefore obtained. However, due to the nature and the precision of the data at hand, further interpretations would not be significant and more information on the volatile constituents would be required.

As regards FuzzyART, it is noteworthy that the results are less stable (see Table IV for results obtained with ρ^{max} between 0.1 and 0.5), but when

Table IV *FuzzyART results obtained with different vigilance parameter values ρ^{max} (learning rate η = 0.01). The numbers in columns 2–5 are the indices of the clusters to which the rose varieties have been assigned. Indices between brackets indicate that the considered variety has been assigned to different clusters for separate runs.*

Rose no.	ρ^{max}= 0.1	ρ^{max}= 0.2	ρ^{max}= 0.3	ρ^{max}= 0.5
1	2	3	4	4
2	2	3	4	4
3	2	3	4	4
4	(2,1)	(1,2,2)	3	3
5	(2,1)	(1,2,2)	3	3
6	(2,1)	(1,2,2)	3	3
7	(2,1)	2	3	3
8	(2,1)	2	3	(3,1,3)
9	(2,1)	2	3	(3,1,3)
10	(1,2)	2	(2,3,3)	(3,1,1)
11	(2,1)	2	3	(1,5,1)
12	(1,2)	2	2	(1,5,1)
13	(1,2)	2	2	(1,5,1)
14	(1,2)	2	2	(1,5,1)
15	(1,2)	2	2	(1,5,1)
16	(1,2)	2	2	(1,5,1)
17	(1,2)	2	2	(1,5,1)
18	(1,2)	2	2	(1,5,1)
19	(1,2)	2	2	(1,5,1)
20	(1,2)	2	2	(1,5,1)
21	(1,2)	(2,3,1)	(2,3,1)	(1,5,5)
22	(2,2)	2	(2,2,3)	(3,1,1)
23	(1,2)	(1,1,3)	(1,1,2)	(2,3,2)
24	(1,2)	(1,1,3)	(1,1,2)	(2,3,2)
25	(1,2)	(1,1,3)	(1,1,2)	(2,3,2)
26	(1,2)	(1,1,2)	(1,1,3)	(5,1,2)
27	(1,2)	(1,3,3)	(1,3,2)	(2,3,2)

comparing the classification obtained at ρ^{max} = 0.3 with that obtained by Flament and coworkers (Flament *et al.*, 1992) and that resulting from the ART-2a at ρ^{max} = 0.6, the same general trends can be observed. However, in that particular case study, the ART-2a results should be preferred for their stability. In the same way, it is interesting to note that HCAs after the same scaling transformation as that used for ART-2a with the average linkage or the Ward's algorithm provide exactly the same results as ART-2a (Table III).

Up to now, as the results obtained by means of ART networks are comparable to those obtained by a visual inspection or HCA, it could be advanced that it is useless to use neural networks when classical methods perform as well. However, one may first note that this case study was chosen for the small size of the data table allowing visual inspection. In addition, the advantage of ART networks over the visual inspection is that by adjusting the ρ^{max} values, it is possible to inspect the structure of a data matrix gradually. In HCA, this kind of interpretation can be made by cutting the dendrograms at different levels. It has also been possible to stress that rose variety 11 could be considered both as an alcohol or as an ester type. The slight divergences observed from one run to another is in fact an advantage in detecting the sensitive points of a classification. Last, as described below, ART-2a and FuzzyART provide a series of numerical values which can be valuable for the interpretation of the results. Indeed, for example, ART-2a and FuzzyART allow one to perform a selection of 'key individuals' in each cluster. Thus, by inspecting the ρ^{calc} values which depict the degree of similarity between an individual and a cluster, it is possible to know which individual is the most characteristic of the considered cluster (i.e., the individual which is closest to the neural weight vector of its cluster). Thus, for the roses, if we consider, for example, the results obtained with ART-2a and ρ^{max} = 0.6, Table V indicates that the 'key rose varieties' (those presenting

Table V *Determination of the four 'key roses' (i.e., most representative) by ART-2a based on highest similarity between the neural weight vectors and the roses belonging to these clusters.*

Rose no.	Cluster membership	Similarity (ρ^{calc})	Key rose
1	4	0.9503	
2	4	0.9659	'key rose 4', SONIA MEILLAND
3	4	0.9551	
4	2	0.9963	'key rose 2', *Rosa rugosa rubra*
5	2	0.9776	
6	2	0.0697	
7	2	0.9234	
8	2	0.8428	
9	2	0.8398	
10	1	0.8752	
11	1	0.7339	
12	1	0.9575	
13	1	0.9876	
14	1	0.9806	
15	1	0.9918	'key rose 1', COCKTAIL 80
16	1	0.9679	
17	1	0.9563	
18	1	0.9550	
19	1	0.9800	
20	1	0.9671	
21	1	0.8316	
22	1	0.8565	
23	3	0.9832	
24	3	0.9963	'key rose 3', YOUKI SAN
25	3	0.9253	
26	3	0.7840	
27	3	0.7755	

Table VI *Comparison of the weight matrix* **W** *with the profiles of the key roses.* w_1 *to* w_4 *are the weight vectors of clusters 1 to 4 respectively (see Table III).*

Variable	Neural weight vector			
	w_1	w_2	w_3	w_4
R–H	0.00	0.00	0.18	0.99
R–OH	0.28	1.00	0.00	0.01
–COOR	0.93	0.00	0.00	0.01
Ar–O	0.25	0.00	0.98	0.14
Roses no.	10–22	4–9	23–27	1–3
Variable	Key rose profile			
	COCKTAIL 80	*R. rugosa rubra*	YOUKI SAN	SONIA MEILLAND
R–H	0.01	0.00	0.09	0.90
R–OH	0.11	1.00	0.00	0.41
–COOR	0.97	0.05	0.08	0.14
Ar–O	0.22	0.01	0.99	0.04

the maximum similarity ρ^{calc}) for clusters 1 to 4 are the varieties 15, 4, 24, and 2, respectively. This is confirmed by the comparison of the neural weight vectors of the different clusters with the profiles of the 'key roses' (Table VI). Note that it could also be possible to inspect the degree of similarity between the roses of a given cluster, with the other clusters (results not shown). The FuzzyART weights (Table VII) for a run with similar results show that weights are no longer 'crisp' numbers as in the case of ART-2a (Table VI). The upper and lower 'fuzzy' bounds of the FuzzyART's neural weights reflect the data ranges of individuals grouped in this particular cluster. The distinct range sizes and their different levels show quantitatively that the data cloud has been approximated by four distinct rectangles of different sizes as illustrated in Figure 4B.

From the weights of an ART-2a or a FuzzyART network, it is also possible to know the general characteristics of the individuals that each cluster contains. Thus, for example, inspection of Table VI indicates that cluster 1 contains rose varieties for which the percentages of esters are high (high weight value). To facilitate the interpretation, the weights can be represented by means of bar-charts. This is especially useful when, for example, the number of variables is high, as it has been recently shown in an ART analysis of spectral or chromatographic data sets (Wienke and Kateman, 1994; Wienke *et al.*, 1994b). As a result, compared with HCA, the possibility to study the weights gives a much more quantitative insight and provides a better understanding of the similarity relationships in a data set.

Last, information could also be extracted from the ART weight matrix by calculating the distance matrix of inter-class or inter-cluster angles (Wienke,

Table VII *Weight matrix* **W** *of a FuzzyART neural network (ρ^{max} = 0.3, η = 0.01, 1350 epochs) for the roses data set (Table II). w_1 to w_4 are the weight vectors of clusters 1 to 4 respectively (see Table IV). Note that the cluster indices are different compared with those of ART-2a but this has no influence on the results.*

Variable	Fuzzy range (lower to upper limit) of neural weight vector			
	w_1	w_2	w_3	w_4
R–H	0.03–0.38	0.00–0.19	0.00–0.20	0.54–0.97
R–OH	0.00–0.55	0.00–0.76	0.84–1.00	0.10–0.41
–COOR	0.01–0.51	0.46–1.00	0.05–0.54	0.10–0.81
Ar–O	0.79–0.99	0.01–0.47	0.00–0.21	0.04–0.25
Roses no.	23–27	10, 12–22	4–9, 11	1–3

1994). Indeed, these data stress the links between the different clusters (results not shown).

CASE STUDY II: OPTIMAL SELECTION OF ALIPHATIC SUBSTITUENTS

The first step of a QSAR study consists in the design of training and testing sets with compounds presenting a maximal information content. In the same way, the research for new biologically active chemicals in medicinal chemistry and agrochemistry requires inspection of several thousands of candidates for which only limited information is available. Due to time and cost problems, it is impossible in practice to test all the possible candidates, so that strategies such as test series design, aimed at selecting the most relevant compounds for biological testing, have to be elaborated. A plethora of publications dealing with this crucial problem can be found in the literature (see for example the review paper of Pleiss and Unger (1990)). Significant advances in the field dealt with the use of clustering and display linear multivariate methods allowing selection of test series by simple visual inspection of 2-D plots summarizing the information content of a matrix of physicochemical properties (e.g., Hansch *et al.*, 1973; Dove *et al.*, 1980; Alunni *et al.*, 1983; van de Waterbeemd *et al.*, 1989; Tosato and Geladi, 1990). More recently, nonlinear approaches such as nonlinear mapping (NLM) (Domine *et al.*, 1994a,b; Devillers, 1995), nonlinear neural mapping (Domine *et al.*, 1996) and genetic algorithms (Putavy *et al.*, 1996) have been proposed, opening new perspectives in the field. In this present case study, our aim was to show that ART networks represented a useful alternative for solving the complex problem of optimal test series selection due to their ability to perform clustering but also the fact that they provide useful information on the clusters formed and their relationships. Under these conditions, ART-2a and FuzzyART have been applied to a data matrix of 103 aliphatic substituents (Table VIII) described by the hydrophobic constant for aliphatic substituents Fr, H-bonding acceptor (HBA) and donor (HBD) abilities, the molar refractivity (MR), and the inductive parameter F of Swain and Lupton, respectively. The data were retrieved from a book of Hansch and Leo (1979). The selection of this data set was directed by the fact that the results could be compared with a series of methods so that the advantages and drawbacks of ART could be more easily evidenced. Indeed, this matrix was first analyzed by means of HCA by Hansch and Leo (1979). In addition, Domine and coworkers (Domine *et al.*, 1994b) proposed the use of NLM for obtaining a single easily interpretable map summarizing the information contained in the original data table.

In the present real-life case study, ART-2a and FuzzyART were tested. However, because the ART-2a results were better and more in accordance with those previously obtained by HCA and NLM than those obtained from FuzzyART, only these are detailed below. For ART-2a, data were scaled by normalizing the data in the range [0, 1] (i.e., using a min/max equation) per column. A quantitative analysis of the network training was performed by running the network with different values of the vigilance parameter and by monitoring the number of clusters formed during learning. Figure 5

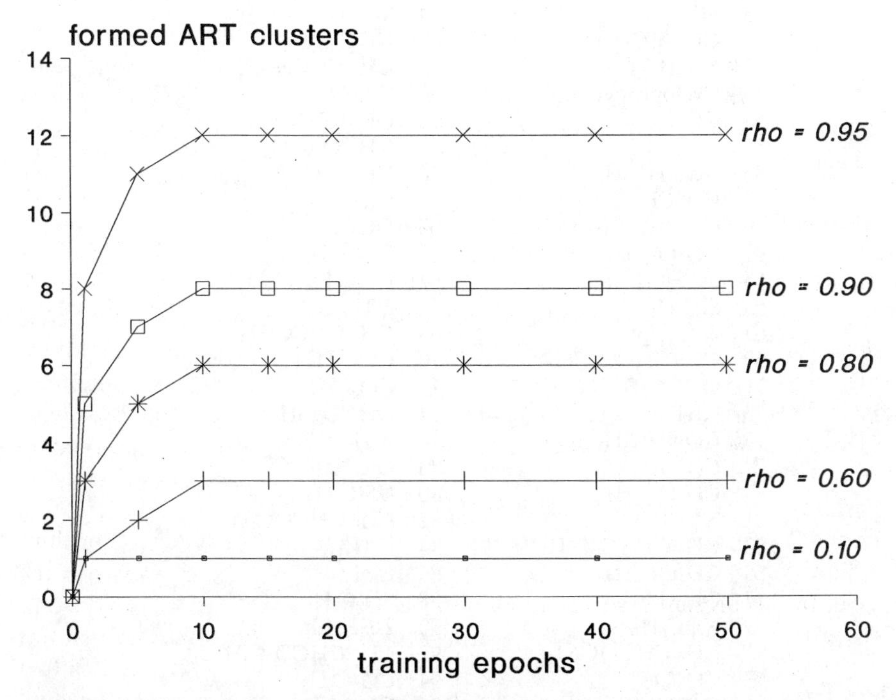

Figure 5 Plot of the number of clusters formed versus number of epochs for an ART-2a neural network trained with the aliphatic data set at various ρ^{max} values.

Table VIII *Aliphatic substituents.*

No.	Substituent	No.	Substituent
1	Br	2	Cl
3	F	4	I
5	NO_2	6	H
7	OH	8	SH
9	NH_2	10	CBr_3
11	CCl_3	12	CF_3
13	CN	14	SCN
15	CO_2^-	16	CO_2H
17	CH_2Br	18	CH_2Cl
19	CH_2I	20	$CONH_2$
21	CH=NOH	22	CH_3
23	$NHCONH_2$	24	OCH_3
25	CH_2OH	26	$SOCH_3$
27	OSO_2CH_3	28	SCH_3
29	$NHCH_3$	30	CF_2CF_3
31	C≡CH	32	CH_2CN
33	$CH{=}CHNO_2$-(trans)	34	$CH{=}CH_2$
35	$COCH_3$	36	$OCOCH_3$
37	CO_2CH_3	38	$NHCOCH_3$
39	$C{=}O(NHCH_3)$	40	CH_2CH_3
41	OCH_2CH_3	42	CH_2OCH_3
43	SOC_2H_5	44	SC_2H_5
45	$CH_2Si(CH_3)_3$	46	NHC_2H_5
47	$N(CH_3)_2$	48	CH=CHCN
49	Cyclopropyl	50	COC_2H_5
51	$CO_2C_2H_5$	52	$OCOC_2H_5$
53	$EtCO_2H$	54	$NHCO_2C_2H_5$
55	$CONHC_2H_5$	56	$NHCOC_2H_5$
57	$CH(CH_3)_2$	58	C_3H_7
59	$OCH(CH_3)_2$	60	OC_3H_7
61	$CH_2OC_2H_5$	62	SOC_3H_7
63	SC_3H_7	64	NHC_3H_7
65	$Si(CH_3)_3$	66	2-Thienyl
67	3-Thienyl	68	$CH{=}CHCOCH_3$
69	$CH{=}CHCO_2CH_3$	70	COC_3H_7
71	$OCOC_3H_7$	72	$CO_2C_3H_7$
73	$(CH_2)_3CO_2H$	74	$NHCOC_3H_7$
75	$CONHC_3H_7$	76	C_4H_9
77	$C(CH_3)_3$	78	OC_4H_9
79	$CH_2OC_3H_7$	80	NHC_4H_9
81	$N(C_2H_5)_2$	82	$CH{=}CHCOC_2H_5$
83	$CH{=}CHCO_2C_2H_5$	84	C_5H_{11}
85	$CH_2OC_4H_9$	86	C_6H_5
87	OC_6H_5	88	$SO_2C_6H_5$
89	NHC_6H_5	90	2-Benzthiazolyl
91	$CH{=}CHCOC_3H_7$	92	$CH{=}CHCO_2C_3H_7$
93	COC_6H_5	94	$CO_2C_6H_5$
95	$OCOC_6H_5$	96	$NHCOC_6H_5$
97	$CH_2C_6H_5$	98	$CH_2OC_6H_5$

Table VIII *continued.*

No.	Substituent	No.	Substituent
99	$CH_2Si(C_2H_5)_3$	100	$CH{=}CHC_6H_5$-(trans)
101	$CH{=}CHCOC_6H_5$	102	Ferrocenyl
103	$N(C_6H_5)_2$		

illustrates, for five different values of the vigilance parameter (ρ^{max}), the number of clusters formed and the number of cycles required for convergence with ART-2a. It is noteworthy that in all cases, convergence was obtained in about 10 epochs. This underlines the ability of ART networks to perform clustering very rapidly. This presents two main advantages. Indeed, the rapid convergence in training allows us to optimize quickly the network parameters and therefore many configurations can be tested for exploratory data analysis.

From Figure 5, it can also be deduced that the dominant clusters are always formed at the very beginning of the training. Indeed, the slopes of the curves shown in Figure 5 are very high at the beginning and then decrease to reach zero very rapidly. Last, the low sensitivity between $\rho^{max} = 0.1$ and $\rho^{max} = 0.6$ which always yield three clusters, indicates that there are basically three strong clusters in the structure of the data set. From a practical point of view, this stability indicates that the clusters are well defined. To compare our results with those previously obtained, the ρ^{max} value was fixed so that 10 clusters could be obtained. This underlines again the advantage of this kind of network which allows a gradual inspection of a data set and also a monitoring of the precision required in the clustering. The appropriate ρ^{max} value for the above constraint was found to be 0.92. The network was run for 30 epochs so as to ensure the stability of the results. It has to be noted that some substituents located at the borderline between two clusters have been found to fall in a neighbor cluster during separate runs but this does not influence the conclusions that are drawn below from one of these runs. Indeed, for the sake of brevity the other runs are not detailed below but they would simply provide complementary information as stressed in the case study on rose varieties. Membership of the 103 substituents to the ten clusters is given in Table IX. For comparison purposes, although the NLM was performed after a different scaling transformation, we have represented, on the nonlinear map of the 103 substituents recently published (Domine *et al.*, 1994b), the membership to the different clusters (Figure 6). It is noteworthy that despite the different preprocessing of the data, the results are generally in agreement. Indeed, the ART-2a clusters allow one to divide the nonlinear map logically into 10 regions of physicochemically similar substituents. Some slight discrepancies with the NLM can be observed in clusters 5 and 10. Indeed, substituent 87, which belongs to cluster 5, is found in the middle of some substituents of cluster 10; this atypical assignment is

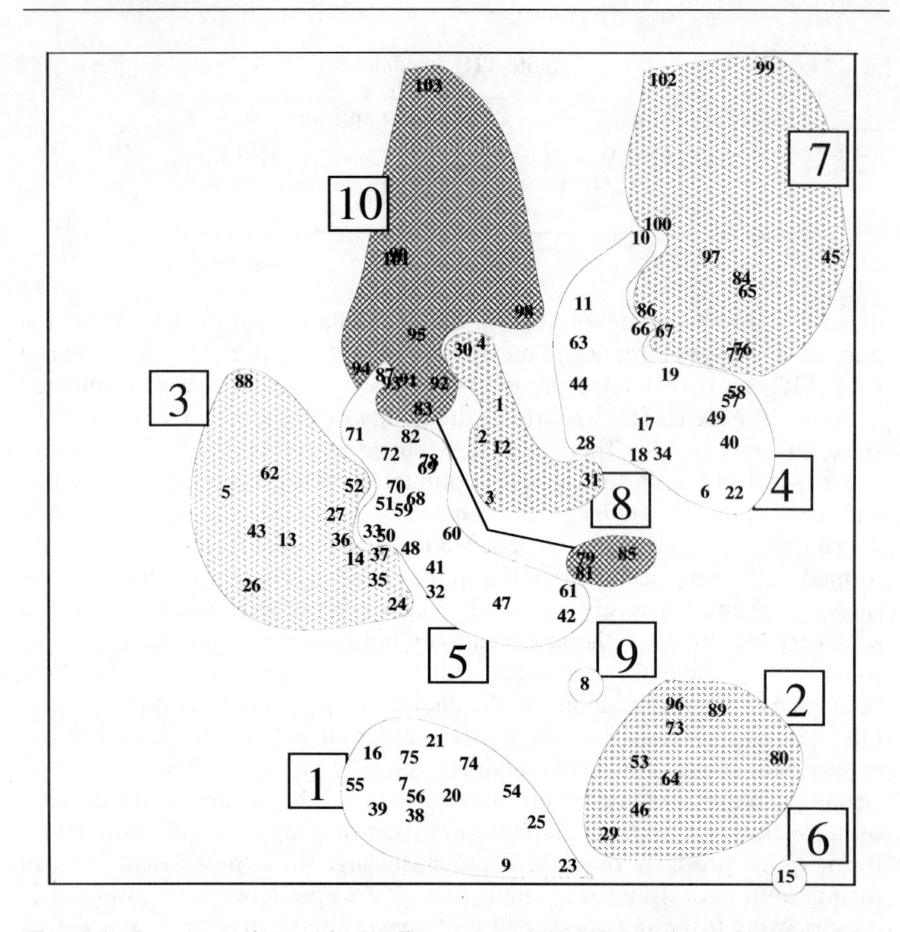

Figure 6 Representation on the nonlinear map of the 103 substituents of their cluster memberships derived by ART-2a. See Table VIII for correspondence between the 'small' numbers and the substituents. The large numbers inside a square correspond to the cluster indices given in Table IX. Cluster 10 is split into two groups linked by a line.

due to its borderline location between the two clusters, as it will be explained later. As regards substituents 79, 81, and 85, a reason for their atypical position might be the different scaling procedures used in the two methods. Compared with NLM and HCA, the atypical character of substituents 8 and 15 is also stressed by the ART-2a results since they are isolated. To go further in the comparison, the key individuals of each cluster derived by ART-2a were determined using the ρ^{calc} values, as explained in the previous case study dealing with rose varieties, and compared to the selection proposed from HCA and NLM in a previous paper (Domine *et al.*, 1994b). The results are summarized in Tables IX and X. Inspection of Table X shows that the

Table IX *Contents of the ten ART-2a derived clusters. The underlined substituents are those proposed in a previous study (Domine* et al., *1994b).*

Cluster no.	Substituents
1	7, 9, 16, 20, 21, 23, 25, 38, 39, 54, 55, 56, 74, 75
2	29, 46, 53, 64, 73, 80, 89, 96
3	5, 13, 14, 24, 26, 27, 35, 36, 37, 43, 52, 62, 88
4	6, 10, 11, 17, 18, 19, 22, 28, 34, 40, 44, 49, 57, 58, 63, 66
5	32, 33, 41, 42, 47, 48, 50, 51, 59, 60, 61, 68, 69, 70, 71, 72, 78, 82, 87
6	15
7	45, 65, 67, 76, 77, 84, 86, 97, 99, 100, 102
8	1, 2, 3, 4, 12, 30, 31
9	8
10	79, 81, 83, 85, 90, 91, 92, 93, 94, 95, 98, 101, 103

numerical values provided by ART-2a, and which guide the selection of the key individuals among the ten clusters, allow one to find similar results to those obtained from HCA and NLM. Indeed, in Table X, it is noteworthy that in all the clusters derived except the sixth one, we find at least one of the previously selected substituents. As regards cluster 6 which contains substituent 15 (i.e., CO_2^-), it has to be considered that HCA also isolated this substituent in a cluster and that it was decided not to select it due to its atypical character. Beside this agreement between the different methods, one advantage of ART-2a is that the preliminary selection of key substituents among each cluster is directed by the ρ^{calc} values (Table X). Indeed, the largest values correspond to the substituents that are the most characteristic of the cluster considered. If ever for any reason, such as difficulty of synthesis or instability in biological systems, a substituent was found not to be adequate, then the ρ^{calc} values can suggest the selection of the second closest substituent and so on. Thus, for example, if in cluster 8, Br (substituent 1) was not satisfactory, ART-2a suggests trying with Cl (substituent 2) or CF_3 (substituent 12), which have the same ρ^{calc} values. Thus, the system is open and freedom is left to the chemist to use any criterion for the selection of an optimal test series. These values can also be used to determine substituents that are at the borderline between two clusters and therefore determine the relative position of each substituent. Thus, for example, it is noteworthy in Figure 6, that substituent 87 is placed in cluster 5 while it is among substituents belonging to cluster 10. Inspection of the ρ^{calc} (data not given) values show that substituent 87 is actually on the borderline between cluster 5 and 10 since its ρ^{calc} values for these two clusters are high and close to each other. As regards substituents 79, 81, and 85, the ρ^{calc} values also suggest a location at the borderline between clusters 5 and 10.

Other useful parameters for the interpretation of ART-2a results and the exploration of a data matrix are the inter-class angles (Table XI). They

Table X *Comparison of test series selected by classical methods (Domine* et al., *1994b) versus those selected by an ART-2a network.* ρ^{calc} *values of the substituents that are closest to the neural weight vector are given between brackets. The* ρ^{calc} *values of the substituents selected by HCA and NLM are also given.*

Cluster no.	Selected by HCA, NLM	Selected by ART-2a
1	7: OH	7: OH (0.998)
2	96: $NHCOC_6H_5$	80: NHC_4H_9 (0.998), 96: $NHCOC_6H_5$ (0.960)
3	13: CN, 24: OCH_3	36: $OCOCH_3$ (0.999), 14: SCN (0.998), 35: $COCH_3$ (0.998), 37: CO_2CH_3 (0.998), 13: CN (0.992), 24: OCH_3 (0.992)
4	6: H	17: CH_2Br (0.999), 34: $CH{=}CH_2$ (0.996), 6: H (0.952)
5	87: OC_6H_5	59: $OCH(CH_3)_2$ (0.999), 60: OC_3H_7 (0.999), 87: OC_6H_5 (0.992),
6		15: CO_2^- (1.000)
7	100: $CH{=}CHC_6H_5$-(trans), 76:C_4H_9	65: $Si(CH_3)_3$ (0.997), 84: C_5H_{11} (0.997), 100: $CH{=}CHC_6H_5$-(trans) (0.992), 76: C_4H_9 (0.990)
8	1: Br	1: Br (0.998), 2: Cl (0.998), 12: CF_3 (0.998), 30: CF_2CF_3 (0.998)
9	8: SH	8: SH (1.00)
10	81: $N(C_2H_5)_2$	95: $OCOC_6H_5$ (0.999), 81: $N(C_2H_5)_2$ (0.992)

provide information on the relationships between the clusters. Thus, the larger the cosine of the angles, the more similar the clusters. From a practical point of view, this allows us to select a substituent in another cluster if ever no appropriate substituents could be found in a given cluster. From Table XI, for example, if one wants to replace a substituent of cluster 1 by a substituent in another cluster, the inter-class angles suggest cluster 2. This is verified on the nonlinear map.

Last, if the derivation of clusters and the selection of key substituents is very useful, it is always also necessary to have information on the clusters formed. This is achieved by inspection of the weights of each cluster which provide an insight into the composition of the clusters in terms of physicochemical properties. Thus, by plotting for the clusters the values of the weights linked to each substituent constant, we obtain physicochemical profiles of the substituents constituting the clusters. In Figures 7–10, four examples of bar-charts corresponding to clusters 1, 2, 3, and 7, respectively, are drawn. Figure 7 indicates that the substituents of cluster 1 are characterized by relatively

Table XI *Cosine of the inter-class angles. The larger the value, the closer the clusters.*

Cluster no.	1	2	3	4	5	6	7	8	9	10
1	1.000	0.918	0.743	0.354	0.735	0.685	0.281	0.402	0.755	0.686
2	0.918	1.000	0.636	0.336	0.683	0.663	0.382	0.226	0.664	0.708
3	0.743	0.636	1.000	0.574	0.989	0.838	0.488	0.639	0.396	0.943
4	0.354	0.336	0.574	1.000	0.611	0.092	0.954	0.915	0.577	0.661
5	0.735	0.683	0.989	0.611	1.000	0.839	0.563	0.616	0.387	0.980
6	0.685	0.663	0.838	0.092	0.839	1.000	0.078	0.118	0.077	0.790
7	0.281	0.382	0.488	0.954	0.563	0.078	1.000	0.764	0.495	0.666
8	0.402	0.226	0.639	0.915	0.616	0.118	0.764	1.000	0.616	0.592
9	0.755	0.664	0.396	0.577	0.387	0.077	0.495	0.616	1.000	0.380
10	0.686	0.708	0.943	0.661	0.980	0.790	0.666	0.592	0.380	1.000

low Fr values; they can accept and donate H-bonds, they possess low MR values and their F values are average. In comparison, the substituents of cluster 2 (Figure 8) generally present higher MR values and lower F values. The principal characteristics of the substituents of cluster 3 (Figure 9) are their H-bond acceptor ability and their relatively high F values. They cannot donate H-bonds. Figure 10 shows that the substituents included in cluster 7 are mainly characterized by their high Fr values and the fact that they cannot accept or donate H-bonds. The same kind of interpretation could be made from the bar-charts of all the other clusters. It therefore becomes obvious that ART-2a does not only cluster the points but it provides numerous numerical data (e.g., weights, ρ^{calc}, intra-class angles) which are useful for the interpretation of the results.

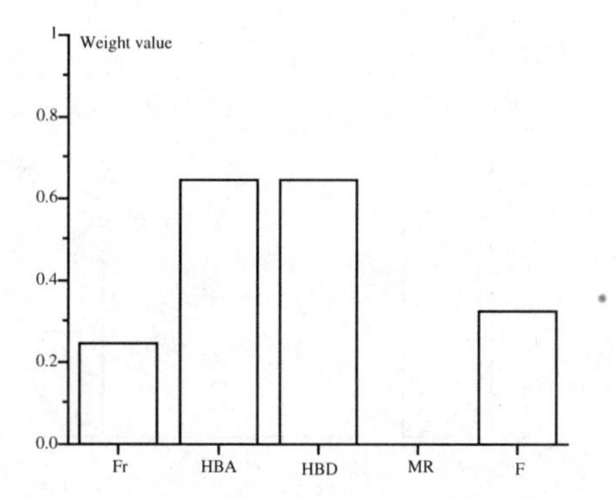

Figure 7 Bar-chart of weight values for cluster 1.

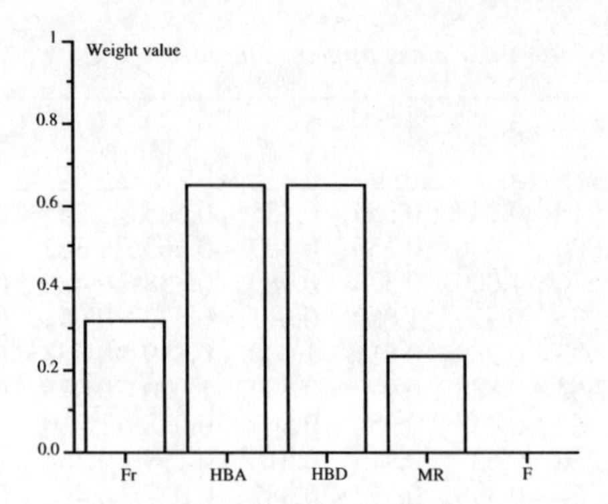

Figure 8 Bar-chart of weight values for cluster 2.

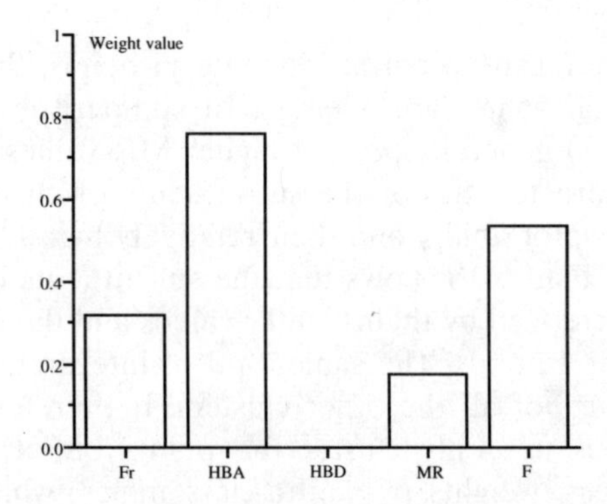

Figure 9 Bar-chart of weight values for cluster 3.

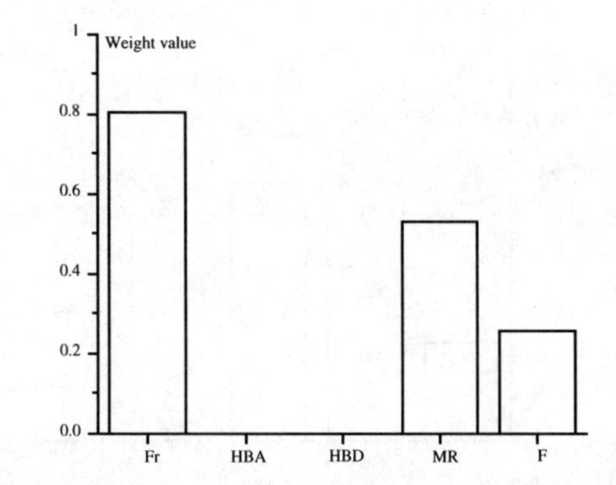

Figure 10 Bar-chart of weight values for cluster 7.

CONCLUSIONS

The present study evidences the heuristic potency of ART networks and more specifically of ART-2a and FuzzyART in computer-aided molecular design. These techniques, which perform nonlinear cluster analyses of data sets, have been shown to be valuable tools for QSAR data analysis. It has been shown that ART networks converge within a few epochs. The size of the data sets used in the present study was rather small. Thus, the gain in speed by using ART compared to classical clustering methods or compared to other neural networks is not really significant. Such gain in speed becomes very interesting for large data sets with many thousands of individuals. ART neural networks offer the possibility of interpretation of the clusters derived in terms of the original variables by means of their weights, which provide the characteristic profiles of the individuals constituting the clusters. Graphical representations, such as bar-charts, are useful complementary tools for visualizing the information condensed in a trained ART neural network. An interesting feature of these networks is also that by setting the vigilance parameter ρ at different levels, it is possible to inspect the structure of the data sets at hand gradually. This is also made possible by the fact that convergence is rapidly obtained. From a practical point of view, ART networks and, more specifically, ART-2a appeared very useful for clustering rose varieties and also for performing optimal test series.

Beside this, however, a few drawbacks can be underlined. Thus, for example, in both case studies, the results obtained with FuzzyART did not show a high stability for separate runs. This is damaging and further studies should be started since the alternative concept of FuzzyART with its ranges of weights could provide more flexibility in the conclusions drawn. FuzzyART also presents the advantage over ART-2a that it can make use of clusters with distinct sizes. However, this instability can in some instances be an advantage since it provides information on the weaknesses of a classification.

ART networks should be considered as new powerful tools in (Q)SAR. They merit further investigations due to the promising results they provide.

REFERENCES

Albano, C., Dunn III, W., Edlund, U., Johansson, B., Norden, B., Sjöström, M., and Wold, S. (1978). Four levels of pattern recognition. *Anal. Chim. Acta* **103,** 429–436.

Alunni, S., Clementi, S., Edlund, U., Johnels, D., Hellberg, S., Sjöström, M., and Wold, S. (1983). Multivariate data analysis of substituent descriptors. *Acta Chem. Scand.* **B 37,** 47–53.

Andrea, T.A. and Kalayeh, H. (1991). Applications of neural networks in quantitative structure–activity relationships of dihydrofolate reductase inhibitors. *J. Med. Chem.* **34,** 2824–2836.

Aoyama, T. and Ichikawa, H. (1991a). Basic operating characteristics of neural networks when applied to structure–activity studies. *Chem. Pharm. Bull.* **39,** 358–366.

Aoyama, T. and Ichikawa, H. (1991b). Obtaining the correlation indices between drug activity and structural parameters using a neural network. *Chem. Pharm. Bull.* **39,** 372–378.

Aoyama, T. and Ichikawa, H. (1991c). Reconstruction of weight matrices in neural networks – A method of correlating outputs with inputs. *Chem. Pharm. Bull.* **39,** 1222–1228.

Aoyama, T. and Ichikawa, H. (1992). Neural networks as nonlinear structure–activity relationship analyzers. Useful functions of the partial derivative methód in multilayer neural networks. *J. Chem. Inf. Comput. Sci.* **32,** 492–500.

Aoyama, T., Suzuki, Y., and Ichikawa, H. (1990). Neural networks applied to quantitative structure–activity relationships analysis. *J. Med. Chem.* **33,** 2583–2590.

Benjamin, C.O., Chi, S.-C., Gaber, T., and Riordan, C.A. (1995). Comparing BP and ART II neural network classifiers for facility location. *Comput. Ind. Engin.* **28,** 43–50.

Bienfait, B. (1994). Applications of high-resolution self-organizing maps to retrosynthetic and QSAR analysis. *J. Chem. Inf. Comput. Sci.* **34,** 890–898.

Borman, S. (1989). Neural network applications in chemistry begin to appear. Neurocomputer systems hold promise for predicting chemical reaction products, drug side effects, protein folding from sequence data. *Sci. Technol.* **April,** 24–28.

Bos, A., Bos, M., and van der Linden, W.E. (1992). Artificial neural networks as a tool for soft-modelling in quantitative analytical chemistry: The prediction of the water content of cheese. *Anal. Chim. Acta* **256,** 133–144.

Bos, M., Bos, A., and van der Linden, W.E. (1993). Data processing by neural networks in quantitative chemical analysis. *Analyst* **118,** 323–328.

Bruchmann, A., Götze, H.-J., and Zinn, P. (1993). Application of Hamming networks for IR spectral search. *Chemom. Intell. Lab. Syst.* **18,** 59–69.

Budzinski, H., Garrigues, P., Connan, J., Devillers, J., Domine, D., Radke, M., and Oudin, J.L. (1995). Alkylated phenanthrene distributions as maturity and origin indicators in crude oils and rock extracts. *Geochim. Cosmochim. Acta* **59,** 2043–2056.

Burke, L.I. (1991). Clustering characterization of adaptive resonance. *Neural Networks* **4,** 485–491.

Cambon, B. and Devillers, J. (1993). New trends in structure–biodegradability relationships. *Quant. Struct.–Act. Relat.* **12,** 49–56.

Carpenter, G.A. and Grossberg, S. (1987). ART 2: Self-organization of stable category recognition codes for analog input patterns. *Appl. Optics* **26,** 4919–4930.

Carpenter, G.A. and Grossberg, S. (1990). ART 3: Hierarchical search using chemical transmitters in self-organizing pattern recognition architectures. *Neural Networks* **3,** 129–152.

Carpenter, G.A. and Grossberg, S. (1991). *Pattern Recognition by Self-Organizing Neural Networks.* The MIT Press, Cambridge, Massachusetts, p. 691.

Carpenter, G.A., Grossberg, S., Markuzon, N., Reynolds, J.H., and Rosen, D.B. (1992). FuzzyARTMAP: A neural network architecture for incremental supervised learning of analog multidimensional maps. *IEEE Trans. Neural Networks* **3,** 698–713.

Carpenter, G.A., Grossberg, S., and Reynolds, J.H. (1991a). ARTMAP: Supervised real-time learning and classification of nonstationary data by a self-organizing neural network. *Neural Networks* **4,** 565–588.

Carpenter, G.A., Grossberg, S., and Rosen, D.B. (1991b). ART 2-A: An adaptive resonance algorithm for rapid category learning and recognition. *Neural Networks* **4,** 493–504.

Carpenter, G.A., Grossberg, S., and Rosen, D.B. (1991c). Fuzzy ART: Fast stable learning and categorization of analog patterns by an adaptive resonance system. *Neural Networks* **4,** 759–771.

Caudell, T.P. (1992). Hybrid optoelectronic adaptive resonance theory neural processor, ART1. *Appl. Optics* **31,** 6220–6229.

Chastrette, M., de Saint Laumer, J.Y., and Peyraud, J.F. (1993). Adapting the structure of a neural network to extract chemical information. Application to structure–odour relationships. *SAR QSAR Environ. Res.* **1,** 221–231.

Chastrette, M., Devillers, J., Domine, D., and de Saint Laumer, J.Y. (1994). New tools for the selection and critical analysis of large collections of data. In, *New Data Challenges in Our Information Age. Thirteenth International CODATA Conference, Beijing, China, October 1992* (P.S. Glaeser and M.T.L. Millward, Eds.). CODATA, Ann Arbor, pp. C29–C35.

Court, J.P., Murgatroyd, R.C., Livingstone, D., and Rahr, E. (1988). Physicochemical characteristics of non-electrolytes and their uptake by *Brugia pahangi* and *Dipetalonema viteae. Mol. Biochem. Parasitol.* **27,** 101–108.

de Saint Laumer, J.Y., Chastrette, M., and Devillers, J. (1991). Multilayer neural networks applied to structure–activity relationships. In, *Applied Multivariate Analysis in SAR and Environmental Studies* (J. Devillers and W. Karcher, Eds.). Kluwer Academic Publishers, Dordrecht, pp. 479–521.

Devillers, J. (1993). Neural modelling of the biodegradability of benzene derivatives. *SAR QSAR Environ. Res.* **1,** 161–167.

Devillers, J. (1995). Display of multivariate data using non-linear mapping. In, *Chemometric Methods in Molecular Design* (H. van de Waterbeemd, Ed.). VCH, Weinheim, pp. 255–263.

Devillers, J. and Cambon, B. (1993). Modeling the biological activity of PAH by neural networks. *Polycyclic Aromatic Compounds* **3 (supp),** 257–265.

Devillers, J. and Domine, D. (1995). Deriving structure–chemoreception relationships from the combined use of linear and nonlinear multivariate analyses. In, *QSAR and Molecular Modelling: Concepts, Computational Tools and Biological Applications* (F. Sanz, J. Giraldo, and F. Manaut, Eds.). J.R. Prous Science Publishers, Barcelona, pp. 57–60.

Devillers, J., Domine, D., and Bintein, S. (1994). Multivariate analysis of the first 10 MEIC chemicals. *SAR QSAR Environ. Res.* **2,** 261–270.

Devillers, J., Domine, D., and Boethling, R.S. (1996). Use of a backpropagation neural network and autocorrelation descriptors for predicting the biodegradation of organic chemicals. In, *Neural Networks in QSAR and Drug Design* (J. Devillers, Ed.). Academic Press, London, pp. 65–82.

Devillers, J. and Karcher, W. (1991). *Applied Multivariate Analysis in SAR and Environmental Studies*. Kluwer Academic Publishers, Dordrecht, p. 530.

Domine, D. and Devillers, J. (1995). Nonlinear multivariate SAR of Lepidoptera pheromones. *SAR QSAR Environ. Res.* **4,** 51–58.

Domine, D., Devillers, J., and Chastrette, M. (1994a). A nonlinear map of substituent constants for selecting test series and deriving structure–activity relationships. 1. Aromatic series. *J. Med. Chem.* **37,** 973–980.

Domine, D., Devillers, J., and Chastrette, M. (1994b). A nonlinear map of substituent constants for selecting test series and deriving structure–activity relationships. 2. Aliphatic series. *J. Med. Chem.* **37,** 981–987.

Domine, D., Devillers, J., Chastrette, M., and Doré, J.C. (1995). Combined use of linear and nonlinear multivariate analyses in structure–activity relationship studies: Application to chemoreception. In, *Computer-Aided Molecular Design. Applications in Agrochemicals, Materials, and Pharmaceuticals* (C.H. Reynolds, M.K. Holloway, and H.K. Cox, Eds.). ACS Symposium Series 589, American Chemical Society, Washington, DC, pp. 267–280.

Domine, D., Devillers, J., Chastrette, M., and Karcher, W. (1993a). Non-linear mapping for structure–activity and structure–property modelling. *J. Chemometrics* **7,** 227–242.

Domine, D., Devillers, J., Chastrette, M., and Karcher, W. (1993b). Estimating pesticide field half-lives from a backpropagation neural network. *SAR QSAR Environ. Res.* **1,** 211–219.

Domine, D., Devillers, J., Garrigues, P., Budzinski, H., Chastrette, M., and Karcher, W. (1994c). Chemometrical evaluation of the PAH contamination in the sediments of the Gulf of Lion (France). *Sci. Total Environ.* **155,** 9–24

Domine, D., Wienke, D., Devillers, J., and Buydens, L. (1996). A new nonlinear neural mapping technique for visual exploration of QSAR data. In, *Neural Networks in QSAR and Drug Design* (J. Devillers, Ed.). Academic Press, London, pp. 223–253.

Dove, S., Streich, W.J., and Franke, R. (1980). On the rational selection of test series. 2. Two-dimensional mapping of intraclass correlation matrices. *J. Med. Chem.* **23,** 1456–1459.

Fang, X., Yu, B., Xiang, B., and An, D. (1990). Application of pyrolysis-high-resolution gas chromatography-pattern recognition to the identification of the Chinese traditional medicine Mai Dong. *J. Chromatogr.* **514,** 287–292.

Feuilleaubois, E., Fabart, V., and Doucet, J.P. (1993). Implementation of the three-dimensional-pattern search problem on Hopfield-like neural networks. *SAR QSAR Environ. Res.* **1,** 97–114.

Flament, I., Debonneville, C., and Furrer, A. (1992). Volatile constituents of roses. Characterization of cultivars based on the headspace analysis of living flower emissions. In, *Bioactive Volatile Compounds From Plants* (R. Teranishi, R.G. Buttery, and H. Sugisawa, Eds.). ACS Symposium Series 525, American Chemical Society, Washington, DC, pp. 269–281.

Forina, M., Armanino, C., Lanteri, S., and Calcagno, C. (1983). Simplified non linear mapping of analytical data. *Anal. Chim.* **73,** 641–657.

Gan, K.M. and Lua, K.T. (1992). Chinese character classification using an adaptive resonance network. *Pattern Recogn.* **25,** 877–882.

Geladi, P. and Tosato, M.L. (1990). Multivariate latent variable projection methods: SIMCA and PLS. In, *Practical Applications of Quantitative Structure–Activity Relationships (QSAR) in Environmental Chemistry and Toxicology* (W. Karcher and J. Devillers, Eds.). Kluwer Academic Publishers, Dordrecht, pp. 171–179.

Grossberg, S. (1976a). Adaptive pattern classification and universal recoding, I: Parallel development and coding of neural feature detectors. *Biol. Cybern.* **23,** 121–134.

Grossberg, S. (1976b). Adaptive pattern classification and universal recoding, II: Feedback, expectation, olfaction, and illusions. *Biol. Cybern.* **23,** 187–203.

Grossberg, S. (1982). *Studies of Mind and Brain.* D. Reidel Publishing Company, Dordrecht, p. 662.

Hansch, C. and Fujita, T. (1964). ρ-σ-π Analysis. A method for the correlation of biological activity and chemical structure. *J. Am. Chem. Soc.* **86,** 1616–1626.

Hansch, C. and Leo, A. (1979). *Substituent Constants for Correlation Analysis in Chemistry and Biology*. John Wiley & Sons, New York, p. 339.

Hansch, C., Unger, S.H., and Forsythe, A.B. (1973). Strategy in drug design. Cluster analysis as an aid in the selection of substituents. *J. Med. Chem.* **16,** 1217–1222.

Henrion, R., Henrion, G., Heininger, P., and Steppuhn, G. (1990). Three-way principal components analysis for multivariate evaluation of round robin tests. *J. Anal. Chem.* **336,** 37.

Ho, C.S., Liou, J.J., Georgiopoulos, M., Heileman, G.L., and Christodoulou, C. (1994). Analog circuit design and implementation of an adaptive resonance theory (ART) neural network architecture. *Int. J. Electronics* **76,** 271–291.

Hohenstein, R. (1994). Classification of neuro-magnetic field patterns using a fuzzy ARTMAP network. Master Student Research Report, Wilhelms-University Münster, Institute Numerical and Instrumental Math/Informatics, Germany.

Hudson, B., Livingstone, D.J., and Rahr, E. (1989). Pattern recognition display methods for the analysis of computed molecular properties. *J. Comput.-Aided Mol. Design* **3,** 55–65.

Jankrift, A. (1993). Classification of neuro-magnetical field patterns by a cascade correlation artificial neural network. Master Thesis, Wilhelms-University Münster, Germany, p. 114.

Kane, J.S. and Paquin, M.J. (1993). POPART: Partial optical implementation of adaptive resonance theory 2. *IEEE Trans. Neural Networks* **4,** 695–702.

Kohonen, T. (1989). *Self-Organization and Associated Memory*. Springer-Verlag, Heidelberg.

Kohonen, T. (1990). The self-organizing map. *Proc. IEEE* **78,** 1464–1480.

Kowalski, B.R. and Bender, C.F. (1972). Pattern recognition. A powerful approach to interpreting chemical data. *J. Am. Chem. Soc.* **94,** 5632–5639.

Kowalski, B.R. and Bender, C.F. (1973). Pattern recognition. II. Linear and nonlinear methods for displaying chemical data. *J. Am. Chem. Soc.* **95,** 686–692.

Lewi, P.J. (1982). *Multivariate Data Analysis in Industrial Practice*. Research Studies Press, Chichester, p. 244.

Lin, C.C.D. and Wang, H.P.B. (1993). Classification of autoregressive spectral estimated signal patterns using an adaptive resonance theory neural network. *Comput. Indus.* **22,** 143–152.

Liu, Q., Hirono, S., and Moriguchi, I. (1992a). Application of functional–link net in QSAR. 1. QSAR for activity data given by continuous variate. *Quant. Struct.–Act. Relat.* **11,** 135–141.

Liu, Q., Hirono, S., and Moriguchi, I. (1992b). Application of functional–link net in QSAR. 2. QSAR for activity data given by ratings. *Quant. Struct.–Act. Relat.* **11,** 318–324.

Livingstone, D.J. (1989). Multivariate quantitative structure–activity relationship (QSAR) methods which may be applied to pesticide research. *Pestic. Sci.* **27,** 287–304.

Livingstone, D.J., Ford, M.G., and Buckley, D.S. (1988). A multivariate QSAR study of pyrethroid neurotoxicity based upon molecular parameters derived by computer chemistry. In, *Neurotox'88: Molecular Basis of Drug and Pesticide Action* (G.G. Lunt, Ed.). Elsevier Science Publishers, Amsterdam, pp. 483–495.

Livingstone, D.J. and Manallack, D.T. (1993). Statistics using neural networks: Chance effects. *J. Med. Chem.* **36,** 1295–1297.

Livingstone, D.J. and Salt, D.W. (1992). Regression analysis for QSAR using neural networks. *Bioorg. Med. Chem. Lett.* **2,** 213–218.

Manallack, D.T. and Livingstone, D.J. (1994). Neural networks and expert systems in molecular design. Neural networks – A tool for drug design. In, *Advanced Computer-Assisted Techniques in Drug Discovery* (H. van de Waterbeemd, Ed.). VCH, Weinheim, pp. 293–318.

Massart, D.L., Vandeginste, B.G.M., Deming, S.N., Michotte, Y., and Kaufman, L. (1988). *Chemometrics – A Textbook*. Elsevier Publishers, Amsterdam, p. 488.

Melssen, W.J., Smits, J.R.M., Rolf, G.H., and Kateman, G. (1993). 2-dimensional mapping of IR spectra using a parallel implemented self-organizing feature map. *Chemom. Intell. Lab. Syst.* **18,** 195–204.

Nicholson, J.K. and Wilson, I.D. (1989). High resolution proton magnetic resonance spectroscopy of biological fluids. *Prog. NMR Spectrosc.* **21,** 449–501.

Peterson, K.L. (1992). Counter-propagation neural networks in the modeling and prediction of Kovats indices for substituted phenols. *Anal. Chem.* **64,** 379–386.

Peterson, K.L. (1995). Quantitative structure–activity relationships in carboquinones and benzodiazepines using counter-propagation neural networks. *J. Chem. Inf. Comput. Sci.* **35,** 896–904.

Pleiss, M.A. and Unger, S.H. (1990). The design of test series and the significance of QSAR relationships. In, *Comprehensive Medicinal Chemistry*, Vol. 4 (C.A. Ramsden, Ed.). Pergamon Press, Oxford, pp. 561–587.

Putavy, C., Devillers, J., and Domine, D. (1996). Genetic selection of aromatic substituents for designing test series. In, *Genetic Algorithms in Molecular Modeling* (J. Devillers, Ed.). Academic Press, London, pp. 243–269.

Rauret, G., Rubio, R., Rius, F.X., and Larrechi, M.S. (1988). Cluster analysis as a tool in the study of groundwater quality. *Intern. J. Environ. Anal. Chem.* **32,** 255–268.

Resch, C. and Szabo, Z. (1994). Category detection in pet image series by adaptive resonance theory (ART) neural networks. *J. Nuclear Med.* **5,** 182.

Rose, V.S., Hyde, R.M., and MacFie, H.J.H. (1990). U.K. usage of chemometrics and artificial intelligence in QSAR analysis. *J. Chemometrics* **4,** 355–360.

Sharaf, M.A., Illman, D.L., and Kowalski, B.R. (1986). *Chemometrics.* John Wiley & Sons, New York, p. 332.

Simon, V., Gasteiger, J., and Zupan, J. (1993). A combined application of two different neural network types for the prediction of chemical reactivity. *J. Am. Chem. Soc.* **115,** 9148–9159.

Sjöström, M. and Kowalski, B.R. (1979). A comparison of five pattern recognition methods based on the classification results from six real data bases. *Anal. Chim. Acta* **112,** 11–30.

Smits, J.R.M., Melssen, W.J., Buydens, L.M.C., and Kateman, G. (1994a). Using artificial neural networks for solving chemical problems. I. Multi-layer feed-forward networks. *Chemom. Intell. Lab. Syst.* **22,** 165–189.

Smits, J.R.M., Melssen, W.J., Buydens, L.M.C., and Kateman, G. (1994b). Using artificial neural networks for solving chemical problems. I. Kohonen self–organizing feature maps and Hopfield networks. *Chemom. Intell. Lab. Syst.* **23,** 267–291.

Sumpter, B.G., Getino, C., and Noid, D.W. (1994). Theory and applications of neural computing in chemical science. *Ann. Rev. Phys. Chem.* **45,** 439–481.

Thomson, J.U. and Meyer, B. (1989). Pattern recognition of the 1H NMR spectra of sugar additols using a neural network. *J. Magnetic Resonance* **84,** 84–93.

Tosato, M.L. and Geladi, P. (1990). Design: A way to optimize testing programmes for QSAR screening of toxic substances. In, *Practical Applications of Quantitative Structure–Activity Relationships (QSAR) in Environmental Chemistry and Toxicology* (W. Karcher and J. Devillers, Eds.). Kluwer Academic Publishers, Dordrecht, pp. 317–341.

Treiger, B., Bondarenko, I., van Espen, P., van Grieken, R., and Adams, F. (1994). Classification of mineral particles by nonlinear mapping of electron microprobe energy dispersive X-ray spectra. *Analyst* **119,** 971–974.

Tusar, M., Zupan, J., and Gasteiger, J. (1992). Neural networks and modelling in chemistry. *J. Chim. Phys.* **89,** 1517–1529.

van de Waterbeemd, H. (1994). *Advanced Computer-Assisted Techniques in Drug Discovery*. VCH, Weinheim.

van de Waterbeemd, H. (1995). *Chemometric Methods in Molecular Design*. VCH, Weinheim, p. 359.

van de Waterbeemd, H., El Tayar, N., Carrupt, P.A., and Testa, B. (1989). Pattern recognition study of QSAR substituent descriptors. *J. Comput.–Aided Mol. Design* **3,** 111–132.

Varmuza, K. (1978). *Pattern Recognition in Chemistry. Lecture Notes in Chemistry.* Springer-Verlag, Berlin, p. 268.

Vriend, S.P., van Gaans, P.F.M., Middelburg, J., and de Nijs, A. (1988). The application of fuzzy *c*-means cluster analysis and non-linear mapping to geochemical datasets: Examples from Portugal. *Appl. Geochem.* **3,** 213–224.

Wessel, M.D. and Jurs, P.C. (1994). Prediction of reduced ion mobility constants from structural information using multiple linear regression analysis and computational neural networks. *Anal. Chem.* **66,** 2480–2487.

Wessel, M.D. and Jurs, P.C. (1995). Prediction of normal boiling points of hydrocarbons from molecular structure. *J. Chem. Inf. Comput. Sci.* **35,** 68–76.

Whiteley, J.R. and Davis, J.F. (1993). Qualitative interpretation of sensor patterns. *IEEE Expert* **April**, 54–63.

Whiteley, J.R. and Davis, J.F. (1994). A similarity-based approach to interpretation of sensor data using adaptive resonance theory. *Comput. Chem. Engng* **18,** 637–661.

Wienke, D. (1994). Neural resonance and adaptation: Towards nature's principles in pattern recognition. In, *Chemometrics: Exploring and Exploiting Chemical Information* (L. Buydens and W. Melssen, Eds.). University Press, University of Nijmegen, pp. 197–222.

Wienke, D. and Buydens, L. (1995). Adaptive resonance theory neural networks – The 'ART' of real-time pattern recognition in chemical process monitoring? *Trends Anal. Chem.* **99,** 1–8.

Wienke, D. and Buydens, L. (1996). An adaptive resonance theory based artificial neural network for supervised chemical pattern recognition (FuzzyARTMAP). Part 1: Theory and basic properties. *Chemom. Intell. Lab. Syst.* **32,** 151–164.

Wienke, D., Gao, N., and Hopke, P.K. (1994). Multiple site receptor modeling with a minimal spanning tree combined with a neural network. *Environ. Sci. Technol.* **28,** 1023–1030.

Wienke, D. and Hopke, P.K. (1994a). Projection of Prim's minimal spanning tree into a Kohonen neural network for identification of airborne particle sources by their multielement trace patterns. *Anal. Chim. Acta* **291,** 1–18.

Wienke, D. and Hopke, P.K. (1994b). Visual neural mapping technique for locating fine airborne particles sources. *Environ. Sci. Technol.* **28,** 1015–1022.

Wienke, D. and Kateman, G. (1994). Adaptive resonance theory based artificial neural networks for treatment of open-category problems in chemical pattern recognition – Application to UV/VIS- and IR-spectroscopy. *Chemom. Intell. Lab. Syst.* **23,** 309–329.

Wienke, D., van den Broek, W., Feldhoff, R., Huth-Fehre, T., Kantimm, T., Quick, L., Melssen, W., Winter, F., Cammann, K., and Buydens, L. (1996a). Adaptive resonance theory based neural network for supervised chemical pattern recognition (FuzzyARTMAP). Part 2: Classification of post-consumer plastics by remote NIR spectroscopy. *Chemom. Intell. Lab. Syst.* **32,** 165–176.

Wienke, D., van den Broek, W., Melssen, W., Buydens, L., Feldhoff, R., Huth-Fehre, T., Kantimm, T., Quick, L., Winter, F., and Cammann, K. (1996b). Comparison of an adaptive resonance theory based artificial neural network with other classifiers for fast sorting of post-consumer plastics by remote NIR sensing with an InGaAs diode detector array. *Anal. Chim. Acta* **317,** 1–16.

Wienke, D., Xie, Y., and Hopke, P.K. (1994). An adaptive resonance theory based artificial neural network (ART-2a) for rapid classification of airbone particles by their scanning electron microscopy image! *Chemom. Intell. Lab. Syst.* **26,** 367–387.

Wienke, D., Xie, Y., and Hopke, P.K. (1995). Classification of airborne particles by analytical scanning electron microscopy imaging and a modified Kohonen neural network (3MAP). *Anal. Chim. Acta* **310,** 1–14.

Wiese, M. and Schaper, K.J. (1993). Application of neural networks in the QSAR analysis of percent effect biological data: Comparison with adaptive least squares and nonlinear regression analysis. *SAR QSAR Environ. Res.* **1,** 137–152.

Willems, P. (1994). The ART neural network models en-lighted: Implementation on sequential and parallel computer systems. Master Student Research Report. Catholic University of Nijmegen, Institute for Informatics.

Wold, S. (1972). Spline-funktioner – Ett nytt verktyg i data-analysen. *Kemisk Tidskrift* **3,** 34.

Wold, S. (1974). Kemometri – Kemi och tillämpad matematik. *Svensk Naturventenskap* 200.

Wunsch, D.C., Caudell, T.P., Capps, C.D., Marks, R.J., and Falk, R.A. (1993a). An optoelectronic implementation of the adaptive resonance theory neural network. *IEEE Trans. Neural Networks* **4,** 673–684.

Wunsch, D.C., Morris, D.J., McGann, R.L., and Caudell, T.P. (1993b). Photo–refractive adaptive resonance theory neural network. *Appl. Optics* **32,** 1399–1407.

Xie, Y., Hopke, P.K., and Wienke, D. (1994). Airbone particle classification with a combination of chemical composition and shape index utilizing an adaptive resonance artificial neural network. *Environ. Sci. Technol.* **28,** 1921–1928.

Zitko, V. (1986). *Multidimensional Data Display by Nonlinear Mapping.* Canadian Technical Report of Fisheries and Aquatic Sciences No. 1428.

Zupan, J. and Gasteiger, J. (1991). Neural networks: A new method for solving chemical problems or just a passing phase? *Anal. Chim. Acta* **248,** 1–30.

Zupan, J. and Gasteiger, J. (1993). *Neural Networks for Chemists. An Introduction.* VCH, Weinheim, p. 305.

Zupan, J., Novic, M., Li, X., and Gasteiger, J. (1994). Classification of multicomponent analytical data of olive oils using different neural networks. *Anal. Chim. Acta* **292,** 219–234.

7 Multivariate Data Display Using Neural Networks

D.J. LIVINGSTONE
ChemQuest, Cheyney House, 19–21 Cheyney St, Steeple Morden, Herts SG8 0LP, UK
Centre for Molecular Design, School of Biological Sciences, University of Portsmouth, Portsmouth, UK

Two neural network techniques for the production of two-dimensional displays of high-dimensional data have been compared to the 'standard' methods of principal component analysis (PCA) and nonlinear mapping (NLM) using two different data sets. The network methods worked well, producing complementary displays to PCA and NLM and, in one case, working better than either of these techniques. A criticism of the ReNDeR method is that compounds may exist which will not be distinguishable from some of the training points used to produce a display. This may be true but it is difficult to assess by experiment and it is proposed that the best way to check this is by use of a training/test set procedure.

KEY WORDS: *pattern recognition; unsupervised learning; overtraining; QSAR.*

INTRODUCTION

Data display, for example plotting graphs, is one of the first techniques that scientists use for data analysis. The success of the approach takes advantage of our exceptional ability in pattern recognition (Livingstone, 1991a) although this is limited to relatively low dimensional problems, i.e., where the number of variables involved is less than, say, five or six (Hudson *et al.*, 1989). One of the simplest methods, the bivariate plot, is particularly easy to produce and interpret but unfortunately, in the case of multivariate data sets, it is also one of the least informative since the information used involves only two variables. The format of a bivariate plot, however, is familiar and thus multivariate techniques which result in two-dimensional displays have considerable appeal (Hyde and Livingstone, 1988).

In, *Neural Networks in QSAR and Drug Design* (J. Devillers, Ed.)
Academic Press, London, 1996, pp. 157–176.
ISBN 0-12-213815-5

Although there are other techniques for the display of multivariate data, e.g., cluster analysis (Livingstone, 1991b) and spectral mapping (Lewi, 1989), this report focuses on multivariate analogues of the bivariate plot and, in particular, neural network techniques for the display of high-dimensional data. The currently used methods for this type of data display may be classified as linear, based on principal component analysis (PCA), and nonlinear. PCA dates back to the beginning of this century (Pearson, 1901) although the modern form of the technique owes much to Hotelling (1933). Bivariate plots from PCA can be used to show relationships between samples (scores plots), between variables (loadings plots) or between both variables and samples (biplot, Gabriel (1971)). Since PCA is such a well known technique it will not be described here at all, further details can be obtained from any of a number of excellent texts (e.g., Jackson (1991)).

Nonlinear techniques such as multidimensional scaling (MDS) and nonlinear mapping (NLM) are a more recent phenonemon. MDS, due to Shepard (1962) and Kruskal (1964), and NLM (Sammon, 1969) are based on similar concepts in that a two-dimensional display is produced in which the N-space interpoint distances are preserved. The map is produced by minimization of an error function such as that shown in Eq. (1) where d_{ij} is the distance between two points, i and j, in the two-dimensional plot and d_{ij}^* is the distance between the same points in N dimensions. Alternatives to this equation and a good description of the technique can be found in the review by Domine and coworkers (1993).

$$E = \sum_{i>j} (d_{ij}^* - d_{ij})^2 / (d_{ij}^*)^\rho \tag{1}$$

One of the earliest examples of the use of NLM in chemistry was reported by Kowalski and Bender (1972, 1973) and it was first used in drug design to visualize the results of compound selection (Goodford *et al.*, 1976). Since then the method has been used to some extent as an 'unsupervised learning' display technique for quantitative structure–activity relationships (QSAR) (de Winter, 1983; Hudson *et al.*, 1989; Clare, 1990; Ford and Livingstone, 1990; Livingstone *et al.*, 1992; Domine *et al.*, 1993) and recent applications have come 'full circle' to use the method to produce compound selection maps for aromatic (Domine *et al.*, 1994a) and aliphatic substituents (Domine *et al.*, 1994b). Nonlinear mapping has the advantage over principal component scores plots in that the method does not impose a linear combination on the variables that go to make up the axes of the plot. It suffers the disadvantage, however, that the axes are unknown, nonlinear combinations of the starting variables whereas the 'structure' of principal component axes can be seen in the loadings of the input variables.

Two types of neural networks have been used to display data for drug design. Rose and coworkers applied an unsupervised learning technique called Kohonen mapping to the analysis of a set of 31 antifilarial antimycin analogues and demonstrated that the technique gave similar results to principal component scores plots, cluster analysis and nonlinear mapping (Rose *et al.*, 1991). Livingstone and coworkers described a technique called ReNDeR and showed its application to three sets of biologically active compounds, including a sub-set of the antimycin analogues reported by Rose *et al.* (1991) (Livingstone *et al.*, 1991). The ReNDeR method is a back-propagation feed forward network which produces a two-dimensional display of a high-dimensional data set. In the terminology of pattern recognition, this technique is 'unsupervised' since the property of interest (activity) is not used to produce the map, but in the terminology of neural networks, this method may be called self-supervised since the input data values themselves are used to guide the training process. The methodology of both of these techniques is described in the next section.

METHODS

Data sets

The 'synthetic' example was taken from a report by Bienfait (1994) on the use of self-organizing maps and consisted of 32 carbonyl containing compounds (originally published by Luce and Govind (1990)) which undergo four different mechanisms of disconnection. The compounds were described by numerical values for substructural descriptors (see Bienfait (1994) for details) at each of six positions in the molecule, as shown in Table I.

The 'QSAR' example involved 26 selective antagonists of the 5-hydroxytryptamine 5-HT_3 receptor. 5-HT_3 antagonists are currently used in the treatment of chemotherapy and radiotherapy induced emesis (Oxford *et al.*, 1992). The compounds are based on the parent structure shown in Table II, the aromatic systems include mono- and bicyclic rings, with and without heteroatoms, and with various substitution patterns. This range of structural variation makes it difficult to treat the analysis of these compounds as a 'standard' QSAR problem and thus they were characterized by building molecular models and calculating properties from these models using a SmithKline Beecham in-house molecular modelling system (Livingstone *et al.*, 1992). Each compound was geometry optimised with molecular mechanics, and the MOPAC (Stewart, 1990) wavefunction was calculated using the PM3 hamiltonian. A common 'core' was defined as the carbon, oxygen and nitrogen atoms of the amide group and the aryl portion of the compounds was treated as a substituent. For each molecule, the program records simple geometric data such as maximum and minimum dimensions in the cartesian axes, calculated log P and molar refractivity using the MEDCHEM algorithms (MEDCHEM software system, Daylight Chemical Information Systems, Inc, Irvine, CA 92713–7821), moments of inertia, and

Table I *Carbonyl compounds, molecular descriptors and disconnection mechanism (taken with permission from Bienfait (1994)).*

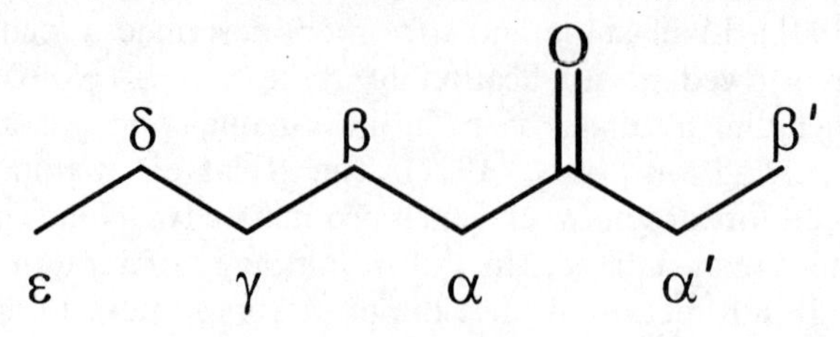

Disconnection	Index	C_ε	C_δ	C_γ	C_β	C_α	$C_{\alpha'}$	$C_{\beta'}$
Aldol-type	1	0	0	0.1	0.36	0.08	0.02	0
	2	0	0	0.1	0.38	0.08	0.1	0
	3	0	0.1	0.08	0.36	0.06	0.02	0
	4	0	0.1	0.08	0.38	0.06	0.1	0
Claisen-type	5	0	0	0.02	0.3	0.08	0.18	0.1
	6	0	0	0.02	0.3	0.08	0.18	0.08
	7	0	0	0.1	0.3	0.06	0.18	0.1
	8	0	0	0.1	0.3	0.06	0.18	0.08
Michael-type	9	0.02	0.3	0.9	0.08	0.08	0.02	0
	10	0.02	0.3	0.9	0.06	0.08	0.02	0
	11	0.02	0.3	0.9	0.08	0.06	0.02	0
	12	0.02	0.3	0.9	0.04	0.08	0.02	0
	13	0.02	0.3	0.9	0.06	0.06	0.02	0
	14	0.02	0.3	0.9	0.04	0.06	0.02	0
enamine-type	15	0.02	0.3	0.08	0.08	0.08	0.02	0
	16	0.02	0.3	0.08	0.08	0.06	0.02	0
	17	0.02	0.3	0.08	0.06	0.08	0.02	0
	18	0.02	0.3	0.06	0.08	0.08	0.02	0
	19	0.02	0.3	0.08	0.06	0.06	0.02	0
	20	0.02	0.3	0.08	0.04	0.08	0.02	0
	21	0.02	0.3	0.06	0.06	0.08	0.02	0
	22	0.02	0.3	0.04	0.08	0.08	0.02	0
	23	0.02	0.3	0.06	0.08	0.06	0.02	0
	24	0.02	0.3	0.06	0.04	0.08	0.02	0
	25	0.02	0.3	0.06	0.06	0.06	0.02	0
	26	0.02	0.3	0.04	0.06	0.08	0.02	0
	27	0.02	0.3	0.08	0.04	0.06	0.02	0
	28	0.02	0.3	0.04	0.08	0.06	0.02	0
	29	0.02	0.3	0.06	0.04	0.06	0.02	0
	30	0.02	0.3	0.04	0.06	0.06	0.02	0
	31	0.02	0.3	0.04	0.04	0.08	0.02	0
	32	0.02	0.3	0.04	0.04	0.06	0.02	0

some wavefunction derived properties such as charges, dipole moment and its components, energies of HOMO and LUMO, etc. In addition to these atom based and whole molecule properties, electrostatic potential, the van der Waals energy of the interaction of a carbon atom and the X, Y and Z components of the electric field were computed at 14 points on the surface of a box which fitted all of the molecules in the set. This gave rise to a total of 116 descriptors (20 whole molecules, 18 for the core atoms, 8 for the substituent and 70 for the grid points) which after removing redundant parameters (a correlation reduction procedure (Livingstone and Rahr, 1989) was used to remove pairwise correlations in excess of 0.7) left a set of 56. Application of a parameter selection routine, SELECT (Kowalski and Bender, 1976), to the compounds described by classified data (activity class 1 or 2) resulted in the identification of 9 important variables as shown in Table II.

Neural networks

Kohonen Mapping

A Kohonen map (Kohonen, 1990) is made up from a two-dimensional array of neurons, each one being connected, *via* a connection weight, to each of the inputs (physicochemical descriptors in the case of a QSAR data set). The set of weights and neurons can be represented by a three-dimensional diagram as shown in Figure 1. In the figure a single layer in the 3-D 'stack'

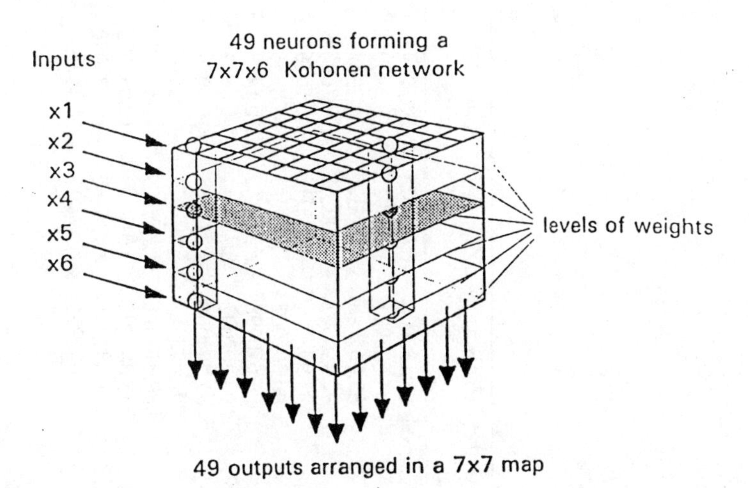

Figure 1 Block representation of a 7 × 7 Kohonen network with 6 input variables. Each layer of the block depicts the connection weights between a particular input variable and all of the neurons in the 7 × 7 plot. The shaded layer shows the weights for input variable 3 with the particular weights for neurons (1,1) and (6,5) picked out as shaded circles (reproduced with permission from Zupan (1994)).

Table II *Calculated molecular descriptors for 5HT₃* antagonists based on the parent structure shown below.*

Compound number	Activity*	CMR	μZ	HOMO	ALP(3)	FZ(4)	VDWE(4)	FY(6)	FZ(9)	FY(11)
1	1	83.156006	–0.023075	–10.125457	0.165465	0.690959	–0.01822	0.146858	–1.331931	–0.571263
2	1	85.682991	1.133915	–10.373082	0.164176	0.825953	–0.067104	–0.095023	–1.814457	–1.311374
3	1	83.169998	–1.009328	–10.291448	0.166767	0.199967	–0.007455	0.131438	–0.261507	1.610443
4	1	92.446007	–1.100368	–10.269679	0.166118	0.350187	–0.092965	–0.355433	–0.178441	3.939417
5	1	93.903	–0.883712	–10.767652	0.165174	0.323102	–0.09015	0.245022	–0.444854	–0.101309
6	1	90.302002	1.681585	–9.896879	0.165143	0.263393	–0.091509	–0.230816	–0.693684	–2.056473
7	1	86.670998	1.509192	–10.141784	0.16502	0.012006	–0.08891	–0.466338	–0.280089	–1.676208
8	1	90.302002	0.374931	–9.85334	0.165122	0.026151	–0.102998	–0.449422	–0.066987	–1.326878
9	1	87.794006	–2.013182	–10.100966	0.166473	0.240704	–0.071487	0.281943	–0.05055	1.565157
10	1	92.19001	–1.59107	–10.354034	0.165545	–0.403645	–0.237743	–0.702534	–0.465868	0.124084
11	1	87.814011	1.266109	–10.571728	0.160491	0.151386	–0.042395	–0.152652	–0.58176	0.474638
12	1	78.888992	–0.01763	–11.056095	0.164764	–0.31937	–0.017028	–0.365937	0.729973	0.701252
13	1	85.682991	–2.126862	–10.373082	0.165448	1.154631	–0.0777	0.106626	–1.773054	–0.115845
14	1	85.682991	–1.260413	–10.457438	0.165508	0.956636	–0.059637	0.089594	–1.630651	–0.489548

Table II *continued.*

Compound number	Activity*	CMR	μZ	HOMO	ALP(3)	FZ(4)	VDWE(4)	FY(6)	FZ(9)	FY(11)
15	1	86.75	0.445967	−10.813911	0.163938	0.051509	−0.013655	−0.101951	0.724446	−0.264842
16	1	86.751007	−0.130124	−10.337708	0.163203	0.342944	−0.054695	0.088702	−1.062614	−1.835011
17	1	81.836998	−0.129692	−10.008447	0.165598	0.304008	−0.08273	0.202748	−0.121693	2.285414
18	1	81.808998	−1.471474	−11.050653	0.164786	0.155182	−0.052069	−0.053286	−0.372007	0.229041
19	2	87.701996	−0.567377	−10.759488	0.163277	0.334046	−0.096817	0.083781	−1.089644	−1.575862
20	2	80.582001	−0.56162	−10.446554	0.164764	0.274729	−0.061855	0.053127	−0.56334	−0.273451
21	2	83.063995	−0.992965	−10.841124	0.163602	0.37626	−0.02528	0.068368	−1.002572	−1.531959
22	2	81.836998	0.665133	−10.383967	0.16327	0.280005	−0.054866	−0.004466	−1.071213	−2.130827
23	2	79.928001	−0.683407	−10.109129	0.166261	0.041216	−0.01325	−0.019164	0.003597	2.001035
24	2	89.233002	−0.955246	−10.55268	0.163597	0.611623	−0.217868	0.116554	−1.201934	−1.457719
25	2	85.49601	−0.464845	−10.503699	0.164757	0.153127	−0.084228	−0.013756	−0.481711	−0.30028
26	2	87.459	−1.887398	−10.830238	0.163055	0.003264	−0.042777	0.031053	−3.251751	−1.093677

*1 = inactive; 2 = active.

CMR Calculated molar refractivity.
μZ Z component of the dipole moment.
HOMO Energy of the highest occupied molecular orbital.
FZ(No) and FY(No) Z and Y components of the electric field at specified (No) grid points.
VDWE(No) The van der Waal's energy of the interaction of a carbon atom at a specified (No) grid point.
ALP(No) The self atom polarizability of the specified atom (No).

represents all the weights associated with a particular input variable. Training a Kohonen map consists of two parts, competitive learning and self-organization. Initially, as with most network training procedures, the connection weights are set to random values. Each pattern (compound) in the training set may be considered to be a vector, **X,** consisting of m values x_i (where there are m physicochemical descriptors in the set); each neuron j in the map is characterized by a weight vector, $\mathbf{W}_j$, consisting of m weights w_{ij}. Euclidean distances, d_{ij}, are calculated between each **X** and each weight vector $\mathbf{W}_j$ by Eq. (2):

$$d_{ij} = \left(\sum_{i=1}^{m} (x_i - w_{ij})^2 \right) 1/2 \quad (2)$$

The neuron with the weight vector $\mathbf{W}_j$ closest to the input pattern **X** is said to be the winning neuron, j^*, and it is updated so that its weight vector, $\mathbf{W}_j^*$, is even closer to the input vector **X**:

$$w_{ij^*}(t+1) = w_{ij^*}(t) + \alpha(t)[x_i - w_{ij^*}(t)] \qquad 0 < \alpha < 1 \quad (3)$$

The terms t and α in Eq. (3) are time and learning rate, respectively, (t and $t + 1$ are successive instants of time) and after each updating the time variable is incremented whereas the learning rate, α, is decreased. This process, which is called a step, is repeated for each of the patterns in the training set; a learning epoch consists of as many steps as there are patterns in the training set.

The competitive learning phase of Kohonen mapping takes no account of the topological relationships between the neurons in the plot, these are updated in an isolated fashion. Self-organization is the second phase of the training process and this is achieved by defining a set of neighbouring neurons, N_{j^*}, as the set of neurons which are topologically close to the 'winning' neuron j^*. The learning algorithm shown in Eq. (3) may be modified so that the neighbouring neurons are updated as well as the j^* neuron:

$$\begin{aligned} w_{ij}(t+1) &= w_{ij}(t) + \alpha(t)\, \gamma(t)\, [x_i - w_{ij}(t)] \\ \gamma(t) &= 1 \quad \forall j \in N_{j^*}(t) \\ \gamma(t) &= 0 \quad \forall j \notin N_{j^*}(t) \end{aligned} \quad (4)$$

The neighbouring neurons are specified by the parameter γ and at first the area of this set is wide but, as training proceeds, the radius of this area is decreased (like the learning rate α) as the time variable t is incremented. The result of this combination of competitive learning and self-organization is to produce a two-dimensional plot in which the data points are arranged according to their similarities in the high-dimensional space defined by the physicochemical parameters in the training set. The similarity between Kohonen mapping and NLM or principal component scores plots is evident.

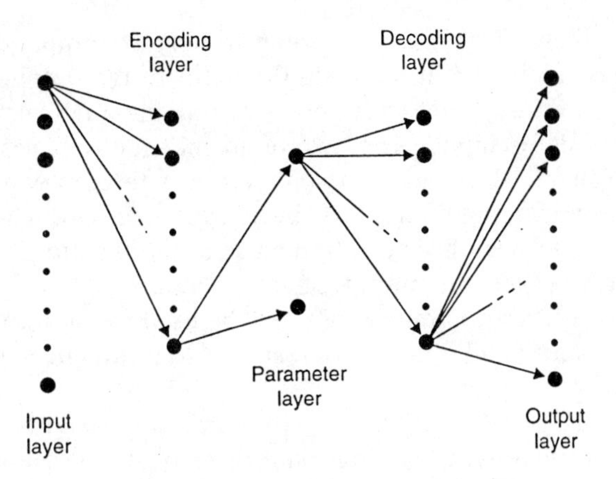

Figure 2 Representation of a ReNDeR backpropagation neural network. The network is fully connected, only a few connections are shown for clarity, see text for details of the layers (reprinted by permission of the publisher from Livingstone *et al.* (1991). Copyright 1991 Elsevier Science Inc.).

ReNDeR Mapping

The acronym, ReNDeR, stands for reversible nonlinear dimension reduction as will be explained below. The neural network used to produce a ReNDeR plot is a feedforward backpropagation network containing, according to terminology (some authors ignore the input layer of neurons in a backpropagation network since these neurons simply act as 'distributors' of the input signals), either four or five layers (Livingstone *et al.*, 1991). A diagram of the ReNDeR network is shown in Figure 2 where it may be seen that it contains an input and output layer and three hidden layers. Each neuron in the input layer corresponds to one of the physicochemical parameters in the training set, in the normal way, and each neuron in the output layer also corresponds to one of the training set descriptors. The number of neurons in the encoding and decoding layers is usually smaller than the number of input parameters but should be of sufficient size to allow a 'reasonable' mapping of the inputs onto the outputs. In the two examples shown here the number of encoding and decoding neurons was 2 (for 7 parameters, hence a 7:2:2:2:7 network) and 3 (a 9:3:2:3:9 network). As is customary with neural network construction the connection weights are set to random values within their allowed ranges at the beginning of network training. Training is carried out by repeated presentation of the input patterns to the network, comparison of the output signals with their target signals and adjustment of the network connection weights to achieve a better 'match' between output and target (Salt *et al.*, 1992). In this case the targets are the values of the input parameters themselves and thus, when training is complete*, the ReNDeR network has succeeded in mapping the input to the output. This may seem a strange procedure, what is the point of training a network to produce the same output as its input? The answer to this question lies in the fact that the central hidden layer contains only two (or three) neurons and thus the high-dimensional input data set has been 'squeezed' through a central 'bottleneck'. Once a network has been trained, each input pattern may be presented to the network and an output signal obtained from each of the central layer neurons. These output signals may be used to produce a two- or three-dimensional plot similar, in many respects, to a nonlinear map or principal component scores plot.

A ReNDeR plot may be used in the same way as an NLM or PC scores plot, it is 'unsupervised learning' pattern recognition in that the activity of the compounds is not used to produce the plot, although this type of network may also be referred to as 'self-supervised' (Salt *et al.*, 1992). Thus far, a ReNDeR plot may not appear to be very different to a nonlinear map and, for that matter, a Kohonen self-organizing map. There is, however, one important feature of a ReNDeR plot which makes it unique. Since the network is symmetrical, connection weights are established which lead **from** the input layer to the parameter layer and also **from** the parameter layer to the output layer. These weights need not be (and probably aren't) equivalent, but they are known, so it is possible to present input signals to the parameter layer neurons and monitor the output from the output neurons. This is equivalent to 'driving' a cursor around in the two-dimensional plot and seeing changes in the individual physicochemical properties which make up the training set N-dimensional data space. A by-product of this reversibility (hence the name of the technique) is the ability to generate mapping errors by projecting a grid of points from N-space to 2-space and then back to N-space. Since the mapping from N-space to 2-space is not 'perfect' the points 'miss' by varying amounts when projected back to N dimensions, and thus error contours may be created and plotted on the two dimensional map (see next section). The error generation is a particular feature of the ReNDeR demonstrator program and is thus only available in this form using this package. In all other respects, the ReNDeR system is a straightforward multilayer feedforward network and it should be possible to produce such a system using most standard neural network packages, although extraction of the X and Y coordinate values from the central neurons may be inconvenient.

*A common problem in the application of neural networks is how to know when training has been carried out for a sufficiently large number of iterations. In the case of these networks any reasonable minimum should be sufficient since the resultant plot can be tested by inspection of mapping errors. For the examples shown here a few thousand complete cycles through the data sets were used.

RESULTS AND DISCUSSION

Bienfait (1994) created a Kohonen self-organizing map for the carbonyl compound data set given in Table I. The training process was run for a total of 20 epochs and at the end of this time the resultant map was clearly split into four distinct areas. Figure 3 shows four of the stages of this training process, including the initial and final plots, where it may be seen that the four different classes of disconnection have been separated by the unsupervised display. Although the classes have been separated, however, it can also be seen that the final map does not display all of the compounds in the set. The four compounds which undergo Claisen-type disconnection, for example, are represented by just a single cell (neuron) on the plot while the six Michael-type compounds are shown in just two cells. Principal component analysis (using the PC package Systat from Systat, Inc., 1800 Sherman Ave., Evanston, IL 60201, USA) of this data set gave two significant (using the often accepted test for 'significance' of an eigenvalue greater than 1 for the components derived from autoscaled data) components which explained approximately 76% of the variance in the data set. The first component

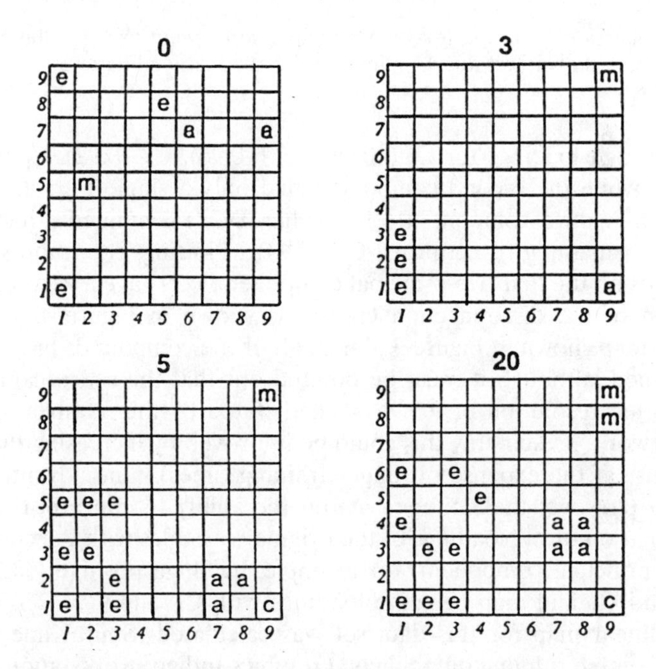

Figure 3 Four stages in the training of a Kohonen self-organizing map for the data shown in Table I. Numbers above each plot indicate the number of training epochs and the letters a, c, m and e represent Aldol, Claisen, Michael and enamine-type disconnections respectively (from Bienfait (1994). Copyright 1994 American Chemical Society).

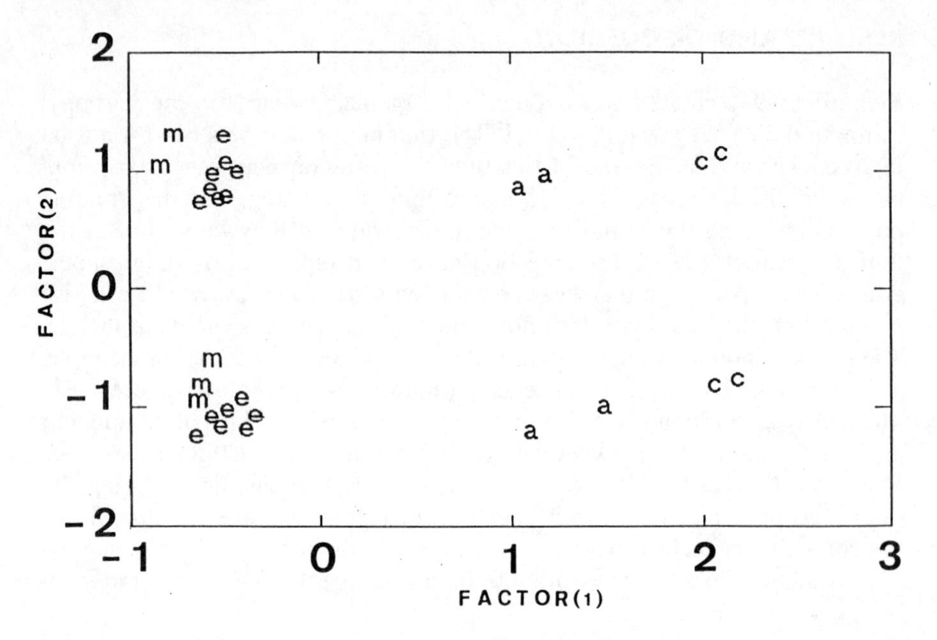

Figure 4 Scores plot on the first two principal component axes for the compounds shown in Table I, points are labelled as for Figure 3.

contained high loadings for C_{δ} (–0.98), C_{ε} (–0.96), $C_{\alpha'}$ (0.92), C_{β} (0.91) and $C_{\beta'}$ (0.82) while the second component had only a single high loading with C_{α} (–0.99). A third component (eigenvalue 0.922) contained a high loading with the remaining descriptor, C_{γ} (0.93). Plotting the scores for the compounds on the first two principal component axes gave a very satisfactory separation of the disconnection classes as shown in Figure 4. Unlike the Kohonen map shown in Figure 3 almost all of the compounds may be clearly distinguished, although it must be pointed out that the self-organizing map has a limited resolution of 9×9 pixels dictated by the number of neurons in the network, presumably this could be improved by increasing the number of neurons (at the expense of longer training times). One advantage of the PC scores plot over the Kohonen map is the ability to interpret the PC axes by examination of the loadings of the original variables with the components. The first principal component, for example, has high positive loadings with $C_{\alpha'}$, C_{β} and $C_{\beta'}$ and high negative loadings with C_{δ} and C_{ε}.

A nonlinear map for this data set was calculated (SmithKline Beecham in-house routine running on a Silicon Graphics Indigo[2] workstation, mapping error 0.027) and is shown in Figure 5. It can be seen from the figure that the compounds fall into distinct groups according to their disconnection type and like the principal component scores plot, each disconnection type appears to form two sub-groups. The compounds which undergo enamine-

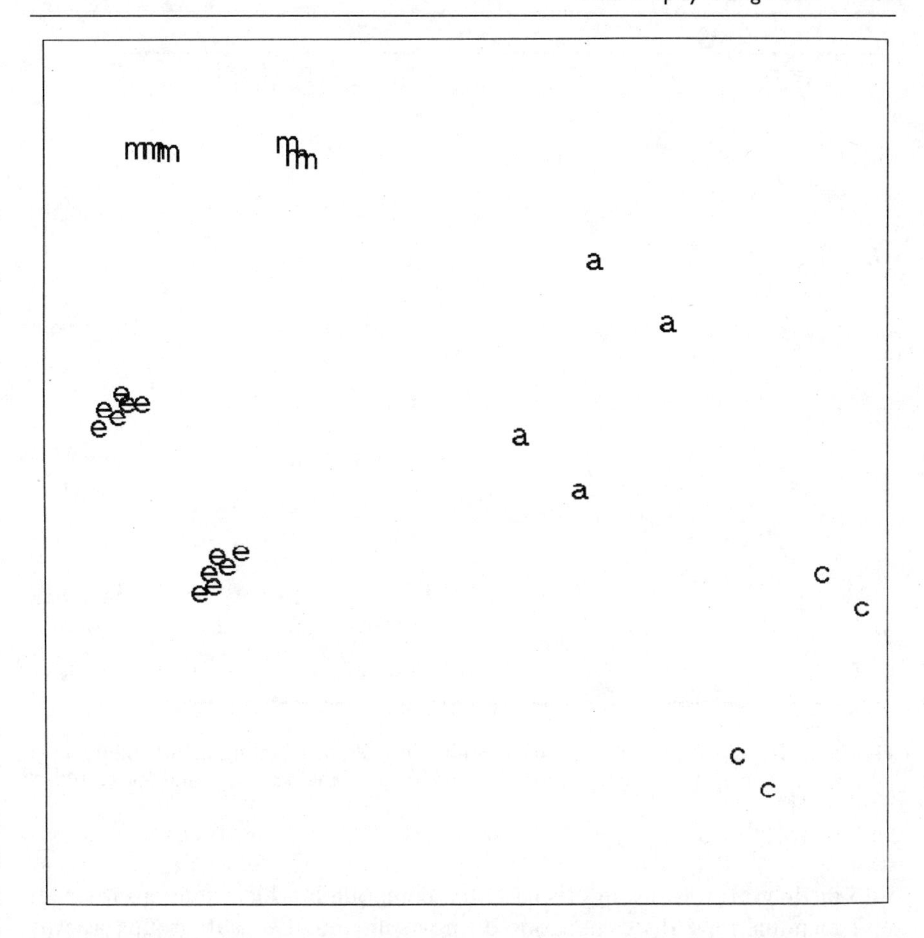

Figure 5 Nonlinear map of the compound set shown in Table I, point labelling is as for Figures 3 and 4, not all compounds of the enamine-type disconnection are shown for clarity.

type disconnections form tighter clusters on the nonlinear map than the PC scores plot and these are separated from the Michael-type disconnections. It is clear that these two types of display are both doing a reasonable job in classifying the compounds according to reaction type, based on quite simple descriptions of chemical structure. It would be interesting to see how well these methods work (and the Kohonen map) using physicochemical properties which describe shape, hydrophobicity, volume, electron distributions, etc.

A ReNDeR plot was also calculated (ReNDeR demonstrator program (supplied by AEA Technology, Harwell, UK) running on a 486 PC) for this data set and is shown in Figure 6. Once again, the compounds are separated

Figure 6 ReNDeR plot of the data from Table I. Point labelling as for Figures 3–5 and, again, some of the enamine-type disconnection labels are omitted for clarity.

into distinct groups but in this case, the compounds which undergo Michael and enamine-type disconnections do not split into two sub-groups as was seen for the NLM and PC scores plots. The Aldol and Claisen-type disconnections do fall into two sub-groups but here two of the sub-groups overlap.

The original report of the ReNDeR technique described its application to three QSAR problems (Livingstone *et al.*, 1991). Since then only one other group (see, however, the chapter by Manallack *et al.* (1996) in this volume) has demonstrated the use of the technique (Good *et al.*, 1993), probably due to the lack of suitable software, although a commercial program does now contain an implementation of ReNDeR (TSAR from Oxford Molecular, Oxford Science Park, Oxford OX4 4GA, UK). Another group, however, has published a comment on the technique concerning the ability of the network to reverse. Reibnegger and coworkers (1993) showed for simple 2-1-2 networks that it was possible to extract the weights from a trained network and use these to create prediction equations for the output neurons. In the case of the network trained on only one input vector, a single equation was obtained, in the case of two input vectors, two equations, and so on. These equations may be represented as lines and any input vector which lies along one of these lines will map onto the same vector as was used to produce the line, meaning that it will be impossible to distinguish this point from the training set vector used to generate the network. This is undoubtedly true and may well lead to confusion if a test set should contain a compound, or compounds, whose parameter values happen to lie on one of the multi-dimensional surfaces (equivalent to the lines shown in this simple example) created by the input vectors of the training set. In the real world of chemistry, however, it is perhaps unlikely that this will happen very frequently, if at all. Of course, as with any analytical method used in QSAR, the best way to test a method is to split a data set into training and test sets and observe how well the technique predicts or classifies the test set compounds.

As a test of the method, and a further exploration of the use of the ReNDeR technique generally, the data shown in Table II for a set of $5HT_3$ antagonists was used to produce some ReNDeR plots. Figure 7 shows a plot of the whole data set where it can be seen that the active compounds (class 2 in the table) are mostly clustered together. The activity classes shown in the table were actually derived from quantitative (ED_{50}) data so it would be worthwhile labelling the points according to activity; the near neighbours of the active set may well be just over the activity threshold used for

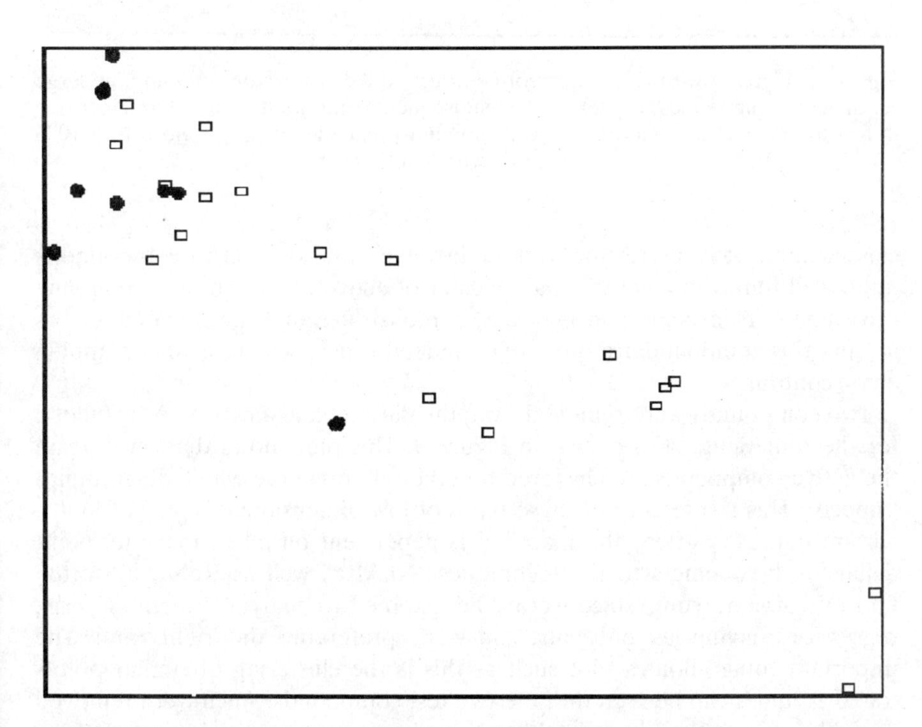

Figure 7 ReNDeR plot of the data from Table II. Filled symbols represent active (class 2) compounds.

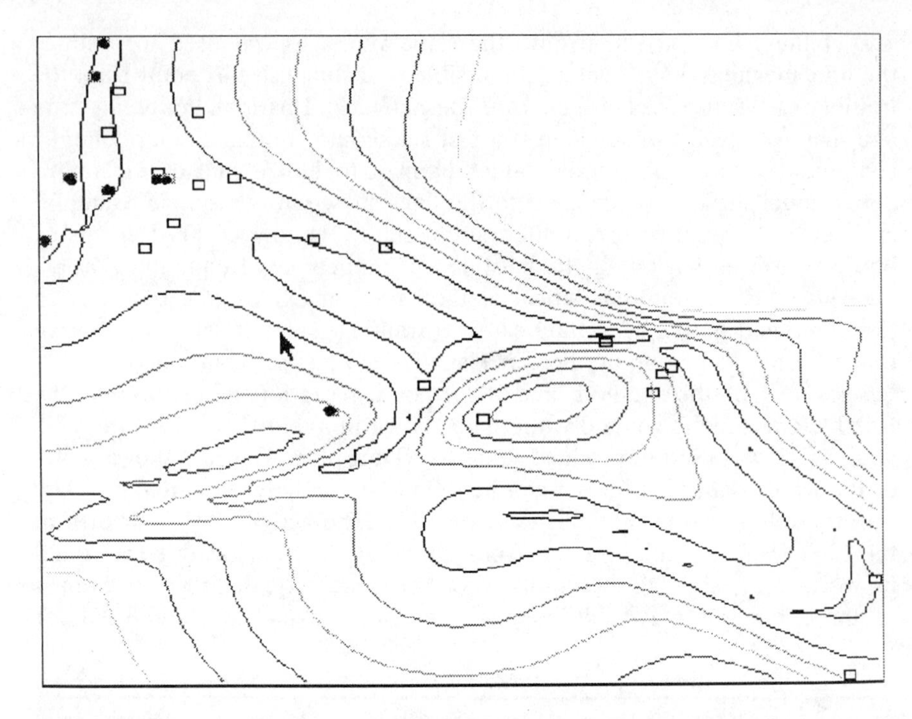

Figure 8 Figure 7 with mapping error contours added. The error bounding the large set of active compounds (top left) is 1%, subsequent contours start at 5% and increase in 5% intervals. The separate active compound (near the cursor) is close to a 10% mapping error contour.

classification. It is interesting to note, however, that one active compound is quite well removed from the main cluster of actives. Generation of mapping error contours, as shown in Figure 8, demonstrates that the mapping errors around this compound are quite high, indeed it lies close to a 10% mapping error contour.

Two compounds were removed from the data set and a new map calculated for the remaining 24 as shown in Figure 9. This plot shows that once again the active compounds are clustered together although the whole diagram has 'flipped'. This is a feature of these types of low-dimensional display of multivariate data sets where the mapping is dependent on all of the inter-point distances. Newcomers to the techniques (NLM as well as ReNDeR) often find this disconcerting, since we are accustomed to analytical methods (e.g., regression) giving us only one 'answer', preferably the right one! The important thing about a plot such as this is the clustering of similar points together and it can be seen that the two test compounds which were removed from the plot both appear in the active cluster, unfortunately one of these compounds is classified as inactive. It would be tempting to try to explain

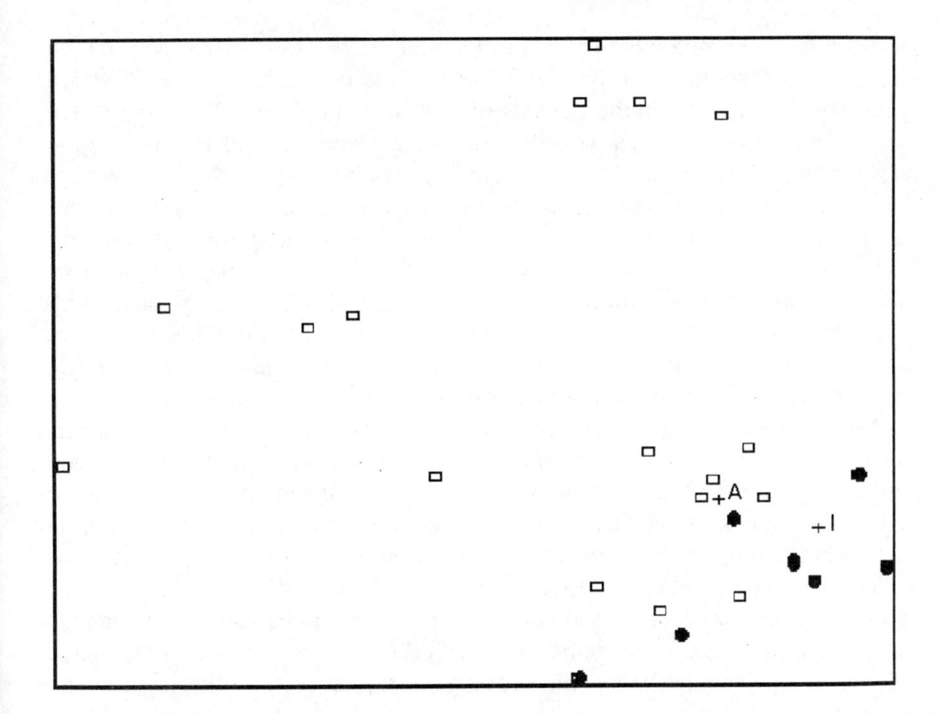

Figure 9 ReNDeR plot created from 24 of the compounds in Table II, the other two (one active, one inactive) are shown plotted as a + with their known activity marked as I or A.

this poor prediction in terms of mapping errors but in fact the region containing these active compounds and the two test compounds lies within a 1% mapping error contour (data not shown). A principal component scores plot, on the first two PC axes, and a nonlinear map was also calculated for this data set but neither display showed a grouping of the active compounds, in both cases the actives were mixed up with the inactives (plots not shown).

CONCLUSIONS

The two neural network techniques for data display have been shown to produce reasonable two-dimensional plots of high-dimensional multivariate data sets. For the data set examined here, the Kohonen self-organizing map separates compounds into chemical reaction classes in a similar fashion to both a principal component scores plot and a nonlinear map. The Kohonen map tends to produce smaller clusters with several compounds occupying a single cell but this may simply be a matter of resolution of the map. The ReNDeR technique also performs well with the 'synthetic' example and appears to be complementary to the PC and NLM plots. In the case of the $5HT_3$ antagonists data set, the ReNDeR method produces a superior plot to either principal component analysis or nonlinear mapping. Only one active compound is not clustered with the remaining seven and this lies in a region of high mapping errors. The criticism that points may exist that cannot be separated from training set points used to create a ReNDeR plot (Reibnegger *et al.*, 1993) is accepted, although it is pointed out that in nature such compounds may not exist. A reduced training set of the $5HT_3$ antagonists was shown to produce a similar clustering of active compounds, with one test compound well predicted and one predicted badly. This, of course, is not a very systematic test of the technique; what is required is to generate many plots with various compounds omitted, but the computational requirements for this are not trivial. The question of testing network performance, however, raises other problems. Is it possible to say that a network has been sufficiently trained? In most applications of neural networks it is usual to calculate some kind of fitting error and use this as a criterion for halting network training. If the network is fitting a model to a set of qualitative or quantitative data, then prediction success may be used as the fitting error. This, however, may lead to good performance in fitting but poor performance in prediction (Manallack *et al.*, 1994). Tetko and coworkers (1995) have proposed a procedure which will prevent neural network models from being overfitted, but the question remains as to how to assess the fitting performance of a method such as ReNDeR.

REFERENCES

Bienfait, B. (1994). Applications of high-resolution self-organizing maps to retrosynthetic and QSAR analysis. *J. Chem. Inf. Comput. Sci.* **34,** 890–898.

Clare, B.W. (1990). Structure–activity correlations for psychotomimetics. 1. Phenylalkylamines: Electronic, volume, and hydrophobicity parameters. *J. Med. Chem.* **33,** 687–702.

de Winter, M.L. (1983). Significant fragment mapping: Lead generation by substructural analysis. In, *Quantitative Approaches to Drug Design* (J.C. Dearden, Ed.). Elsevier, Amsterdam, pp. 286–287.

Domine, D., Devillers, J., and Chastrette, M. (1994a). A nonlinear map of substituent constants for selecting test series and deriving structure–activity relationships. 1. Aromatic series. *J. Med. Chem.* **37,** 973–980.

Domine, D., Devillers, J., and Chastrette, M. (1994b). A nonlinear map of substituent constants for selecting test series and deriving structure–activity relationships. 2. Aliphatic series. *J. Med. Chem.* **37,** 981–987.

Domine, D., Devillers, J., Chastrette, M., and Karcher, W. (1993). Nonlinear mapping for sructure–activity and structure–property modelling. *J. Chemom.* **7,** 227–242.

Ford, M.G. and Livingstone, D.J. (1990). Multivariate techniques for parameter selection and data analysis exemplified by a study of pyrethroid neurotoxicity. *Quant. Struct.–Act. Relat.* **9,** 107–114.

Gabriel, K.R. (1971). The biplot graphic display of matrices with application to principal component analysis. *Biometrika* **58,** 453–467.

Good, A.C., So, S.S., and Richards, W.G. (1993). Structure–activity relationships from molecular similarity matrices. *J. Med. Chem.* **36,** 433–438.

Goodford, P.J., Hudson, A.T., Sheppey, G.C., Wootton, R., Black, M.H., Sutherland, G.J., and Wickham, J.C. (1976). Physicochemical–activity relationships in asymmetrical analogues of methoxychlor. *J. Med. Chem.* **19,** 1239–1247.

Hotelling, H. (1933). Analysis of a complex of statistical variables into principal components. *J. Educ. Psychol.* **24,** 417–441, 498–520.

Hudson, B., Livingstone, D.J., and Rahr, E. (1989). Pattern recognition display methods for the analysis of computed molecular properties. *J. Comput.-Aided Mol. Design* **3,** 55–65.

Hyde, R.M. and Livingstone, D.J. (1988). Perspectives in QSAR: Computer chemistry and pattern recognition. *J. Comput.-Aided Mol. Design* **2,** 145–155.

Jackson, J.E. (1991). *A User's Guide to Principal Components.* John Wiley & Sons, New York.

Kohonen, T. (1990). The self-organizing map. *Proc. IEEE* **78,** 1464–1480.

Kowalski, B.R. and Bender, C.F. (1972). Pattern recognition. A powerful approach to interpreting chemical data. *J. Am. Chem. Soc.* **94,** 5632–5639.

Kowalski, B.R. and Bender, C.F. (1973). Pattern recognition. II. Linear and nonlinear methods for displaying chemical data. *J. Am. Chem. Soc.* **95,** 686–693.

Kowalski, B.R. and Bender, C.F. (1976). An orthogonal feature selection method. *Pattern Recognition* **8,** 1–4.

Kruskal, J.B. (1964). Multidimensional scaling by optimizing goodness of fit to a nonmetric hypothesis. *Psychometrika* **29,** 1–27.

Lewi, P.J. (1989). Spectral map analysis: Factorial analysis of contrasts, especially from Log ratios. *Chemom. Intell. Lab. Syst.* **5,** 105–116.

Livingstone, D.J. (1991a). Pattern recognition methods in rational drug design. In, *Molecular Design and Modeling: Concepts and Applications.* Part B. *Methods in Enzymology*, Vol. 203 (J.J. Langone, Ed.). Academic Press, San Diego, pp. 613–638.

Livingstone, D.J. (1991b). Applications of cluster analysis in molecular modelling and drug design. *Anal. Proc.* **28,** 247–248.

Livingstone, D.J. and Rahr, E. (1989). Corchop – An interactive routine for the dimension reduction of large QSAR data sets. *Quant. Struct.-Act. Relat.* **8**, 103–108.

Livingstone, D.J., Evans, D.A., and Saunders, M.R. (1992). Investigation of a charge-transfer substituent constant using computational chemistry and pattern recognition techniques. *J. Chem. Soc. Perkin Trans.* **2,** 1545–1550.

Livingstone, D.J., Hesketh, G., and Clayworth, D. (1991). Novel method for the display of multivariate data using neural networks. *J. Mol. Graph.* **9,** 115–118.

Luce, H.H. and Govind, R. (1990). Neural network applications in synthetic organic chemistry: I. A hybrid system which performs retrosynthetic analysis. *Tetrahedron Comput. Methodol.* **3,** 213–237.

Manallack, D.T., Ellis, D.D., and Livingstone, D.J. (1994). Analysis of linear and nonlinear QSAR data using neural networks. *J. Med. Chem.* **37,** 3758–3767.

Manallack, D.T., Gallagher, T., and Livingstone, D.J. (1996). Quantitative structure–activity relationships of nicotinic agonists. In, *Neural Networks in QSAR and Drug Design* (J. Devillers, Ed.). Academic Press, London, pp. 177–208.

Oxford, A.W., Kilpatrick, G.J., and Tyers, M.B. (1992). Ondansetron, and related 5-HT_3 antagonists: Recent advances. In, *Progress in Medicinal Chemistry* (G.P. Ellis, and D.K. Luscombe, Eds.). Elsevier, Amsterdam, p. 239.

Pearson, K. (1901). On lines and planes of closest fit to systems of points in space. *Phil. Mag. Ser. B* **2,** 559–572.

Reibnegger, G., Werner-Felmayer, G., and Wachter, H. (1993). A note on the low-dimensional display of multivariate data using neural networks. *J. Mol. Graph.* **11,** 129–133.

Rose, V.S., Croall, I.F., and MacFie, H.J.H. (1991). An application of unsupervised neural network methodology (Kohonen topology-preserving mapping) to QSAR analysis. *Quant. Struct.–Act. Relat.* **10,** 6–15.

Salt, D.W., Yildiz, N., Livingstone, D.J., and Tinsley, C.J. (1992). The use of artificial neural networks in QSAR. *Pest. Sci.* **36,** 161–170.

Sammon, J.W. (1969). A nonlinear mapping for data structure analysis. *IEEE Trans. Comput.* **C-18,** 401–409.

Shepard, R.N. (1962). The analysis of proximities: Multidimensional scaling with an unknown distance function. *Psychometrika* **27,** 125–139, 219–246.

Stewart, J.J.P. (1990). MOPAC: A semiempirical molecular orbital program. *J. Comput.-Aided Mol. Design* **6,** 67–84.

Tetko, I.V., Livingstone, D.J., and Luik, A.I. (1995). Neural network studies. 1. Comparison of overfitting and overtraining. *J. Chem. Inf. Comput. Sci.* **35**, 826–833.

Zupan, J. (1994). Introduction to artificial neural network (ANN) methods: What they are and how to use them. *Acta Chimica Slovenica* **41,** 327–352.

8 Quantitative Structure–Activity Relationships of Nicotinic Agonists

D.T. MANALLACK[1]*, T. GALLAGHER[2], and
D.J. LIVINGSTONE[3,4]
[1]Chiroscience Limited, Cambridge Science Park, Milton Road, Cambridge CB1 4WE, UK
[2]School of Chemistry, University of Bristol, Cantock's Close, Bristol BS8 1TS, UK
[3]ChemQuest, Cheyney House, 19–21 Cheyney Street, Steeple Morden SG8 OLP, Hertfordshire, UK
[4]Centre for Molecular Design, School of Biological Sciences, University of Portsmouth, Portsmouth, UK

To assist in the search for novel therapies to treat Alzheimer's disease, we have examined a series of potent, structurally diverse, nicotinic agonists to develop a pharmacophore model for these compounds. This model consists of three primary pharmacophore points and a lipophilic region. Two of these three points represent locations for a charged nitrogen atom and the ring centroid of a pyridine ring or the carbon atom of a carbonyl group. The third point is a dummy atom indicating the location of an atom in the 'receptor' *which interacts with either the pyridyl nitrogen or carbonyl oxygen of the nicotinic agonist.*

The quantitative structure–activity relationships (QSARs) of two series of nicotinic agonists were examined using calculated physicochemical properties and multivariate statistical techniques using the pharmacophore model for molecular alignment. Analysis of the dataset for the first series of nicotinic analogues using ReNDeR neural networks and principal components analysis suggested that biological activity was associated with six molecular properties measuring steric parameters. A QSAR model developed for the second series of compounds possessed poor predictive abilities. This study has demonstrated the usefulness of ReNDeR neural networks for multivariate data analysis and has shown the complementarity of combining pharmacophore and 3D-QSAR techniques.

---

*Author to whom all correspondence should be addressed.

In, *Neural Networks in QSAR and Drug Design* (J. Devillers, Ed.)
Academic Press, London, 1996, pp. 177–208.
ISBN 0-12-213815-5

KEY WORDS: *nicotinic agonists; neural networks; quantitative structure–activity relationships; multivariate analysis; computer-aided molecular design.*

INTRODUCTION

Recent research in the field of Alzheimer's disease, senile dementia of Alzheimer type (AD/SDAT) has pointed to the use of nicotinic agonists as a potential therapeutic treatment for this and other types of dementia. AD/SDAT is a debilitating neurodegenerative disorder which progressively reduces cognitive function in the elderly and there is considerable evidence implicating an involvement of CNS cholinergic systems (for reviews see Court and Perry, 1994; Gopalakrishnan and Sullivan, 1994).

In normal aging, changes occur to various brain transmitter systems (Reisine *et al.*, 1978; McGeer *et al.*, 1984) including decreases in the number of nicotinic acetylcholine receptors in the CNS (Giacobini, 1990; Court and Perry, 1994). These changes can be related to memory deficits in normal aged humans and animals (Drachman and Leavitt, 1974; Bartus *et al.*, 1982). Losses of cholinergic receptors in AD/SDAT is more marked, however, as has been demonstrated in a number of studies (Perry *et al.*, 1981; Perry, 1986). Significant reductions are not seen with muscarinic acetylcholine receptors (Whitehouse *et al.*, 1988) while nicotinic binding sites in the entorhinal cortex and subicular formation suffer substantial losses even from the early stages of AD/SDAT (see Court and Perry, 1994).

Exposure of neonatal brains to nicotine has been shown to reduce the number of nicotinic binding sites (Navarro *et al.*, 1989). Conversely, smoking in later life appears to offer some protection against Parkinson's and AD/SDAT (Shahi and Moochhala, 1991; van Duijn and Hofman, 1991) as the number of nicotinic receptors in human and experimental animals is increased (Wonnacott, 1990). Clinical trials administering nicotine to AD/SDAT patients proved to be encouraging as improvements were seen in a number of measurable behaviour paradigms (Newhouse *et al.*, 1988; Sahakian *et al.*, 1989; Jones *et al.*, 1992). Studies examining the memory capabilities of aged animals provide good evidence for the cognitive enhancing effects of nicotine (Cregan *et al.*, 1989). This has also been studied in mice genetically engineered to lack the neuronal nicotinic acetylcholine β2 receptor protein (Picciotto *et al.*, 1995). Nicotine was shown to enhance memory after acute administration, however, the effects of long term exposure are unknown (Picciotto *et al.*, 1995). All this evidence, including neuroregenerative effects of nicotine (Jason *et al.*, 1989; Owman *et al.*, 1989), suggest that compounds of this type may be used for treating AD/SDAT. Unfortunately, nicotine stimulates a number of receptor subtypes (Heinemann *et al.*, 1990) as well as possessing unwanted side effects (e.g., cardiotoxicity). Using rational drug design techniques, there is a real possibility that the development of a nicotinic agonist possessing the appropriate pharmacological profile can be achieved and be progressed as a therapeutic agent for AD/SDAT.

To assist in this development, molecular modelling and quantitative structure–activity relationship (QSAR) studies will be able to make significant contributions. The first study which attempted to describe the pharmacophoric elements of the nicotinic binding site is still referred to today (Beers and Reich, 1970). Their model consisted of a cationic centre (e.g., an ammonium nitrogen) and an atom bearing unshared electrons (e.g., a carbonyl oxygen) which could accept a hydrogen bond. The Beers–Reich distance (5.9 Å) refers to the separation between the centre of charge and the van der Waals surface of the H-bond acceptor through a plane defined by the centre of charge and a line through the carbonyl bond. An angle is also defined (120°) by the centre of charge, the centre of the H-bond acceptor and the line through the carbonyl bond directed to the H-bond donor (Beers

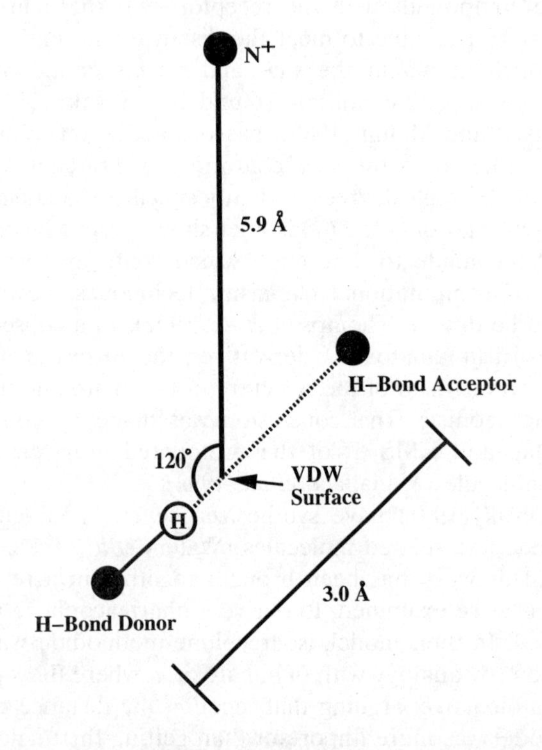

Figure 1 The Beers–Reich (1970) pharmacophore model. Diagram showing the relationship of the charged nitrogen atom and H-bond acceptor of the nicotinic agonist to the H-bond donor group in the receptor. The distance from the H-bond acceptor atom to its van der Waals (VDW) surface was assumed to be 1.4 Å and the distance to the H-bond donor was defined as 3 Å.

and Reich, 1970; Figure 1). This model has since been discussed at length and modified by several groups.

Using distance geometry techniques, Sheridan and coworkers (1986) were able to refine the Beers–Reich model employing 4 semi-rigid agonists and antagonists. Both the Beers–Reich and Sheridan models have been criticised, as a number of weak compounds fit very well to the pharmacophores (Gund and Spivak, 1991). Many factors will affect compound activity, including receptor fit, as judged by these pharmacophore models. Physicochemical properties and kinetics also play major roles in compound efficacy which are related to receptor occupancy. To address the problems of the above pharmacophores, a model was proposed containing a centre of positive charge and an area of planarity (Barlow and Johnson, 1989); an idea supported by the suggestion that agonists were interacting with a key phenylalanine residue in the receptor (Cockcroft *et al.*, 1990). Hacksell and Mellin (1989) used (+)-anatoxin-*a* and some newly identified nicotinic agonists to propose a pharmacophore which could address the stereochemical interactions of these compounds with the receptor. (+)-Anatoxin-*a* has caused some controversy as it is able to meet the distance and angle requirements of the Beers–Reich model in the *s-cis* and *s-trans* enone conformations, respectively, but never both conditions (Gund and Spivak, 1991). It has been argued by Hacksell and Mellin (1989), based on receptor-excluded volume, that the *s-trans* rather than the *s-cis* conformer of (+)-anatoxin-*a* provides an explanation of the high degree of enantiospecificity associated with this potent ligand (note that only the (+) isomer shows potent nicotinic activity). An effort has been made to determine which conformer was most likely from a variety of computational chemistry techniques, however, no firm conclusion could be drawn (Thompson *et al.*, 1992). In a subsequent QSAR study of a series of (+)-anatoxin-*a* derivatives, the *s-trans* arrangement was used based on a comparison of the structure of (+)-anatoxin-*a* to (–)-cytisine (a rigid nicotinic agonist). This conclusion was made by considering both molecular overlap and RMS fit of the protonated nitrogen and carbonyl atoms of each molecule (Manallack *et al.*, 1994).

Gund and Spivak (1991) have synthesized several analogues of isoarecolone methiodide and related molecules (Waters *et al.*, 1988; Spivak *et al.*, 1989a,b, 1991). This work has been brought together in a review in which these compounds were examined, to derive a pharmacophore model (Gund and Spivak, 1991). In their model, isoarecolone methiodide was used in the *s-cis* configuration (by analogy with (+)-anatoxin-*a* where the *s-cis* conformer was regarded as bioactive, arguing that meeting the distance criteria of the Beers–Reich model was more important than getting the angle correct) and each molecule was compared to this compound using the Beers–Reich distance, electrostatic potential (ESP) maps and conformational energy as fit criteria. The final model listed a number of requirements for good activity including electrostatic similarity and steric fit to the template molecule.

While these models have many features in common, controversy still exists regarding (+)-anatoxin-*a* and related compounds. Given the renewed interest in the potential use of nicotinic agonists for treating AD/SDAT, we have embarked on the determination of a pharmacophore model to encompass a range of structurally diverse nicotinic agonists. This model will employ both steric and electrostatic fitting criteria as well as using methods which remove (to a certain extent) the unintended bias that molecular modellers often introduce from their preconceived ideas. It is our hope that some of these controversies may be extinguished with the ultimate aim of using the model to assist in future drug design efforts. One aspect of the use of such a model is to direct 3D-QSAR studies. In this study we have examined two sets of nicotinic compounds to illustrate multivariate QSAR methods, including the use of neural networks for dimension reduction. Refinement of these models, both pharmacophore and QSAR, will no doubt take place as specific pharmacology data becomes available for the different nicotinic receptor subtypes.

METHODS

Pharmacophore model

To generate a pharmacophore for nicotinic agonists we used a method which is able to compare molecular surface properties *via* gnomonic projections. The software package ANACONDA (Oxford Molecular Ltd, Oxford Science Park, Oxford) employs this method allowing the interactive comparison of pairs of compounds which can suggest superimposition modes for molecules and thus lead to the generation of a pharmacophore model(s). Briefly, surface properties of a molecule (e.g., electrostatic potential [ESP] or van der Waals radius) are projected onto the surface of a sphere from a defined centre of interest (Chau and Dean, 1987). The spheres of two molecules may then be compared by either manual manipulation of one sphere relative to another or by automatically computing differences through all possible relative orientations. This latter choice produces a contour map showing regions of similarity and minimum points on the map may be examined in further detail (typically the global minimum) by extracting the molecular structures and examining their fit *via* a molecular graphics package. Molecular volume overlap was then used to choose between minima if more than one reasonable orientation was suggested. Unfortunately, it is not within the scope of this study to describe the theory behind the generation of sphere points or contour maps and we refer the reader to Chau and Dean (1987) for details.

Molecular choice was based primarily on structural diversity and good potency at the nicotinic receptor. In addition, compounds were chosen which had limited conformational flexibility containing one or less rotatable bonds and possessing two clearly identifiable pharmacophoric elements (i.e.,

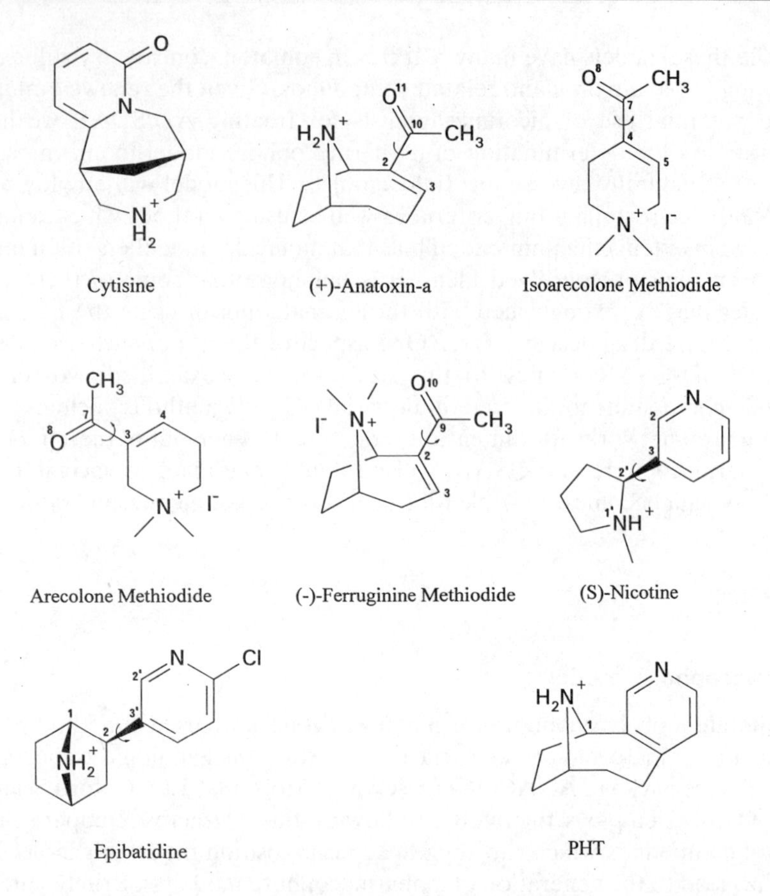

Figure 2 Structures of the seven compounds used to derive the pharmacophore model. The structure of (1*R*)-PHT is also shown.

compounds like tetramethyl ammonium were not considered). Seven compounds were selected consisting of (+)-anatoxin-*a*, isoarecolone methiodide, arecolone methiodide, (–)-ferruginine methiodide, (–)-cytisine, (S)-(–)-nicotine and (2*R*)-epibatidine (Figure 2, see Table I for activities). Both isomers of epibatidine are active at the nicotinic receptor (in contrast to anatoxin-*a*) and in this study we chose to use the naturally occuring (2*R*)-isomer. Construction of each molecule and conformational analysis of compounds with rotatable bonds was performed using SYBYL (Tripos Inc., St Louis, USA). Potential derived partial charges were calculated using the semi-empirical molecular orbital package MOPAC using the key words ESP and MNDO (Stewart, 1990). Compounds were modelled using a positively charged nitrogen for pharmacophore generation as this was the species expected to be in solution at physiological pH.

Table I *Diverse nicotinic agonists.*

Compound number*	Name	Relative potency†	Activity category‡
1	(+)-anatoxin-*a*	110	Potent
2	(1*R*)-PHT§	100	Potent *(test)*
3	isoarecolone methiodide	50	Potent
7	dihydroisoarecolone methiodide	9.1	Potent
8	arecolone methiodide	8.6	Potent
10	1-methyl-4-acetyl-piperazine methiodide	4.6	Potent (*test*)
11	(–)-ferruginine methiodide	3.3	Potent
14	isoarecoline methiodide	2.7	Potent
15	1-methyl-4-acetyl-piperazine methiodide	2.6	Potent
17	arecoline methiodide	1.3	Moderate
18	cytisine	1.1	Moderate
19	*S*-(–)-nicotine methiodide	1.05	Moderate
20	1-methyl-4-carbamyl-1,2,3,6-tetrahydropyridine methiodide	0.77	Moderate
21	muscarone	0.77	Moderate
22	*S*-(–)-nicotine	0.56	Moderate (*test*)
23	isoarecolone	0.48	Moderate
24	1-methyl-1,2,3,6-tetrahydropyridine-4-methanol methiodide	0.35	Moderate
26	*N*-methyl cytisine	0.26	Moderate
30	1-methyl-4-carbamylpiperazine methiodide	0.15	Moderate
31	(–)-norferruginine	0.14	Moderate (*test*)
33	isoarecoline	0.09	Weak (*test*)
35	nornicotine	0.069	Weak
36	1-methyl-4-carbamylpiperidine methiodide	0.052	Weak (*test*)
37	(–)-ferruginine	0.04	Weak
41	octahydro-2-methyl-*trans*-(1*H*)-isoquinolone methiodide	0.015	Weak
43	arecoline	0.011	Weak
44	1-methyl-4-piperidone oxime methiodide	0.0055	Weak
45"	(2*R*)-epibatidine	¶	Potent (*test*)

*The numbering scheme used in this table corresponds to the numbers used in the original table of data from Gund and Spivak (1991).

†The potency is quoted relative to the reference compound carbamylcholine determined in frog rectus abdominus muscles. This data was collated from several studies and the original references may be found in Gund and Spivak (1991).

‡Activity categories were chosen using the following scheme:
Potent, relative potency > 2.0;
Moderate, 2.0 ≥ relative potency ≥ 0.1;
Weak, relative potency < 0.1.

§While the (1*R*) isomer of PHT has been used in this QSAR study, the biological data is given for the racemate.

"(2*R*)-Epibatidine was not included in the original table and has been nominally given the number **45**.

¶The activity of epibatidine has been quoted as having a potency of about 40 pM at neuronal nicotinic acetylcholine receptors (see Gopalakrishnan and Sullivan, 1994). It has also been shown to be 100 times more potent than nicotine in inducing lethality in mice (see Gopalakrishnan and Sullivan, 1994).

(–)-Cytisine was employed as the template molecule for the molecular comparison studies using ANACONDA. While it may be argued that comparing each compound to a single template molecule may not be ideal, we felt that the rigid structure and high potency of (–)-cytisine made it suitable for this purpose. Following the comparison of ESP surfaces with ANACONDA, all seven molecules were then superimposed in their final orientations relative to (–)-cytisine using SYBYL. Coordinates representing the location of the protonated nitrogen and the carbonyl group (or ring centroid and pyridine nitrogen) with an associated dummy atom representing a point of interaction with the receptor are then derived by averaging these positions for the seven compounds to produce the pharmacophore model. Additional features of the pharmacophore model may then be identified, such as regions containing lipophilic groups. Finally, both enantiomers of PHT were analysed using the methodology above to predict the active enantiomer. While PHT is often drawn in the literature as the (1*R*)-isomer (Figure 2), a prediction made by Hacksell and Mellin (1989), this compound has yet to be either resolved or synthesized in enantiomerically pure form. As a consequence, the nicotinic potency (if any) of the individual enantiomers has yet to be determined.

QSAR

To exploit the pharmacophore model, suitable sets of compounds were sought for analysis using 3D-QSAR techniques. Two series were found for this study, consisting of 21 nicotine analogues (Lin *et al.*, 1994) and 28 diverse nicotinic agonists (Gund and Spivak, 1991).

Nicotine Analogues

A series of pyrrolidine modified nicotine analogues was recently described by Lin and coworkers (1994) with biological data relating to their affinity for rat central nervous system nicotinic acetylcholine receptors. A small number of compounds were excluded from this analysis as data were only available for the racemate. The subset of compounds used in this study consisted of nicotine and 20 analogues with substituents in the 4' and 5' positions of the pyrrolidine ring (Table II) ranging in potency from 1.15 to 3353 nM. Each compound was constructed and geometry optimised using SYBYL, employing Gasteiger–Huckel partial charges. Using the conformation of (S)-(–)-nicotine determined for the pharmacophore model, each analogue was altered to this configuration and minimized. In most cases this configuration did not differ markedly from that of (S)-(–)-nicotine, although for a few compounds the conformation of the pyrrolidine ring was altered to a lower energy conformer to relieve any steric strain caused by ring substituents. Each compound was then superimposed on to (S)-(–)-nicotine using 3 atoms in the pyridine ring and the pyrrolidine nitrogen. Typically,

Table II *Nicotinic analogues.*

Compound*	R_1	R_2	Binding affinity (nM)[†]	Activity category‡
A ((*S*)-(–)-nicotine)	H	H	1.15 ± 0.4	Potent
B	(β)-OH	H	27.6 ± 0.8	Potent (*test*)
C	(β)-OMe	H	36.6 ± 0.8	Potent
D	(β)-OMs	H	363.6 ± 17.9	Moderate
E	(α)-CN	H	82.0 ± 2.0	Potent
F	(β)-OAc	H	102.9 ± 16.7	Moderate
G	(β)-CH_2OH	H	157.8 ± 7.4	Moderate
H	(α)-CH_2OH	H	294.3 ± 11.0	Moderate
I	(β)-CH_2F	H	11.1 ± 1.9	Potent
J	(β)-CH_2CN	H	52.0 ± 2.9	Potent
K	(β)-CH_2SMe	H	492.8 ± 19.2	Moderate (*test*)
L	(β)-CH_2OMe	H	510.0 ± 46.6	Moderate
M	(β)-Me	H	4.23 ± 0.28	Potent
N	(β)-Et	H	50.2 ± 1.1	Potent
O	(β)-CH_2Ph	H	119.4 ± 18.5	Moderate
P	H	(β)-Me	34.9 ± 1.9	Potent
Q	H	(α)-Me	1205.3 ± 34.6	Weak
R	H	(β)-*n*-Bu	125.2 ± 4.7	Moderate
S	H	(α)-*n*-Bu	1381.4 ± 209.0	Weak
T	H	(β)-Ph	1242.3 ± 12.4	Weak
U	H	(α)-Ph	3353.5 ± 196.9	Weak (*test*)

*This data table originates from the study by Lin and coworkers (1994), however a number of compounds from this original study were not included as data was only available for the racemate.

[†]Binding affinities are given as the K_i ± SEM.

‡Activity categories were chosen using the following scheme:
Potent, activity < 100 nM;
Moderate, 100 nM $\leq$ activity $\leq 1{,}000$ nM;
Weak, activity $> 1{,}000$ nM.

this resulted in an RMS fit of the 4 atoms less than 0.1 Å. In those cases where the conformation of the 5 membered ring was altered no major change was observed in the location of the pyrrolidine nitrogen nor in the direction of the nitrogen lone pair vector.

Compounds were split into three activity categories: potent, moderate and weak nicotinic activity. Three representative test compounds were chosen using the hierarchical cluster analysis facility within the 3D-QSAR package, TSAR (Oxford Molecular Ltd); one from each of the three activity categories. TSAR was also used to perform neural network and multivariate data analysis.

Diverse Nicotinic Agonists

The second series of compounds we chose to investigate were collated into a table in the review by Gund and Spivak (1991). Molecular choice was based on similar criteria to those compounds selected to derive the pharmacophore model. From the original set of 44 compounds a subset of 26 compounds was examined for QSAR analysis. In addition to these compounds, the active enantiomer of PHT predicted from the pharmacophore model was used as a test compound. Similarly, (2*R*)-epibatidine was also included as a test compound as it represents a recently identified and highly potent nicotinic agonist with a novel structure (see, Spande *et al.*, 1992; Qian *et al.*, 1993; Senokuchi *et al.*, 1994). Each compound was constructed and minimized as described above using SYBYL (Tripos Ltd) and MOPAC, MNDO ESP charges. The same superimposition strategy used for the pharmacophore model was applied to generate the final set of conformations for QSAR analysis. Once again compounds were split into three activity categories (potent, moderate and weak) and seven compounds were held back as a test set (chosen using hierarchical cluster analysis to obtain two representatives from each class. (2*R*)-Epibatidine was chosen as one of the representatives of the 'potent' class and the seventh test compound was (1*R*)-PHT). Molecular similarities were generated using the ASP program within TSAR which was also used for data analysis.

ASP

ASP allows the calculation of molecular similarity between sets of molecules. These similarity matrices have been shown to be of use in a growing number of QSAR studies using this technique (Hopfinger, 1980, 1983; Burt *et al.*, 1990; Good *et al.*, 1993a,b). The original idea of determining a similarity index to compare two molecules was introduced by Carbo and coworkers (1980).

Carbo Index
$$R_{AB} = \frac{\int P_A P_B \, dv}{\left(\int P_A^2 dv\right)^{1/2} \left(\int P_B^2 dv\right)^{1/2}} \quad (1)$$

where molecular similarity R_{AB} is derived from the structural properties P_A and P_B of the two compounds being compared. The denominator in this equation provides a normalising factor for the numerator which is a measure of property overlap. The similarity index therefore takes a value in the range 0–1. Typically the structural property P is either ESP, electrostatic field, shape or combinations of these properties.

A problem arises if two molecules have electron densities which correlate, as this leads to similarity indices tending towards unity. Hodgkin and Richards (1987) addressed this problem by increasing the sensitivity of the Carbo index. It was shown that if $P_A = nP_B$, then R_{AB} equals unity. This lead to the definition of the Hodgkin index:

Hodgkin Index
$$H_{AB} = \frac{2\sum_{i=1}^{N} (P_A P_B)}{\sum_{i=1}^{N} P_A^2 + \sum_{i=1}^{N} P_B^2} \quad (2)$$

such that if $P_A = nP_B$, then $H_{AB} = 2n/(1 + n^2)$. This index was employed in this study.

Neural Networks and Multivariate Statistical Methods

Neural networks have been applied in the chemistry field for nearly a decade and the first report of their use in QSAR was published over 5 years ago (Aoyama *et al.*, 1990). Problems associated with regression and discriminant data analysis using neural networks have been described (Manallack and Livingstone, 1994) and a number of suggestions have been investigated to overcome these (see Manallack and Livingstone, 1994). The ReNDeR (reversible nonlinear dimension reduction) technique was introduced by Livingstone *et al.* (1991) as a method of using neural networks for dimension reduction (Figure 3). This method does not suffer from the problems of overfitting which is a potential danger when networks are used for regression and discriminant analysis. As the ReNDeR technique is used to reproduce, at the output layer, the same data presented to the input layer, it effectively squeezes the data through the parameter layer (Figure 3). ReNDeR may therefore be termed an 'unsupervised learning' technique.

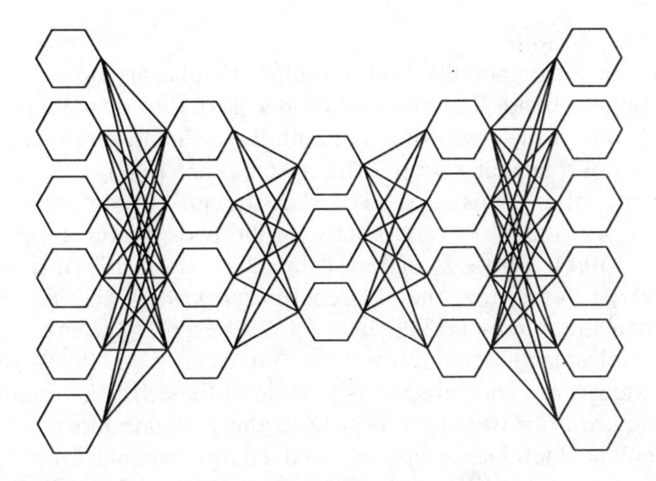

Figure 3 Example of a ReNDeR neural network used for dimension reduction. This network is fully connected and contains 6 units (hexagons) in the input layer and 3 units in the central parameter layer. Following training the values of the units in the parameter layer are used for plotting purposes (i.e., these units represent the *X*, *Y* and *Z* coordinates).

Any clustering of activity categories should consequently not be due to chance, although this possibility cannot be completely ruled out. The ReNDeR method is implemented within TSAR and has been modified to increase the speed of training. Training proceeds by presenting each row of data to the network in turn and the error (true vs. predicted) is minimised over 2000 cycles. The initial weights and biases of the network may be selected using a Monte-Carlo simulation in order to achieve a better starting set. Whilst training proceeds, the user is presented with a graph of RMS deviation vs. cycle number. This can be used to determine when network training has converged. TSAR uses feedforward backpropagation and a description of this technique may be found in the following references (Rumelhart and McClelland, 1988; Hertz *et al.*, 1991; Katz *et al.*, 1992).

As a comparison to the neural network dimension reduction procedure, principal components analysis (PCA) was employed. PCA is also used to reduce the dimensionality of large datasets to provide simple two- or three-dimensional plots whilst retaining as much information from the original dataset as possible. A description of the PCA method and other multivariate techniques can be found in the following references (Chatfield and Collins, 1986; Livingstone, 1991, 1995). PCA is implemented within TSAR.

RESULTS

Pharmacophore model

Conformational Analysis

Six of the seven compounds used to define the pharmacophore contained a single rotatable bond. The conformational flexibility of (+)-anatoxin-*a* has been studied previously showing that both the *s-cis* and *s-trans* enone conformations are of a similar energy (Thompson *et al.*, 1992). Comparisons with the structure of (–)-cytisine showed that the *s-trans* conformation of (+)-anatoxin-*a* gave the best overlap between the two structures using a torsion angle (τ 3,2,10,11, Figure 2) of 160° (Manallack *et al.*, 1994). In the present study, SYBYL found that the *s-cis* conformer was 1.4 kcal mol^{-1} higher in energy than the *s-trans* (τ 3,2,10,11 = 151.4°, Figure 2), agreeing with the previous molecular modelling results (Thompson *et al.*, 1992). Both enone conformations were compared to (–)-cytisine (Table III) confirming that the *s-trans* conformer provided the best fit to the template molecule.

Isoarecolone methiodide also showed energy minima for the *s-cis* and *s-trans* conformations (*s-trans* was lower in energy by 0.25 kcal mol^{-1}). Table III shows that there is no significant difference in the fit to (–)-cytisine between the two enone conformers of isoarecolone methiodide. The final choice of using the *s-trans* conformer (τ 5,4,7,8 = –155.5°, Figure 2) in this study was based on (+)-anatoxin-*a* assuming that the location of the

Table III *Comparison of molecular fit to (–)-cytisine.*

Compound	Torsion angle*	ANACONDA RMS fit†	% Volume overlap‡	ASP shape index§
(+)-anatoxin-a *s-trans*[1]	151.4	0.316	78.6	0.903
(+)-anatoxin-a *s-cis*	30.9	0.351	69.4	0.832
isoarecolone methiodide *s-trans*[1]	–155.5	0.410	79.8	0.883
isoarecolone methiodide *s-cis*	8.0	0.409	75.3	0.842
arecolone methiodide *s-trans*[1]	158.9	0.410	79.1	0.892
arecolone methiodide *s-trans*	–15.8	0.437	79.4	0.892
(–)-ferruginine methiodide *s-trans*[1]	157.9	0.386	76.5	0.933
(–)-ferruginine methiodide *s-cis*	–19.0	0.515	58.1	0.730
(*S*)-(–)-nicotine[1]	–58.8	0.341	78.8	0.905
(*S*)-(–)-nicotine	117.1	0.561	54.0	0.616
(2*R*)-epibatidine[1]	4.3	0.311	62.1	0.773
(2*R*)-epibatidine	171.5	0.644	48.1	0.603
(1*R*)-PHT	–	0.302	71.8	0.857
(1*S*)-PHT	–	0.571	61.1	0.739

*Torsion angle definitions are shown in Figure 2.
†As a measure of similarity, the ANACONDA program calculates the RMS difference value for the ESP gnomonic spheres of (–)-cytisine and each test compound (a low value for this parameter indicates a good match between the gnomonic spheres).
‡Volume overlap is determined by dividing the volume of overlap with (–)-cytisine, with the volume of the test molecule itself. The superimposition orientations were derived from ANACONDA.
§The ASP values quoted here give the steric similarity value to (–)-cytisine using the overlap orientation derived from ANACONDA.
Conformers indicated with the symbol[1] were ultimately used in the determination of the pharmacophore model.

C=C double bond plays a role in ligand binding. The 1,2,3,6-tetrahydropyridine ring of isoarecolone methiodide was modelled in a half chair conformation (Spivak *et al.*, 1986). In this arrangement, the charged N atom in Figure 2 will lie above the plane of the page. This choice was also based on a preliminary comparison of isoarecolone methiodide with the structure of (–)-cytisine to maximize the fit between the H-bond accepting groups and the positively charged nitrogen centres.

A similar strategy was employed for arecolone methiodide to choose the conformation used in further modelling work (Table III). The *s-trans* isomer of arecolone methiodide was 0.6 kcal mol^{-1} lower in energy (τ 4,3,7,8 = 158.9°, Figure 2) and once again a half chair conformation was used for the 1,2,3,6-tetrahydropyridine ring (charged N atom lies above the plane of the page,

Figure 2). The structure of (–)-ferruginine methiodide is similar to that of (+)-anatoxin-*a* and was modelled in a similar manner; the *s-trans* conformer was used for pharmacophore determination (*s-trans* was 0.3 kcal mol^{-1} lower in energy than the *s-cis* conformer, τ 3,2,9,10 = 157.9°, Figure 2). The results of the fit of the *s-cis* and *s-trans* enone conformers of (–)-ferruginine methiodide to (–)-cytisine clearly show the better fit of the *s-trans* conformer (Table III).

The previous molecules investigated in this chapter have all contained conjugated α,β-unsaturated ketone systems resulting in low energy conformations corresponding to the *s-trans* and *s-cis* isomers (as expected). Rotation of the bond τ 2,3,2',1' (Figure 2) of (*S*)-(–)-nicotine showed that there were two low energy minima with a barrier to rotation of 2.6 kcal mol^{-1}. Again a comparison of these low energy conformers were made to (–)-cytisine to decide on the likely bioactive configuration for use in developing the pharmacophore model (Table III). The conformer chosen (τ 2,3,2',1' = –58.8°, Figure 2) corresponds to the NMR solution conformation (Pitner *et al.*, 1978) where the pyridine and pyrrolidine rings are arranged perpendicular to each other. In addition, the pyrrolidine ring is in an envelope conformation and the *N*-methyl moiety and pyridine rings are oriented *trans* with respect to each other. Rotation of the pyridine ring of (2*R*)-epibatidine relative to the 7-azabicyclo[2.2.1]heptane ring system (τ 1,2,3',2', Figure 2) gave two low energy conformations with a barrier to rotation of 2.7 kcal mol^{-1}. A preliminary check of (2*R*)-epibatidine to the structure of (–)-cytisine showed that a good fit could be obtained using a high energy conformation. Since (2*R*)-epibatidine is highly potent it was decided to compare both low energy conformers to (–)-cytisine (Table III). This comparison clearly shows that the minimum energy structure corresponding to τ 1,2,3',2' = 4.3° (Figure 2) provides the superior fit. We have made the assumption that if the bioactive conformer was high in energy, then this would be reflected in a lower potency hence the modelling has used one of these minimum energy structures.

In this study, (–)-cytisine (Figure 2) was modelled in its lowest energy conformation. The piperidine ring of (–)-cytisine adopts a chair conformation.

ANACONDA

The ANACONDA program was used to compare the ESP gnomonic surfaces of (–)-cytisine and the six other pharmacophore compounds (Table III). This resulted in suggested overlaps where the charged nitrogen atoms were closely superimposed. In addition, the H-bond accepting groups were also located together, thus demonstrating the usefulness of this technique to produce sensible molecular superimpositions. Moreover, very good spatial overlap between the compounds and a region containing many lipophilic groups was also identified (Figures 4A and 4B). Dummy atoms representing a point of interaction on the receptor were attached to the carbonyl oxygen at a

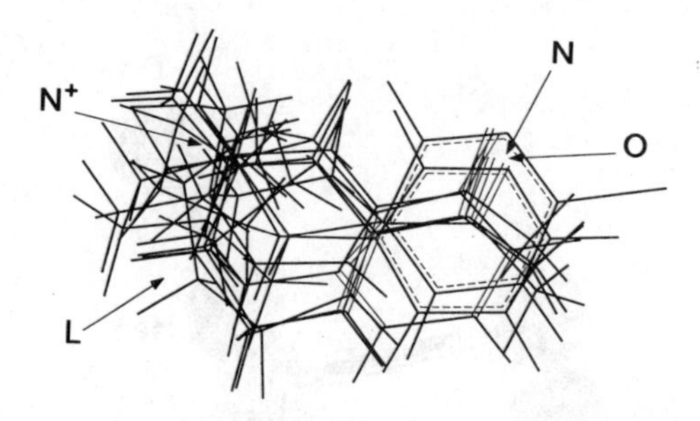

Figure 4A Plot of the seven pharmacophore compounds superimposed together. Positions of the charged nitrogen atoms (N^+), lipophilic site, pyridyl nitrogens (N) and carbonyl oxygens (O) have been indicated.

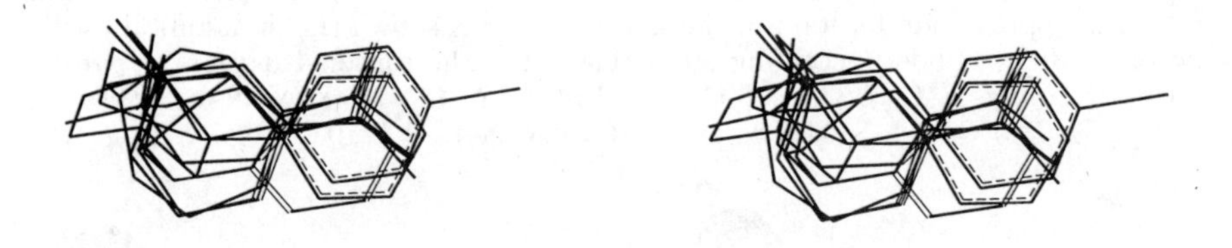

Figure 4B Stereo plot of the seven pharmacophore compounds. Hydrogen atoms have been removed for clarity.

distance of 3 Å along the carbon oxygen vector of this group. Similarly, a dummy atom was attached 3 Å from the nitrogen atom of the pyridine rings along the ring centroid, pyridyl nitrogen vector. The positions of the seven charged nitrogens (**N**), seven dummy atoms (**D**) and the five carbonyl carbons plus two pyridine ring centroids (**C**) were averaged to produce the final pharmacophore model. Figures 5A and 5B illustrate the disposition of these three averaged positions in space and indicates the approximate location of the lipophilic region (**L**). The lipophilic region was located close to the 3',4' carbon atoms of the *S*-(–)-nicotine pyrrolidine ring. To compare the present pharmacophore to previously described models it is instructive to examine the distances and angles between the three key pharmacophoric points (Figures 5A and 5B). **N–C** = 3.94 Å; **C–D** = 4.14 Å; **N–D** = 6.48 Å; **N–L** = 2.50 Å; **C–L** = 4.07 Å; **D–L** = 6.70 Å. Of these measurements the most important relate to the Beers–Reich model. The angle of 107° for **N–C–D** compares well to previous studies. However, to be able to compare the Beers–Reich distance the **C** to **D** distance was reduced to 1.4 Å (i.e., van der Waals radii) resulting in a Beers–Reich distance of 5.4 Å.

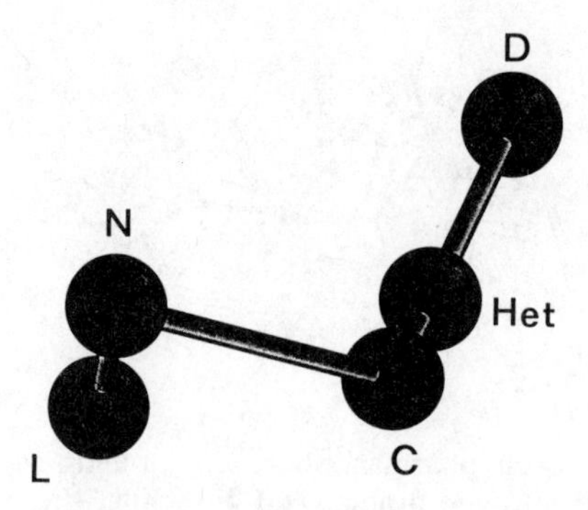

Figure 5A Diagram of the pharmacophore model showing the locations of the charged nitrogen (**N**), carbonyl carbon/pyridyl ring centroid (**C**) and the dummy atom (**D**). The approximate location of the lipophilic site is shown (**L**), in addition to the position of the H-bond accepting atom (**Het**) (i.e., the carbonyl oxygen or pyridyl nitrogen atom). **N–C** = 3.94 Å; **C–D** = 4.14 Å; **N–D** = 6.48 Å; **N–L** = 2.50 Å; **C–L** = 4.07 Å; **D–L** = 6.70 Å; **N–C–D** = 107°.

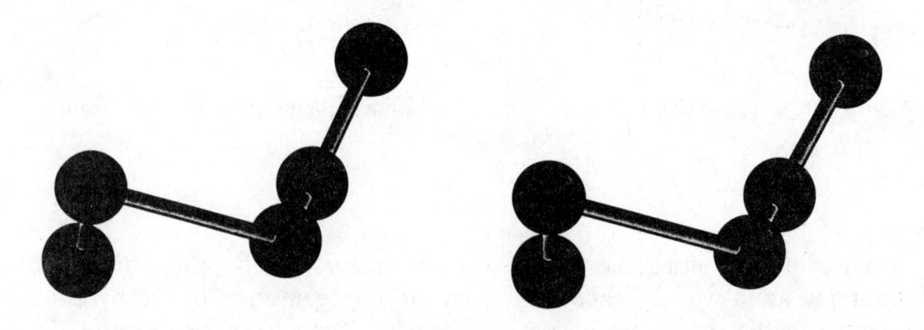

Figure 5B Stereo plot of the pharmacophore model.

QSAR

Nicotine Analogues

Lin and coworkers (1994) briefly discussed the effects of pyrrolidine ring substitution on nicotinic activity for the compounds (A–U) listed in Table II. They concluded that in general, steric factors predominated in affecting nicotinic potency. As a consequence, this study chose a number of parameters relating to steric properties. A total of 15 parameters were calculated for the 21 nicotine analogues listed in Table II consisting of:

- moments of inertia in the *X*, *Y* and *Z* directions (*IX*, *IY* and *IZ*);
- principal ellipsoid axes (*RX*, *RY*, *RZ*);
- ellipsoidal volume;

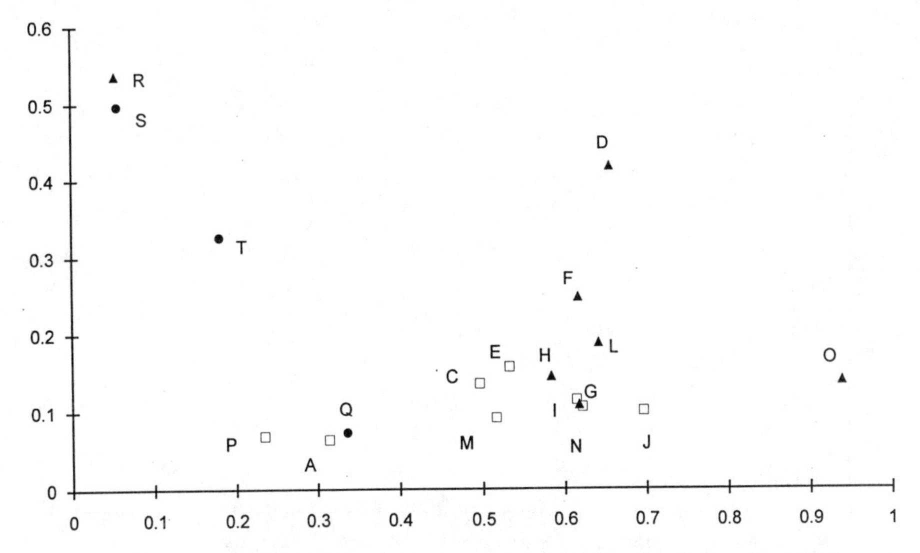

Figure 6 Neuron plot of the 18 training compounds of the nicotine analogues dataset. (□) Potent, (▲) moderate and (●) weak compounds. See Table II for compound lettering.

- molecular volume;
- shape indices K1, Kα1, K2, Kα2, K3, Kα3;
- flexibility (ϕ) (Hall and Kier, 1992).

A number of these properties were highly correlated ($R^2 > 0.9$) with other properties and were removed to leave six representative parameters for further analysis. Figure 6 shows the neuron plot of the 18 training compounds (Table II) using a 6,3,2,3,6 ReNDeR neural network (where each number represents the number of processing units in each layer of the network). This network took 70 seconds to train on an Indigo 2 Extreme Silicon Graphics workstation. Clustering of activity categories can be seen with only a few compounds not falling in the appropriate region. For example, compound Q (Table II) fell within the region associated with active compounds; however, this is not surprising based on its structure, which is similar to (*S*)-(–)-nicotine itself. This would indicate that there are strict steric requirements in the position adjacent to the 4'-α position. Indeed if this compound is fitted into the pharmacophore model, the 4'-α-methyl substituent extends beyond the volumes occupied by the eight pharmacophore compounds. Obviously, the parameters chosen for this QSAR analysis will be insufficient to describe compound Q unless additional properties are included for this sterically sensitive region of the receptor (perhaps an indicator variable).

Despite this, the good separation of categories (Figure 6) means that this approach can be used for prediction *via* the introduction of the three

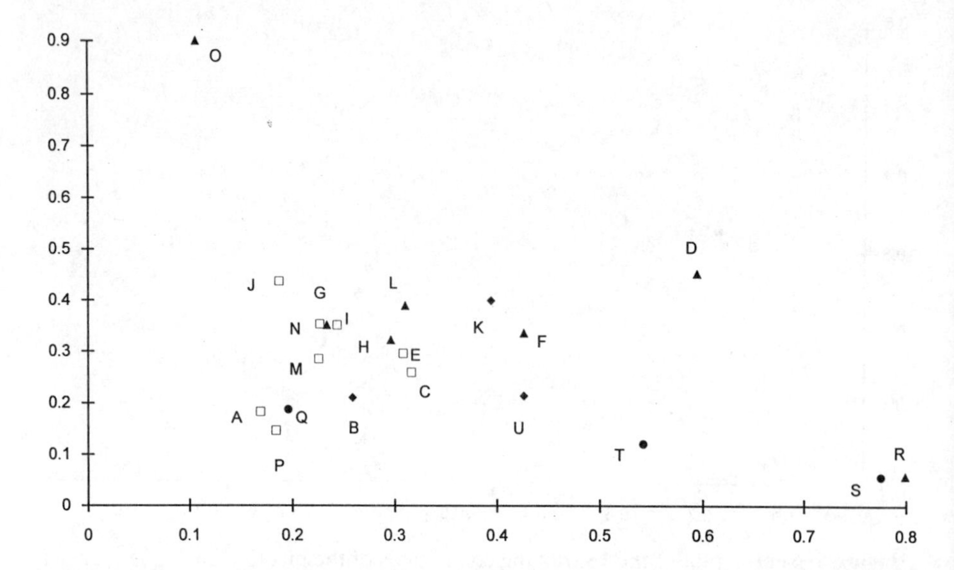

Figure 7 Neuron plot of the compounds in the nicotine analogues dataset. (□) Potent, (▲) moderate, (●) weak and (◆) test compounds. See Table II for compound lettering.

remaining test compounds (B, K, and Q, Table II). Figure 7 shows the neuron plot including the three test compounds. It should be noted that the plot needs to be recalculated when additional compounds are included. Once again, it is clear that there is clustering of activity categories, and at least two (B and K) of the three test compounds fall within areas occupied by their respective categories. The third compound, U, falls adjacent to two other weak compounds but is also located approximately equidistant from the other two categories. This therefore demonstrates that this model is useful for predictive purposes. Interestingly, the location of the weak compounds has changed from the original training plot and there are subtle differences in the locations of all compounds.

Using PCA to examine the 18 training compounds (Table II) also resulted in clustering of the three activity categories (data not shown). The associations of compounds in the plot of PC1 vs. PC2 were remarkably similar to those found in the neuron plot (Figure 8). Introduction of the test compounds to this analysis gave the same clustering pattern as the ReNDeR analysis. Test compounds B and K fell in the correct regions while the third (U) fell between these clusters (Figure 8).

Diverse Nicotinic Agonists

The same approach for determining molecular superimpositions, as applied to the generation of the pharmacophore model, was used for the diverse nicotinic

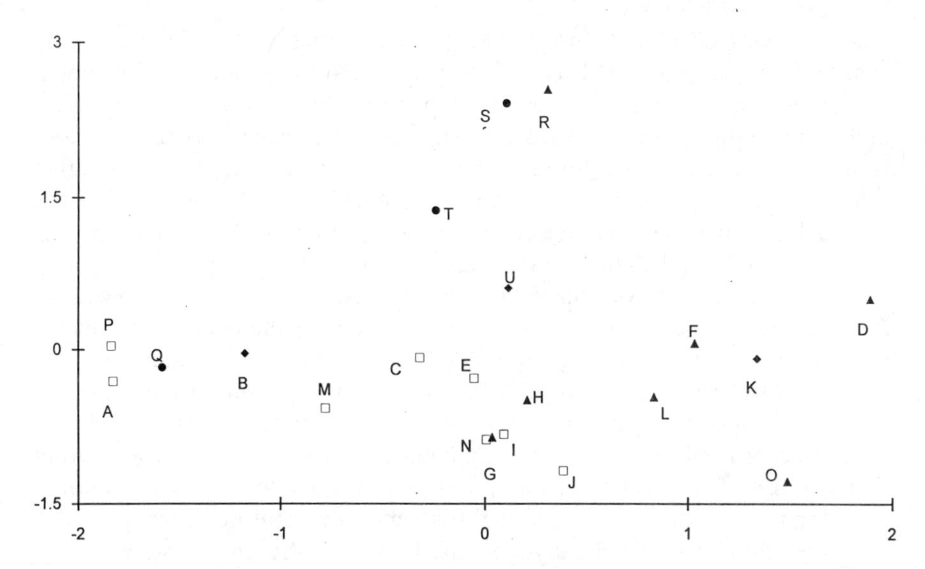

Figure 8 Principal components plot of the nicotine analogues dataset derived from 6 variables. The first two principal components explained over 88% of the variance of the dataset. (□) Potent, (▲) moderate, (●) weak and (◆) test compounds. See Table II for compound lettering.

agonists dataset (compounds 1–45, Table I). Indeed, the conformations of the compounds used in the pharmacophore model were employed for structurally related compounds. For example, compound **20** employed the same conformation as compound **3** (Table I). Typically, for structurally unrelated compounds, a local minimum conformation (often the lowest minimum found in the search) would be chosen which was within 1 kcal mol^{-1} of the global minimum. Reasonable overlap of the charged nitrogen and H-bond accepting group of the compound with the corresponding groups in (–)-cytisine was an important criterion if more than one conformer was considered. This was a relatively simple task, as the majority of the structurally unrelated compounds differed only slightly with any of the seven pharmacophore compounds. Muscarone (**21**), which has two rotatable bonds, was modelled in its global minimum energy conformation which gave an excellent overlap with (–)-cytisine using the pharmacophore groups described above.

Following this preliminary conformational work, each compound was compared using the program ANACONDA to produce the final orientations for superimposition. Again, if more than one reasonable superimposition was suggested for a given compound, superior molecular overlap with (–)-cytisine was used to decide between these. Each compound in its final configuration was then imported into TSAR for QSAR analysis.

Since the compounds in Table I represent several different structural classes, it was decided to use molecular similarities as the descriptors for analysis. The program ASP is able to produce *N* by *N* matrices of molecular similarities by comparing each compound in turn with the others in the table. Interestingly, when electrostatic similarity was calculated between each compound these values differed from unity by only 0.01. In other words they were considered to be extremely similar to each other. As clear differences should be seen with this dataset, the question arose to how well the ASP program could handle charged molecules. To overcome the problem, a negatively charged acetate was placed in the same location in each file to counter the positive charge on the nitrogen (the location of the acetate was placed so that each oxygen was located 3.5 Å from the hydrogens attached to the charged nitrogen of (–)-cytisine. This position was maintained throughout for each compound so that it would have a constant effect). It was felt that as the receptor site presumably has a counter charge for the nitrogen atom, then the use of the acetate group in our analyses may mimic this compound receptor interaction. This technique was very successful in allowing ASP to generate numbers which demonstrated clear differences between the compounds.

Using a trial and error approach it was found that the use of combined shape and ESP similarities provided a more reasonable QSAR model for the training compounds than using either the separate shape or ESP indices alone. A total of 28 properties (combined shape, ESP similarities) was generated and 6 of these were removed as they were found to be highly correlated ($R^2 > 0.9$) with other properties. In addition, the similarity indices for the test compounds were also removed to analyze the training set. A neuron plot generated using a 15,5,3,5,15 ReNDeR network for the training compounds (Table I) is shown in Figure 9. Using this particular orientation (it should be noted that this plot was derived from three-dimensional data and the original coordinates were rotated to produce the view in Figure 9) it can be seen that there is a cluster of active compounds. Unfortunately, when this was repeated by including the test compounds (Table I), they did not fall in the appropriate regions of the plot (Figure 10). This may suggest that the original cluster of compounds was merely a fortuitous association which had no predictive usefulness. In comparison to the neuron plots, PCA also showed a clustering of active compounds for the training set (data not shown) and inclusion of the test compounds also resulted in these being misassigned (Figure 11).

DISCUSSION

This study has employed a computational chemistry approach with molecular modelling to develop a nicotinic agonist pharmacophore model. This model adequately accounts for the activity of various structurally diverse nicotinic

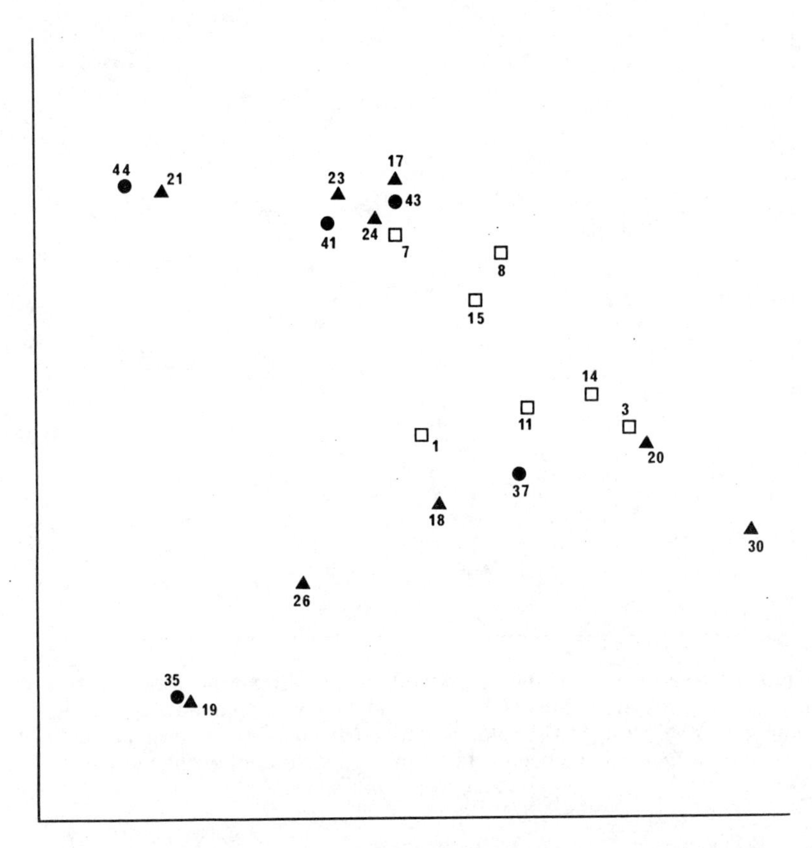

Figure 9 Neuron plot of the 21 training compounds of the diverse nicotinic agonists dataset. (□) Potent, (▲) moderate and (●) weak compounds. See Table I for compound numbering. This plot is derived from three-dimensional data and the original coordinates have been rotated to produce the following view. The *X*, *Y* and *Z* axes are not shown.

agonists and gives the location of both primary and secondary binding sites in the receptor. Furthermore, the model clarifies a number of minor controversies regarding the binding mode of some nicotinic agonists. A QSAR study of a series of nicotine analogues demonstrated that activity could be accounted for using steric physicochemical properties.

Pharmacophore model

One aim of this study was to define a pharmacophore model using computational methods which take away some of the decision-making from the molecular modeller. Ideally this would mean giving a software package the structures and activities of many nicotinic agonists and allowing it to derive a model. Such techniques are present (e.g., Catalyst–Molecular Simulations

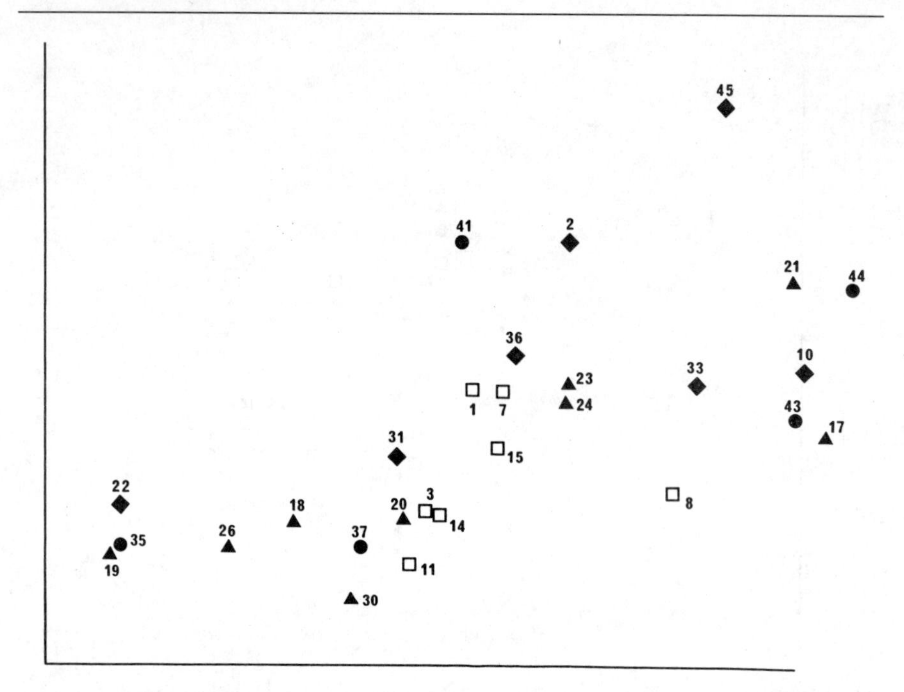

Figure 10 Neuron plot of the compounds in the diverse nicotinic agonists dataset. (□) Potent, (▲) moderate, (●) weak and (◆) test compounds. See Table I for compound numbering. This plot is derived from three-dimensional data and the original coordinates have been rotated to produce the following view. The *X*, *Y* and *Z* axes are not shown.

Inc., Massachusetts, USA), however, in the absence of such methods, modellers must draw on their experience and, if possible, previously published attempts to define a model. Unfortunately, as scientists, we can often overlook potential solutions to a problem by avidly following our pet theories. Although considerable input to the description of this present model was dependent on the choice of conformers for some of the compounds, the final superimpositions were derived using electrostatic alignments thus partially taking away modeller bias at the last step. Of course, final acceptance of any model usually compels the modeller to apply the 'bloody obvious' test, which refers back to what they consider is logical based on the SAR data.

One of the advantages of using ANACONDA to align each compound is that this should reproduce what the receptor will see of an approaching molecule, i.e., the electrostatic surface of the molecule (Marshall and Naylor, 1990). Moreover, this allows the model to be derived without using dummy atoms attached to the oxygen along the carbon–oxygen vector of the carbonyl group (or ring centroid–pyridyl nitrogen vector) as in the Beers–Reich model (1970), (cf. Sheridan *et al.*, 1986). Indeed, it has been shown that interaction with a carbonyl group does not occur preferentially along the carbon–oxygen

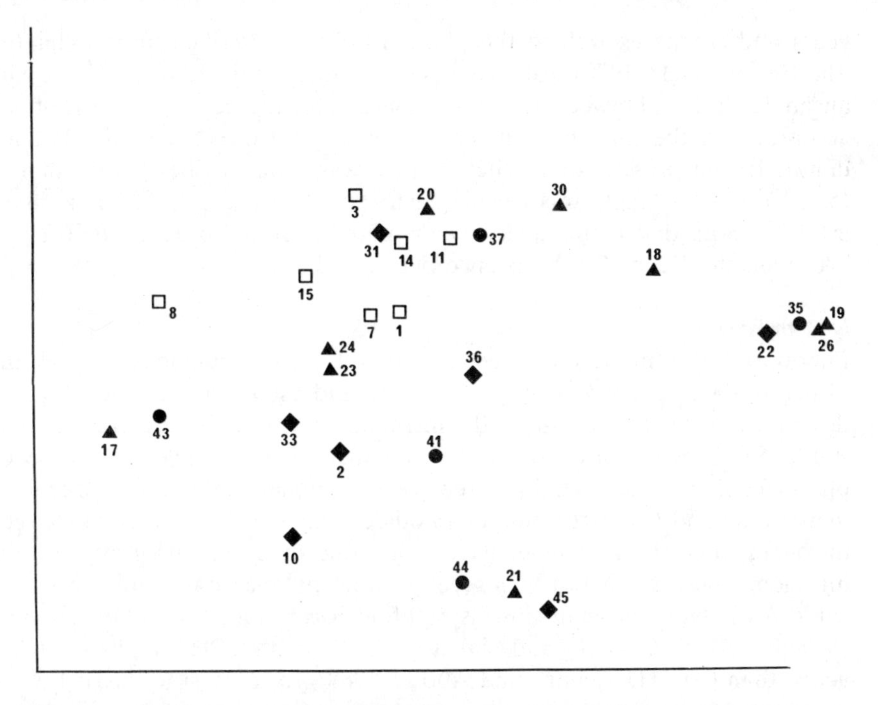

Figure 11 Principal components plot of the diverse nicotinic agonists dataset derived from 20 variables. The first three principal components explained over 79% of the variance of the dataset. (□) Potent, (▲) moderate, (●) weak and (◆) test compounds. See Table I for compound numbering. This plot is derived from three-dimensional data and the original coordinates have been rotated to produce the following view. The *X*, *Y* and *Z* axes are not shown.

bond but in a cone of possible interacting points (Tintelnot and Andrews, 1989). Thus, (+)-anatoxin-*a*, which does not fit the Beers–Reich distance (Gund and Spivak, 1991) in the *s-trans* arrangement, is still able to interact with the appropriate point in the receptor through a different location on its cone of points to the one directly along the carbon–oxygen vector. While these dummy points have been used in this present model, they were used merely to identify a common point which all the molecules could interact with, as well as simplifying the description of the final model by the use of a single point. The present model also considered molecular overlap as an important criterion for predicting common binding modes. Figure 4B illustrates the high degree of overlap between the seven compounds. The other important aspect of this model was in the identification of a region of lipophilicity, an aspect of ligand/pharmacophore characterization that has not been described previously.

It would be remiss at this point not to mention the usefulness and longevity of the Beers–Reich (1970) model. This model has set the scene for over 25 years and compares well to the pharmacophore described in this chapter. The Beers–Reich (1970) distance was originally calculated to be 5.9 Å with an angle of 120° between the centre of charge, the centre of the H-bond acceptor and the line through the carbonyl bond directed to the H-bond donor. In the present study, the distance was found to be slightly shorter (5.4 Å) and the angle was calculated to be more acute (107°). It is interesting to note that if the angle in our model is adjusted within SYBYL to 120° then the Beers–Reich distance becomes 5.8 Å.

N-Methylation

Pharmacophore models can be considered to be simplifications of the available data to allow them to be easily understood and applied to drug design problems. In the case of the nicotinic pharmacophore there are some subtle SARs which are not easily explained. For example, of the seven pharmacophore compounds, three (isoarecolone methiodide, arecolone methiodide and (–)-ferruginine methiodide), have an *N,N*-dimethyl charged nitrogen, while the remainder have either one or no methyl groups on the nitrogen. These methyl groups greatly affect potency but there are no set rules. *N,N*-Dimethyl anatoxin-*a* is 500-fold less potent than (+)-anatoxin-*a* (Wonnacott *et al.*, 1991) and similarly, (±)-*N*-methyl-PHT is 200-fold less active than (±)-PHT (Kanne and Abood, 1988). In contrast to this, (–)-ferruginine methiodide is 80-fold more potent than ferruginine and isoarecolone methiodide is 100-fold more potent than isoarecolone (Gund and Spivak, 1991; Table I). It is possible to consider that (–)-ferruginine methiodide is a conformationally restricted analogue of arecolone methiodide. If one is able to use the ester mimic arecoline methiodide for comparison, then this compound is 100-fold more potent than arecoline itself. It may appear that two classes of compounds exist depending on the nature of the charged nitrogen atom, however, the activities of S-(–)-nicotine and (–)-cytisine are not greatly affected by the addition of a methyl group; 2-fold increase and a 4-fold decrease in activity, respectively (Table I). One can only presume that subtle steric effects influence activity greatly; however, this is difficult to understand when the structural similarity of (+)-anatoxin-*a* and (–)-ferruginine are considered.

The assumption applied to the comparisons above is that all these compounds are binding in the same way. There are dangers associated with this assumption. Firstly, the nicotinic binding site is thought to contain two anionic groups about 11 Å apart, each of which may interact with cationic groups in an agonist (Sheridan *et al.*, 1986). Located 5 Å away from each of the anionic groups is an H-bond donor which can interact with an electronegative atom in the ligand (see Sheridan *et al.*, 1986). The possibility exists that agonists may interact with one or other half of these binding sites. The other danger with the above assumption has been shown from X-ray crystal structure evidence of enzyme inhibitor complexes. Inhibitors can bind in different conformations or to new sites in a protein (Mattos and Ringe, 1993). In the review by Mattos and Ringe (1993) the authors note that minor changes, such as the addition of a small moiety (e.g., a methyl group), can cause radical changes to the mode of binding.

A second assumption implicit in this study is that the receptor protein itself does not change its conformation on binding of the ligand and that this response of the receptor is the same for all ligands. Experience should tell us not to be surprised about different binding modes, as we do not fully understand the processes which take place on binding of ligands to receptors (Mattos and Ringe, 1993). Clearly, these are issues which deserve further attention.

Miscellaneous SAR

Another interesting SAR issue which is relevant to compounds using a carbonyl group to interact with the H-bond donating group in the receptor, is the requirement of a methyl group adjacent to the carbonyl moiety. The desmethyl analogue of (+)-anatoxin-*a*, anatoxinal, is 250 times weaker than the parent compound. A similar, albeit less marked, 7-fold drop in potency has been shown for the hydroxy analogues of isoarecolone methiodide and desmethyl-isoarecolone methiodide. As the electrostatics of the parent and desmethyl compounds look very similar, then this methyl group may be enhancing potency in a number of ways: (a) this group may be located in a specific pocket; (b) the methyl may affect the conformational freedom of the carbonyl group; or (c) may donate electron density to the carbonyl group (Waters *et al.*, 1988). Again, these subtle changes are difficult to describe in a simple pharmacophore model. Changes in electronic structure on binding of a ligand to a receptor are not easily modelled or predicted, and these induced alterations may play a vital role in maintaining agonist activity.

Many nicotinic agonists possess an α,β-unsaturated ketone system where the carbon–carbon double bond is beneficial for activity; dihydroanatoxin (as a mixture of α and β diastereomers) is 7 times less potent than (+)-anatoxin-*a* (Wonnacott *et al.*, 1991). A similar drop in potency is seen by comparing compounds **3** and **7** (Table I). Waters and coworkers (1988) put forward the idea that π-orbital overlap in the conjugated enone system populated a ground state identical with that of the pharmacophore. In saturated systems, conformational freedom, in addition to a small cost in energy required to place the carbonyl in the correct orientation, may explain the lower activity (Waters *et al.*, 1988). The other possibility is that the double bond may interact with a π-orbital system in the binding site. Indeed, the double bonds of the pharmacophore compounds are located together and are coplanar with the pyridine ring systems of the other agonists (Figures 4A and 4B). This provides evidence for a possible π interaction with the binding site and thus a requirement for planarity. Cockcroft and coworkers (1990) have suggested that a key phenylalanine residue interacts with agonists on binding to the nicotinic receptor.

In this study we have successfully demonstrated our approach to the development of a nicotinic agonist pharmacophore model. Considerable efforts have been made over the years to refine the methods for developing binding site models of biological macromolecules. This is a challenging task, but one that is necessary in the drug discovery cycle to rationalize diverse SAR information (Marshall, 1993). To help address some of the fine SAR details we have turned to the use of QSAR and 3D-QSAR methods in an attempt to develop models which could explain the large changes in activity from small molecular alterations. The beauty of multivariate techniques is that we are able to reduce *N*-dimensions of data to something that can be easily displayed and interpreted. If our molecular descriptors are suitable, then we should be able to understand more of the available SAR.

QSAR

The successful use of ReNDeR neural networks for QSAR analyses has been demonstrated in previous studies (Livingstone *et al.*, 1991; Good *et al.*, 1993a,b). In this study we have also shown the utility of this technique to demonstrate the importance of steric properties in a series of nicotine analogues. The technique also complements the results obtained from PCA and could be considered to be examining the same dataset from a different point of view. QSAR researchers often use different methods to examine the same dataset and the ReNDeR implementation in TSAR is fast enough for this to be a viable addition to their armoury.

In the original paper, Lin and coworkers (1994) also recognised the sensitivity of the binding site to the steric properties of the pyrrolidine ring substituents. They also noted that electronic properties played a role in binding affinity. In the model presented in this study, improvements could be expected if additional electronic descriptors are included in the analysis. This may help to separate compounds such as G from N and I (Table II) which on steric grounds alone, are similar to each other, but differ in their electronic structure. It is interesting to compare our previous study of the QSAR of anatoxin analogues, which highlighted a number of electronic properties as being important for discriminating activity categories (Manallack *et al.*, 1994). Unfortunately, the substituents in each of the two series of compounds studied are not located in the same position. Although not planned, inclusion of electronic properties in the nicotine agonists dataset, would be a worthwhile study.

In constrast to the success of the QSAR analysis of the nicotine analogues dataset, investigations into the diverse nicotinics dataset did not yield useful predictive models. The analysis of the latter dataset does, however, raise an important point. The training set of compounds produced an interesting cluster which appeared to indicate that a possible model had been developed (Figure 9). By itself, a researcher may have drawn conclusions on this analysis but these would have been untested. The test compounds clearly showed that the model was not predictive (Figures 10 and 11). The original cluster therefore, may have been a chance association. This highlights the need for test compounds in QSAR studies, be it for multiple linear regression, or for multivariate techniques which have been said to be less susceptible to chance effects. Further work is therefore needed to develop a useful QSAR for this dataset, and one variable which could be looked at includes the method for calculating partial charges (Burt *et al.*, 1990).

A criticism which must be levelled at this work is the use of an acetate counter charge. As mentioned previously, one of the assumptions we have used, is that the binding site is rigid as simulated by our acetate ion. Small changes in the location of the counter charge will affect the molecular similarity calculations. Ideally, improvements to the ASP program to handle charged molecules is needed.

The use of molecular similarities for QSAR analyses has shown to be a successful method (Good *et al.*, 1993a; Seri-Levy *et al.*, 1994). Considerable research is still being done in this area to provide general methods for drug research. Questions such as, 'How similar is a seagull to a jet airliner?' are sometimes bandied about to illustrate that people will respond differently to this question. In other words, we are talking about molecular similarity in terms of molecular frameworks, molecular formulae, steric surfaces, etc. (Dean, 1993). In practice, the needs of the researcher often require special solutions, thus making it difficult to provide guidelines for the use of molecular similarities in 3D-QSAR. ASP is able to calculate a number of similarity measures and it is up to the analyst to determine the optimum method. Benigni and coworkers (1995) recently compared similarity matrices with classical QSAR descriptors and found that there was no advantage in using molecular similarities. Again, it needs to be pointed out that it is difficult to generalize using these techniques and that other data sets may be suited to analysis using similarity matrices.

One additional problem that has not been alluded to so far, is that the biological data for many of the compounds used in this study have come from different pharmacological assays. Many subtypes exist for nicotinic receptors and thus pharmacophore studies will benefit in the future with the availability of data from single receptor subtypes. Refinement of pharmacophore models for each subtype will hopefully facilitate the development of subtype specific compounds to help in the treatment of AD/SDAT.

In conclusion, we have demonstrated the use of a method which is able to align molecules using gnomonic surfaces generated from electrostatic potentials. The pharmacophore we have described here allows more flexibility for superimposition modes than by using dummy atoms attached to pharmacophore groups. Our model not only agrees with previously published work but extends these to include a lipophilic region. ReNDeR neural network analysis of a series of nicotine analogues using the alignment suggested by the pharmacophore model was used successfully to predict the activities of a set of test compounds.

Acknowledgement

The authors would like to thank Dr Robert A. Scoffin for his help and advice with the neural network analyses.

REFERENCES

Aoyama, T., Suzuki, Y., and Ichikawa, H. (1990). Neural networks applied to structure–activity relationships. *J. Med. Chem.* **33,** 905–908.

Barlow, R.B. and Johnson, O. (1989). Relations between structure and nicotine-like activity: X-ray crystal structure analysis of (–)-cytisine and (–)-lobeline hydrochloride and a comparison with (–)-nicotine and other nicotine-like compounds. *Br. J. Pharmacol.* **98,** 799–808.

Bartus, R.T., Dean, R.L., Beer, B., and Lippa, A.S. (1982). The cholinergic hypothesis of geriatric memory dysfunction. *Science* **217,** 408–417.

Beers, W.H. and Reich, E. (1970). Structure and activity of acetylcholine. *Nature* **228,** 917–922.

Benigni, R., Cotta-Ramusino, M., Giorgi, F., and Gallo, G. (1995). Molecular similarity matrices and quantitative structure–activity relationships: A case study with methodological implications. *J. Med. Chem.* **38,** 629–635.

Burt, C., Huxley, P., and Richards, W.G. (1990). The application of molecular similarity calculations. *J. Comput. Chem.* **11,** 1139–1146.

Carbo, R., Leyda, L., and Arnau, M. (1980). An electron density measure of the similarity between two compounds. *Int. J. Quantum Chem.* **17,** 1185–1189.

Chatfield, C. and Collins, A.J. (1986). *Introduction to Multivariate Analysis.* Chapman and Hall, London.

Chau, P.L. and Dean, P.M. (1987). Molecular recognition: 3D surface structure comparison by gnomonic projection. *J. Mol. Graph.* **5,** 97–100.

Cockcroft, V.B., Osguthorpe, D.J., Barnard, E.A., and Lunt, G.G. (1990). Modeling of agonist binding to the ligand-gated ion channel superfamily of receptors. *Proteins* **8,** 386–397.

Court, J.A. and Perry, E.K. (1994). CNS nicotinic receptors. Possible therapeutic targets in neurodegenerative disorders. *CNS Drugs* **2,** 216–233.

Cregan, E., Ordy, J.M., Palmer, E., Blosser, J., Wengenack, T., and Thomas, G. (1989). Spatial working memory enhancement by nicotine of aged Long Evans rats in the T-maze. *Soc. Neurosci. Abstr.* **15,** 2952.

Dean, P.M. (1993). Molecular similarity. In, *3D QSAR in Drug Design: Theory, Methods and Applications* (H. Kubinyi, Ed.). ESCOM Science Publishers, Leiden, pp. 150–172.

Drachman, D.A. and Leavitt, J. (1974). Human memory and the cholinergic system. *Arch. Neurol.* **30,** 113–121.

Giacobini, E. (1990). Cholinergic receptors in human brain: Effects of aging and Alzheimer disease. *J. Neurosci. Res.* **27,** 548–560.

Good, A.C., Peterson, S.J., and Richards, W.G. (1993a). QSAR's from similarity matrices. Technique validation and application in the comparison of different similarity evaluation methods. *J. Med. Chem.* **36,** 2929–2937.

Good, A.C., So, S., and Richards, W.G. (1993b). Structure–activity relationships from molecular similarity matrices. *J. Med. Chem.* **36,** 433–438.

Gopalakrishnan, M. and Sullivan, J.P. (1994). Targeting nicotinic cholinergic receptors. *Drug News Perspect.* **7,** 444–448.

Gund, T.M. and Spivak, C.E. (1991). Pharmacophore for nicotinic agonists. *Meth. Enzymol.* **203,** 677–693.

Hacksell, U. and Mellin, C. (1989). Stereoselectivity of nicotinic receptors. *Prog. Brain Res.* **79,** 95–100.

Hall, L.H. and Kier, L.B. (1992). The molecular connectivity chi indexes and kappa shape indexes in structure-property modeling. In, *Reviews in Computational Chemistry* (K.B. Lipkowitz and D.B. Boyd, Eds.). VCH, pp. 367–382.

Heinemann, S., Boulter, J., Deneris, E., Connolly, J., Duvoisin, R., Papke, R., and Patrick, J. (1990). The brain nicotinic acetylcholine receptor gene family. *Prog. Brain Res.* **86,** 195–203.

Hertz, J., Krogh, A., and Palmer, R.G. (1991). *Introduction to the Theory of Neural Computation.* Addison-Wesley Publishing Co., Santa Fe, CA.

Hodgkin, E.E. and Richards W.G. (1987). Molecular similarity based on electrostatic potential and electric field. *Int. J. Quantum Chem. Quantum Biol. Symp.* **14,** 105–110.

Hopfinger, A.J. (1980). Investigations of DHFR inhibition by Bakers triazines based upon molecular shape analysis. *J. Am. Chem. Soc.* **102,** 7196–7206.

Hopfinger, A.J. (1983). Theory and analysis of molecular potential energy fields in molecular shape analysis: A QSAR study of 2,4-diamino-5-benzylpyrimidines as DHFR inhibitors. *J. Med. Chem.* **26,** 990–996.

Jason, A.M., Fuxe, K., Agnati, L.F., Jasson, A., Bjeklke, B., Sundstrom, E., Anderson, K., Harfstrand, A., Goldstein, M., and Owman, C. (1989). Protective effects of chronic nicotine treatment on lesioned nigrostriatal dopamine neurons in the male rat. *Prog. Brain Res.* **79,** 257–265.

Jones, G.M.M., Sahakian, B.J., Levy, R., Warburton, D.M., and Gray, J.A. (1992). Effects of acute subcutaneous nicotine on attention, information processing, and short term memory in Alzheimer's disease. *Psychopharmacol.* **108,** 485–494.

Kanne, D.B. and Abood, L.G. (1988). Synthesis and biological characterization of pyridohomotropanes. Structure–activity relationships of conformationally restricted nicotinoids. *J. Med. Chem.* **31,** 506–509.

Katz, W.T., Snell, J.W., and Merickel, M.B. (1992). Artificial neural networks. *Meth. Enzymol.* **210,** 610–636.

Lin, N.-H., Carrera, G.M.Jr., and Anderson, D.J. (1994). Synthesis and evaluation of nicotine analogues as neuronal nicotinic acetylcholine receptor ligands. *J. Med. Chem.* **37,** 3542–3553.

Livingstone, D.J. (1991). Pattern recognition methods in rational drug design. *Meth. Enzymol.* **203,** 613–638.

Livingstone, D.J. (1995). *Data Analysis for Chemists: Applications to QSAR and Chemical Product Design.* Oxford University Press, Oxford.

Livingstone, D.J., Hesketh, G., and Clayworth, D. (1991). Novel method for the display of multivariate data using neural networks. *J. Mol. Graph.* **9,** 115–118.

Manallack, D.T., Ellis, D.D., Thompson, P.E., Gallagher, T., and Livingstone, D.J. (1994). Quantitative structure–activity relationships of (+)-anatoxin-a derivatives. *Nat. Prod. Lett.* **4,** 121–128.

Manallack, D.T. and Livingstone, D.J. (1994). Neural Networks – A tool for drug design. In, *Methods and Principles in Medicinal Chemistry. Volume 3. Advanced Computer-Assisted Techniques in Drug Discovery* (H. van de Waterbeemd, Ed.). VCH, Cambridge, pp. 293–318.

Marshall, G.R. (1993). Binding-site modeling of unknown receptors. In, *3D QSAR in Drug Design: Theory, Methods and Applications* (H. Kubinyi, Ed.). ESCOM Science Publishers, Leiden, pp. 80–116.

Marshall, G.R. and Naylor, C.B. (1990). Use of molecular graphics for structural analysis of small molecules. In, *Comprehensive Medicinal Chemistry. Volume 4.* (C. Hansch, P.G. Sammes, J.B. Taylor, and C.A. Ramsden, Eds.). Pergamon Press, Oxford, pp. 432–458.

Mattos, C. and Ringe, D. (1993). Multiple binding modes. In, *3D QSAR in Drug Design: Theory, Methods and Applications* (H. Kubinyi, Ed.). ESCOM Science Publishers, Leiden, pp. 226–254.

McGeer, P.L., McGeer, E.G., Suzuki, J., Dolman, G.E., and Nagai, T. (1984). Aging, Alzheimer's disease, and the cholinergic system of the basal forebrain. *Neurology* **34,** 741–745.

Navarro, H.A., Seidler, F.J., Eylers, J.P., Baker, F.E., Dobbins, S.S., Lappi, S.E., and Slotkin, T.A. (1989). Effects of prenatal exposure on development of central and peripheral cholinergic neurotransmitter systems. *J. Pharmacol. Exp. Ther.* **251,** 894–900.

Newhouse, P.A., Sunderland, Y., Tariot, P.N., Blumhardt, C.L., Weingartner, H., and Mellow, A. (1988). Intravenous nicotine in Alzheimer's disease: A pilot study. *Psychopharmacol.* **95,** 171–175.

Owman, C., Fuxe, K., Jason, A.M., and Kahrstrom, J. (1989). Studies of protective actions of nicotine on neuronal and vascular functions in the brain of rats: Comparison between sympathetic noradrenergic and mesostriatal dopaminergic fiber system, and the effect of a dopamine agonist. *Prog. Brain Res.* **79,** 267–276.

Perry, E.K. (1986). The cholinergic hypothesis – Ten years on. *Br. Med. Bull.* **42,** 63–69.

Perry, E.K., Blessed, G., Tomlinson, B.E., Perry, R.H., Crow, T.J., Cross, A.J., Dockray, G.J., Dimaline, R., and Arredui, R. (1981). Neurochemical activities in human temporal lobe related to aging and Alzheimer-type changes. *Neurobiol. Aging* **2,** 251–256.

Picciotto, M.R., Zoli, M., Lena, C., Bessis, A., Lallemand, Y., LeNovere, N., Vincent, P., Pich, E.M., Brulet, P., and Changeux, J.P. (1995). Abnormal avoidance learning in mice lacking functional high-affinity nicotine receptor in the brain. *Nature* **374,** 65–67.

Pitner, T.P., Edwards, W.B. III, Bassfield, R.L., and Whidby, J.F. (1978). The solution conformation of nicotine. A H^1 and H^2 nuclear magnetic resonance investigation. *J. Am. Chem. Soc.* **100,** 246–251.

Qian, C., Li, T., Shen, T.Y., Libertine-Garahan, L., Eckman, J., Biftu, T., and Ip, S. (1993). Epibatidine is a nicotinic analgesic. *Eur. J. Pharmacol.* **250,** R13–R14.

Reisine, T.D., Yamamura, H.I., Bird, E.D., Spokes, E., and Enna, S.J. (1978). Pre- and postsynaptic neurochemical alterations in Alzheimer's disease. *Brain Res.* **159,** 477–481.

Rumelhart, D.E. and McClelland, J.L. (1988). *Parallel Distributed Processing. Volume 1.* MIT Press, Cambridge, Massachusetts.

Sahakian, B., Jones, G., Levy, R., Gray, J., and Warburton, D. (1989). The effects of nicotine on attention, information processing, and short term memory in patients with dementia of Alzheimer type. *Br. J. Psychiatry* **154,** 797–800.

Senokuchi, K., Nakai, H., Kawamura, M., Katsube, N., Nonaka, S., Sawaragi, H., and Hamanaka, N. (1994). Synthesis and biological evaluation of (±)-epibatidine and the congeners. *Synlett* 343–344.

Seri-Levy, A., West, S., and Richards, W.G. (1994). Molecular similarity, quantitative chirality, and QSAR for chiral drugs. *J. Med. Chem.* **37,** 1727–1732.

Shahi, G.S. and Moochhala, S.M. (1991). Smoking and Parkinson's disease – A new perspective. *Rev. Environ. Health* **9,** 123–136.

Sheridan, R.P., Nilakantan, R., Dixon, J.S., and Venkataraghavan, R. (1986). The ensemble approach to distance geometry: Application to the nicotinic pharmacophore. *J. Med. Chem.* **29,** 899–906.

Spande, T.F., Garraffo, H.M., Edwards, M.W., Yeh, H.J.C., Pannell, L., and Daly, J.W. (1992). Epibatidine: A novel (chloropyridyl)azabicycloheptane with potent analgesic activity from an Ecuadoran poison frog. *J. Am. Chem. Soc.* **114,** 3475–3478.

Spivak, C.E., Gund, T.M., Liang, R.F., and Waters, J.A. (1986). Structural and electronic requirements for potent agonists at a nicotinic receptor. *Eur. J. Pharmacol.* **120,** 127–131.

Spivak, C.E., Waters, J.A., and Aronstam, R.S. (1989a). Binding of semirigid nicotinic agonists to nicotinic and muscarinic receptors. *Mol. Pharmacol.* **36,** 177–184.

Spivak, C.E., Waters, J.A., Yadav, J.S., Shang, W.C., Hermsmeier, M., Liang, R.F., and Gund, T.M. (1991). (±)-Octahydro-2-methyl-*trans*-5 (1*H*)-isoquinolone methiodide – A probe that reveals a partial map of the nicotinic receptors recognition site. *J. Mol. Graph.* **9,** 105–110.

Spivak, C.E., Yadav, J.S., Shang, W.C., Hermsmeier, M., and Gund, T.M. (1989b). Carbamyl analogues of potent nicotinic agonists: Pharmacology and computer assisted molecular modeling study. *J. Med. Chem.* **32,** 305–309.

Stewart, J.J.P. (1990). MOPAC: A semiempirical molecular orbital program. *J. Comput.-Aided Mol. Design* **4,** 1–105.

Thompson, P.E., Manallack, D.T., Blaney, F.E., and Gallagher, T. (1992). Conformational studies on (+)-anatoxin-a and derivatives. *J. Comput.-Aided Mol. Design* **6,** 287–298.

Tintelnot, M. and Andrews, P. (1989). Geometries of functional group interactions in enzyme-ligand complexes: Guides for receptor modelling. *J. Comput.-Aided Mol. Design* **3,** 67–84.

van Duijn, C.M. and Hofman, A. (1991). Relation between nicotine intake and Alzheimer's disease. *Brit. Med. J.* **302,** 1491–1494.

Waters, J.A., Spivak, C.E., Hermsmeier, M., Yadav, J.S., Liang, R.F., and Gund, T.M. (1988). Synthesis, pharmacology, and molecular modelling studies of semirigid, nicotinic agonists. *J. Med. Chem.* **31,** 545–554.

Whitehouse, P.J., Martino, A.M., Wagster, M.V., Price, D.L., Mayeux, R., Atack, J.R., and Kellar, K.J. (1988). Reduction in [^{3}H]nicotine acetylcholine binding in Alzheimer's disease and Parkinson disease. *Neurology* **38,** 720–723.

Wonnacott, S. (1990). The paradox of nicotinic acetylcholine receptor upregulation by nicotine. *Trends Pharmacol. Sci.* **11,** 216–219.

Wonnacott, S., Jackman, S., Swanson, K.L., Rapoport, H., and Albuquerque, E.X. (1991). Nicotinic pharmacology of anatoxin analogues. II. Side chain structure–activity relationships at neuronal nicotinic ligand binding sites. *J. Pharmacol. Exp. Ther.* **259,** 387–391.

9 Evaluation of Molecular Surface Properties Using a Kohonen Neural Network

S. ANZALI[1], G. BARNICKEL[1], M. KRUG[1],
J. SADOWSKI[2], M. WAGENER[2], and J. GASTEIGER[2]*
[1]*E. Merck, Department of Medicinal Chemistry/Drug Design, 64271 Darmstadt, Germany*
[2]*Computer-Chemie-Centrum, University Erlangen-Nürnberg, 91052 Erlangen, Germany*

Based on an application of a Kohonen neural network, a transformation of 3D-molecular surfaces into 2D-Kohonen maps can be realized. The trained neurons of the Kohonen maps are then colored according to the molecular electrostatic potential (MEP) values on the van der Waals surfaces. A template approach is presented for the comparison of the shape and the MEP values of a molecule with that of a reference structure. Applications of these methods are illustrated with datasets of ryanodines, cardiac glycosides and steroids. The results indicate that the Kohonen neural network may be a useful tool for the investigation of large datasets of molecules and for the fast and accurate comparison of molecular electrostatic potentials and shapes within a series of compounds.

KEY WORDS: *Kohonen neural network; self-organizing map; template approach; SAR; ryanodines; cardiac glycosides.*

INTRODUCTION

A variety of approaches based on shape analysis has been used for years by chemists to study structure–activity relationships. The molecular electrostatic potential and the shape are the most important properties often used in QSAR studies. The handling of molecular shape similarity is a complex field

*Author to whom all correspondence should be addressed.

In, *Neural Networks in QSAR and Drug Design* (J. Devillers, Ed.)
Academic Press, London, 1996, pp. 209–222.
ISBN 0-12-213815-5

for the development of drug design methods. The need to quantify the similarity between the shapes, and the question of how to decide on an appropriate molecular conformation from the entire space of allowed conformations, are the basis of many methods in drug design.

Based on an application of a Kohonen neural network (Kohonen, 1982) we present here a method for the comparison of shapes by a template approach.

In a series of papers the methodology and the recognition of molecular surface properties projected into a Kohonen map was described (Gasteiger and Li, 1994; Gasteiger *et al.*, 1994a). Details of the method for mapping molecular electrostatic potentials (MEP) on surfaces by a Kohonen network have also been given (Gasteiger *et al.*, 1994b). In this present paper the use of a Kohonen map in SAR and drug design is described. The methods will be illustrated by a set of examples, which show the different possibilities for application.

MATERIALS AND METHODS

The structures of cardiac glycosides were built from connection tables using the program CORINA (Sadowski *et al.*, 1992). The structures of ryanodines were obtained from crystal structure data (Cambridge Crystallographic Database, Cambridge Crystallographic Data Center, Cambridge, UK). The lack of molecular coordinates from crystal structures for some ryanodine derivatives was overcome by using the SYBYL fragment library (Tripos Associates, St Louis, MO, SYBYL 6.0, USA).

The conformational analysis of the side groups was carried out using the SYBYL SEARCH option at 10° intervals per full cycle rotation.

Charge distributions were calculated by the Gasteiger-Hückel method (Gasteiger and Marsili, 1980; Purcell and Singer, 1967). The MEP is calculated for the van der Waals surface by Coulomb's law using the program SURFACE (Sadowski, unpublished results, 1993). A Kohonen net is trained by random-sampling of points on a van der Waals surface as input using the KMAP simulator (Li *et al.*, unpublished results, 1993). The trained neurons were then colored according to the MEP values at these points.

The parameters of the network were: size = 50×50, step $(\Delta) = 400$, learning rate $(\eta_0) = 0.3$, span $(\sigma_0) = 15$, reduction speed for learning rate $(\lambda_\eta) = 0.95$, reduction speed for span $(\lambda_\sigma) = 1$, training steps = 20 000. The topology of the network was a torus. The main advantage of the toroidal network (compared to a rectangular network) is that there is no boundary on a torus, and therefore all neurons in the net are treated equally and the result is a map without boundaries which may be circularly shifted in vertical or horizontal directions.

Also, depending on the random initialization of the Kohonen network and the way the data points are presented to the network, as well as on the sites chosen for making the cuts into the torus, the Kohonen maps can be oriented in a variety of ways. Through a toroidal network, the maps can be shifted, mirrored and rotated with respect to each other to achieve a similar position of their patterns.

η_0, σ_0, λ_η, λ_σ, and step are parameters which have to be chosen according to the individual tasks. The η_0 determines the initial maximal modification rate, the σ_0, called the learning span, determines the initial size of neighborhood around the winner neuron in which the connection weights will be modified. The parameters step, λ_η, λ_σ together determine the speed of the convergence of the training process. The choice of proper training parameters depends generally on the target object, the net configuration and the individual requirements. It demands some experience. To determine the proper values for these parameters, several molecules were examined. For each molecule, 50 Kohonen maps were trained. The experiment was repeated with different sets of parameters and it was discovered that the 50 Kohonen maps obtained for every molecule, with the above selected parameters, show solid constant patterns. Nevertheless, the results obtained with other proper parameters show very little variation with the above parameters.

KOHONEN NETWORK

The Kohonen neural network, as a self-organizing network, can be used to generate a nonlinear projection of objects from a higher-dimensional space onto a lower-dimensional surface. This method enables a decrease in dimension while conserving the topology of the information as much as possible.

The method for obtaining two-dimensional maps of molecular surfaces consists of training the Kohonen network with a set of several thousands of the three Cartesian coordinates of points spotted at random on a molecular van der Waals surface (in our cases 20 000 points).

The Cartesian coordinates of a point from such a space were used as the input. Thus, when selecting a Kohonen network with 50×50 neurons, there are $3 \times 50 \times 50 = 7500$ weights. The architecture of the Kohonen network is shown in Plate 1.

As the size of the network is much smaller than the size of the dataset, a neuron is normally excited by several points of the dataset. In this case, the weights of the neuron obtain the average value of the information of those points that excite it. The information on the molecular surface (in our case, the individual electrostatic values at the points on the van der Waals surface) is not used in the learning process.

During processing by unsupervised learning, the points having similar coordinates are put close together by the network into the same or adjacent neurons. This is shown by a neighborhood kernel within the Kohonen map simulator.

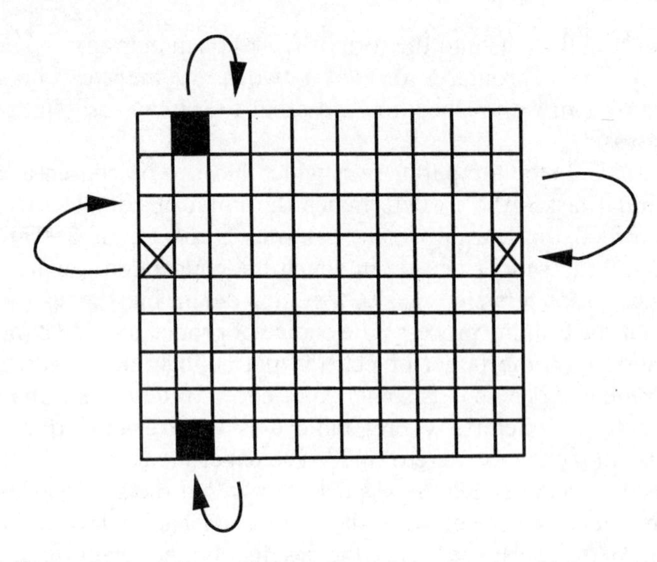

Figure 1 Plane obtained by making two cuts into a torus, which can be shifted into any direction. Thus, the two neurons marked by a cross are direct neighbors as are the two neurons shown as black squares.

In order to have a plane without beginning and without end, the neurons placed at the outskirts of the square shown in Figure 1 are made the neighbors of the ones located on the other side. This leads to the surface of a torus. For visualization, this torus is cut along two perpendicular lines and then the torus surface is spread into a plane. Since these cuts can be made at arbitrary lines, the maps can be shifted into any direction.

TEMPLATE APPROACH

The binding of a ligand to a receptor is a combination of steric and electrostatic factors. Using the template approach both factors can be considered. The basis of the template approach is an analysis of the shapes of molecules and the quantification of a shape-similarity or -dissimilarity within a series of compounds using a reference molecule. A reference network is trained with coordinates of points from the van der Waals surface of the most active and rigid compound. Then the analogs are filtered through this network and thus produce comparative networks. The dissimilarity of the comparative maps to the reference map can then be deduced by the size of the blank areas (empty neurons) in these maps. The larger the difference in shape between the reference molecule and the compared molecule, the more empty neurons (blank areas) are obtained in the comparative map. A preceding

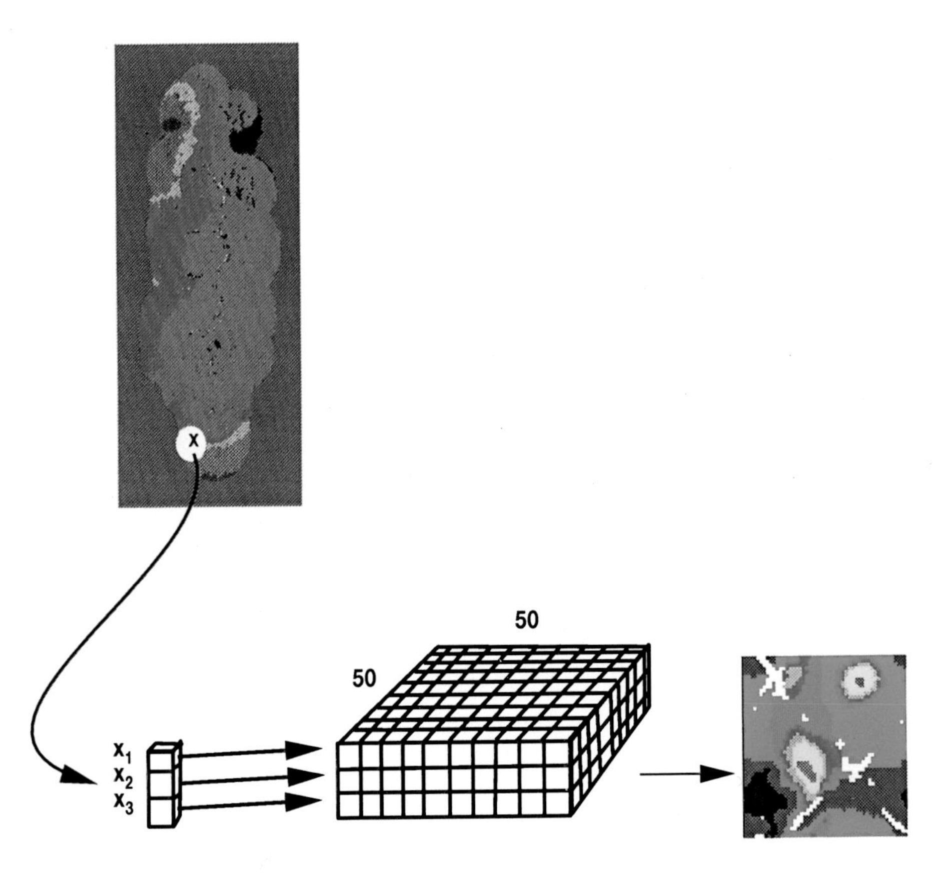

Plate 1. Mapping of a molecular surface into a Kohonen network leading to a map of the molecular electrostatic potential.

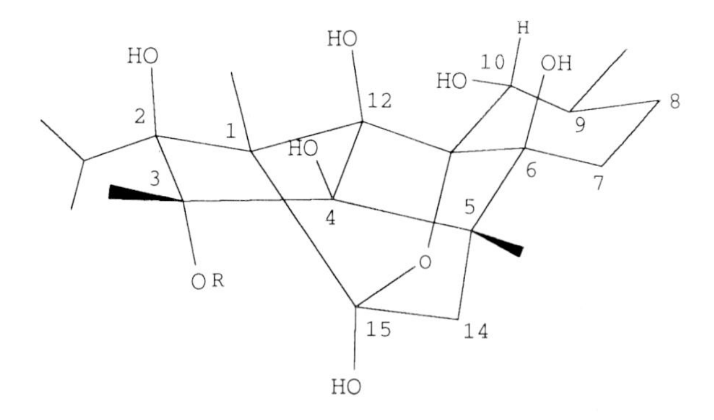

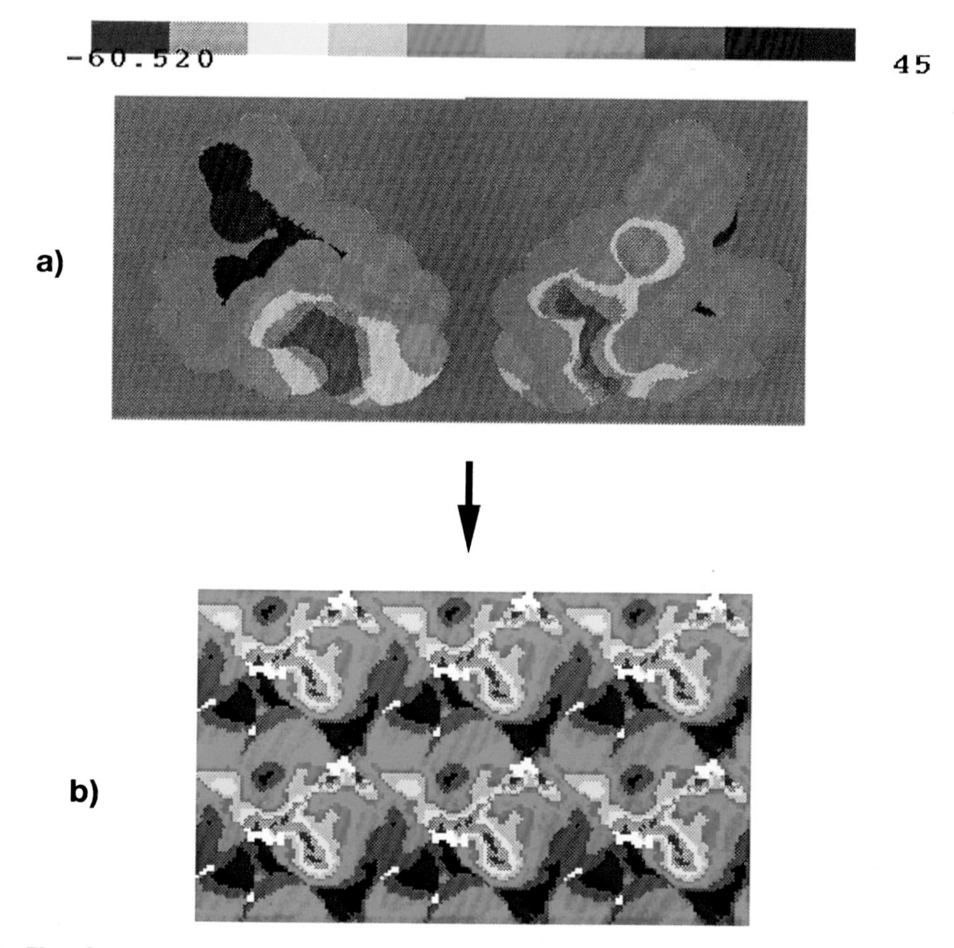

Plate 2. (a) Parallel projection of the electrostatic potential on the van der Waals surface of ryanodine. (b) The corresponding Kohonen map (replication of the map is replicated three times in horizontal and twice in vertical direction 2 x 3).

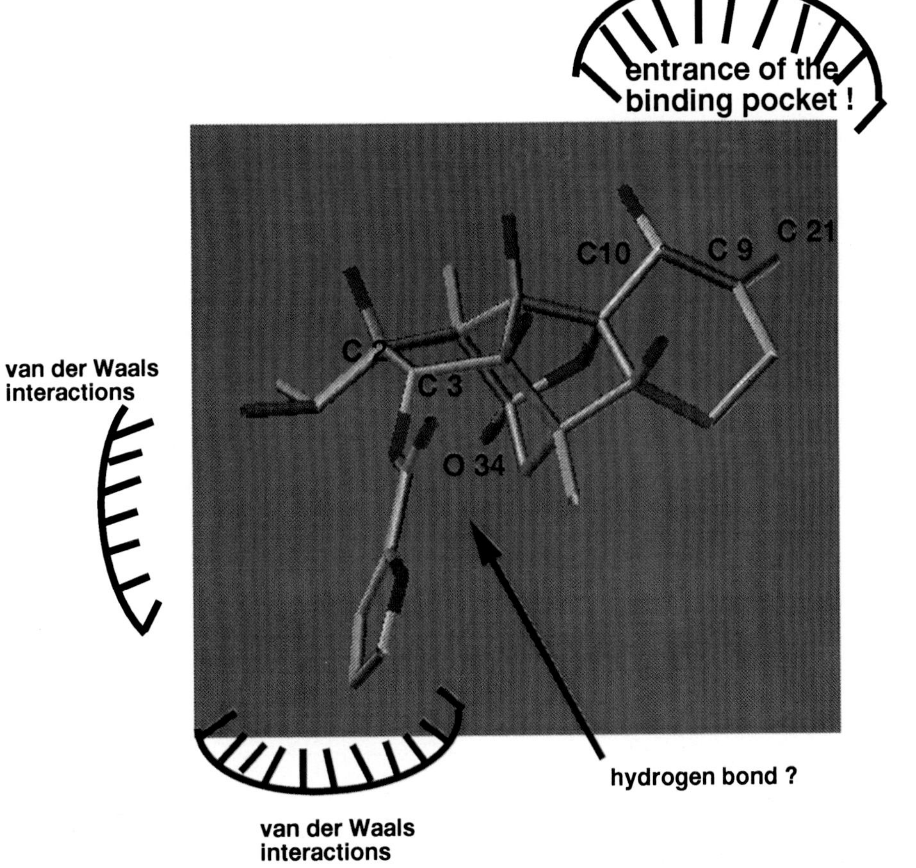

Plate 3. The hypothesis about the ryanodine binding site proposed by Welch *et al.* (1994).

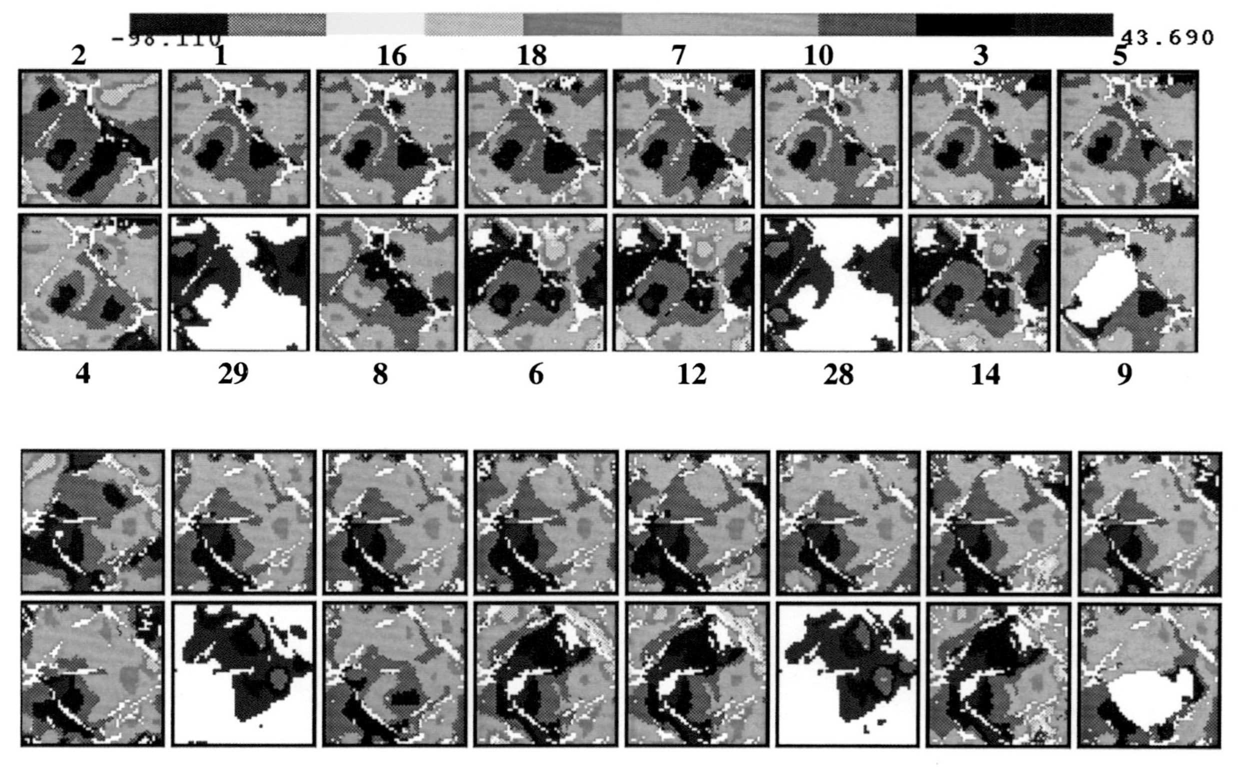

Plate 4. The trained Kohonen maps of the ryanodine compound's series. The maps are ordered with decreasing binding affinities according to the ryanodine receptor. The sets of A and B are obtained from two local conformers for the orientation of the pyrrole ring. The upper nine maps of molecules show similar MEP patterns and molecular shape. The seven further maps show more dissimilarity of the MEP patterns and/or the molecular shape to the reference molecule.

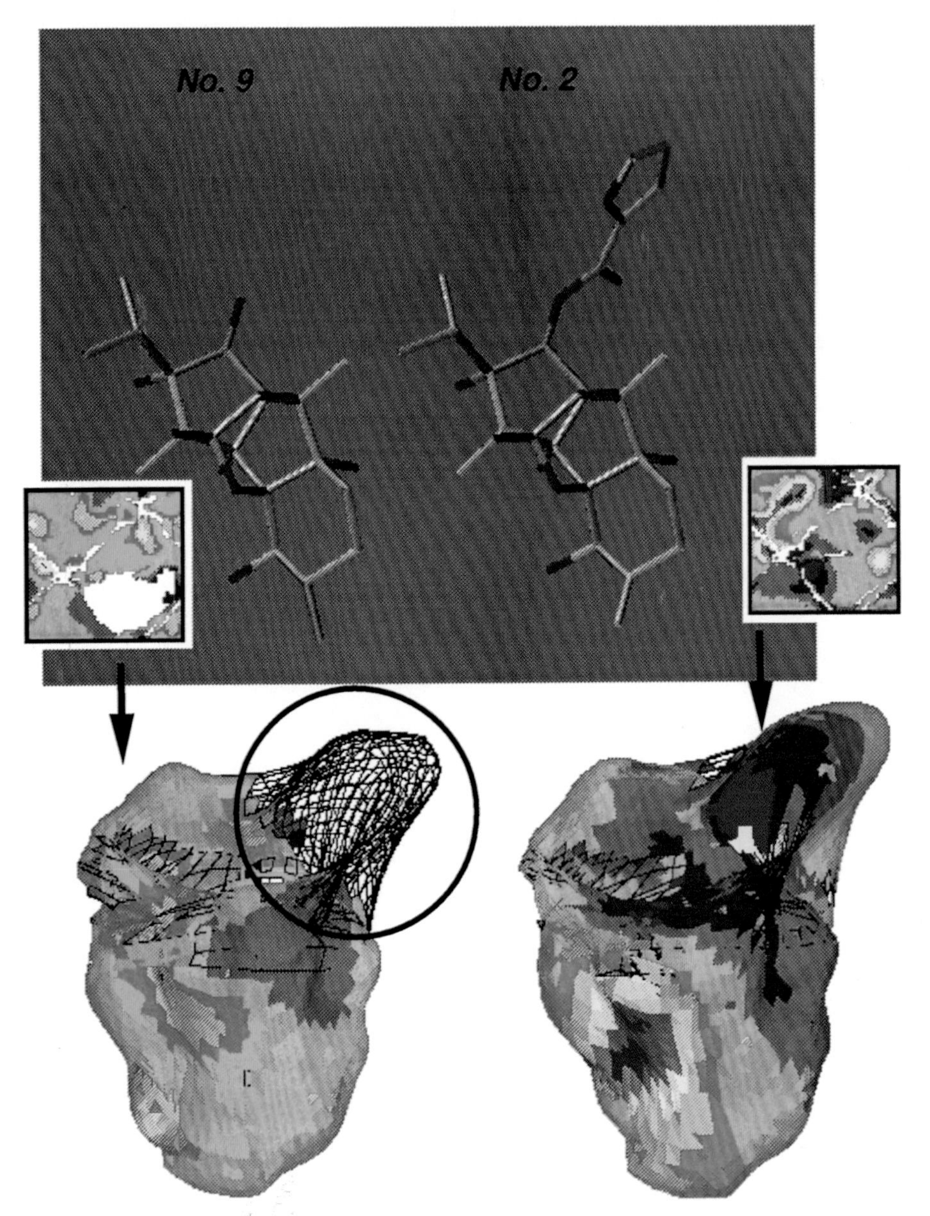

Plate 5. Backprojection of the maps of ryanadol (no.: 9, Table I) and dehydroryanodine (no.: 2, Table I) onto the 3D-space. The area where the two molecules differ in shape is indicated by an open mesh.

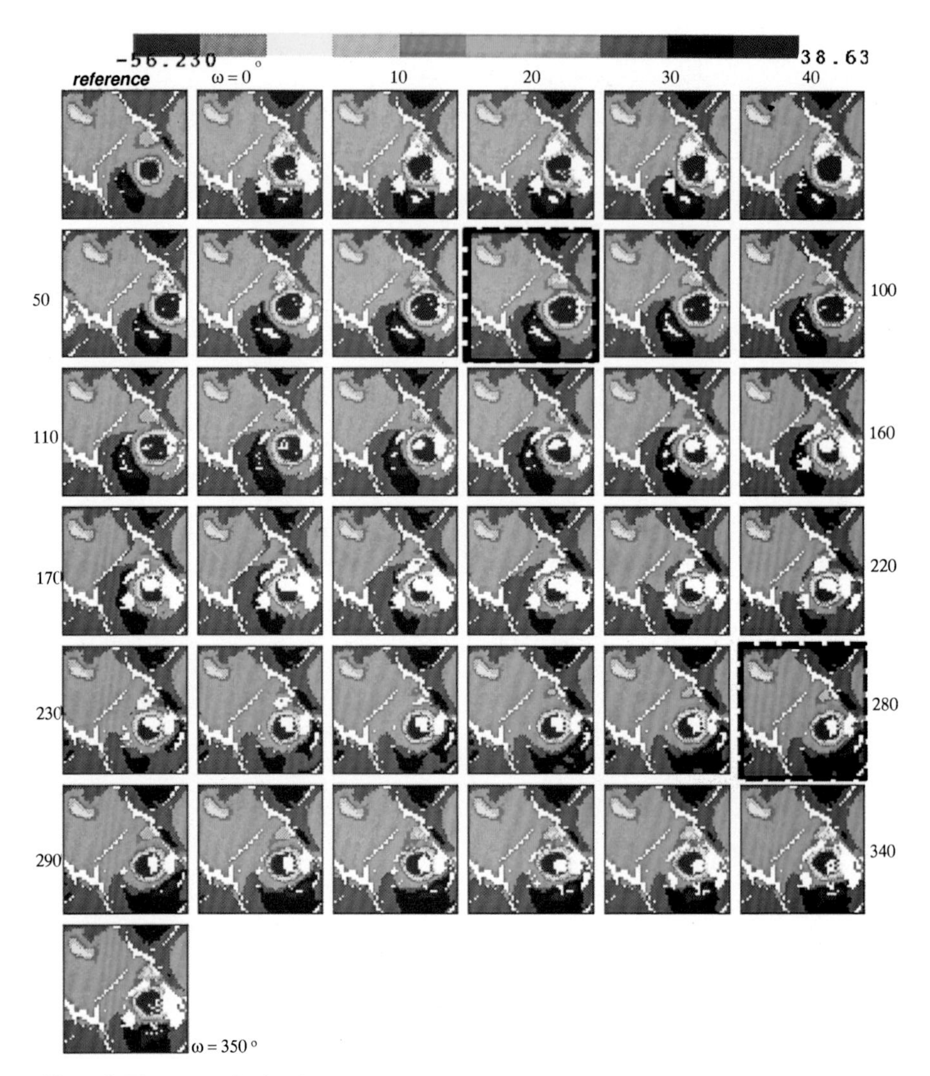

Plate 6. The maps obtained using the template approach. The first map is trained by the van der Waals surface of bufalin as the reference molecule and the others show the maps of 36 conformers of digitoxigenin taken from a rotation of 0° to 350°. Outlined are two maps that show the least number of empty neurons.

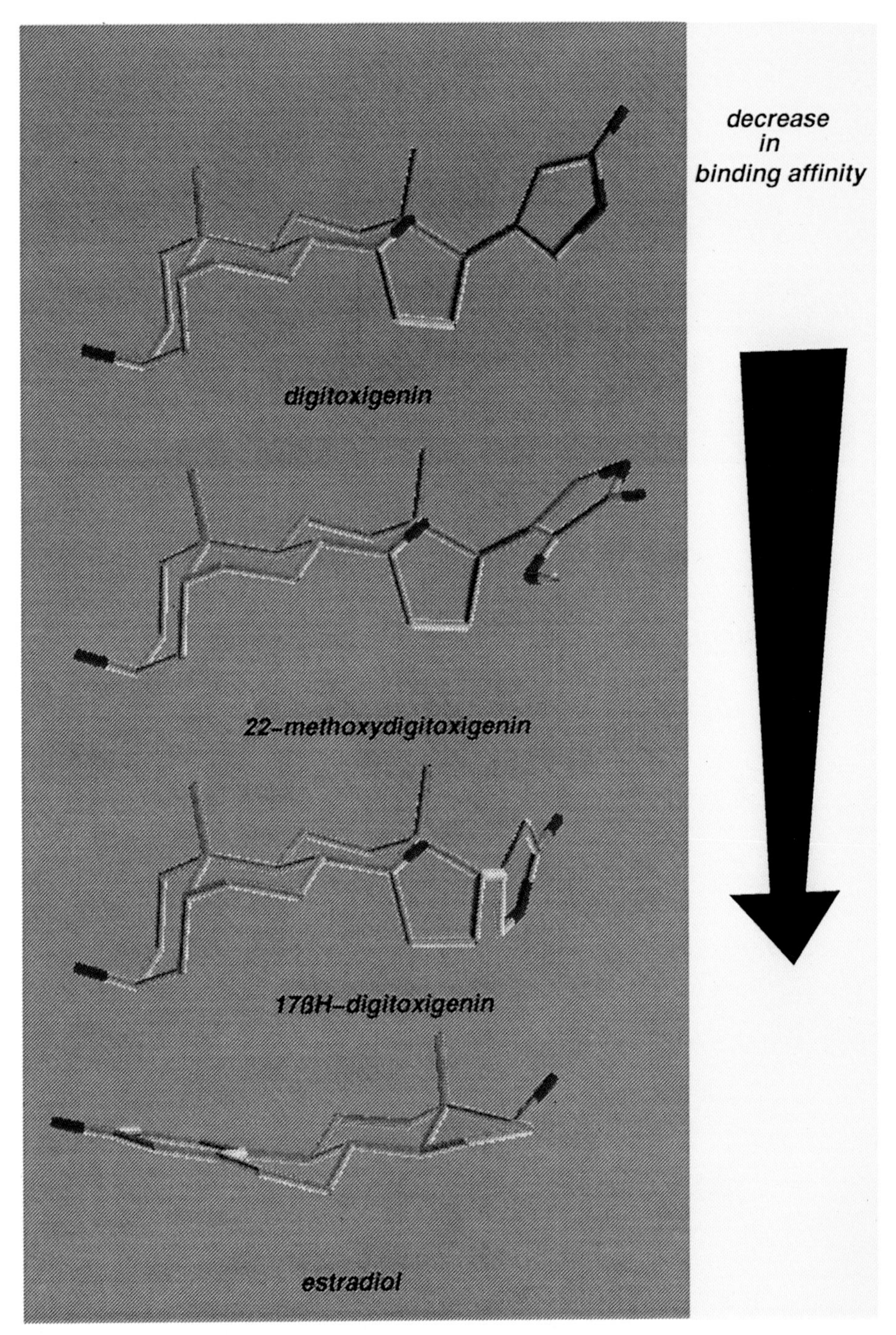

Plate 7. Four compounds of the cardiac glycosides series are ordered according to the decreasing binding affinities of digitalis receptor.

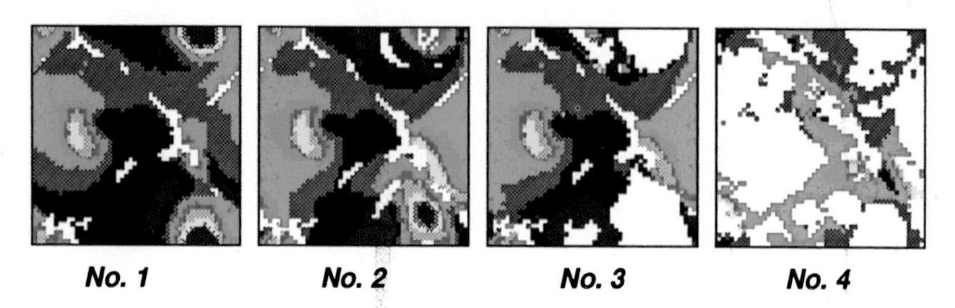

compound	N_e	S_{AB}	Activity
No. 1(reference)			+ + +
No. 2	316	272.6	+ +
No. 3	669	184.3	+−
No. 4	1586	44.5	−

Plate 8. The obtained Kohonen maps of cardiac glycosides from Plate 7 with their evaluated empty neurons (N_e) and the similarity index (S_{AB}).

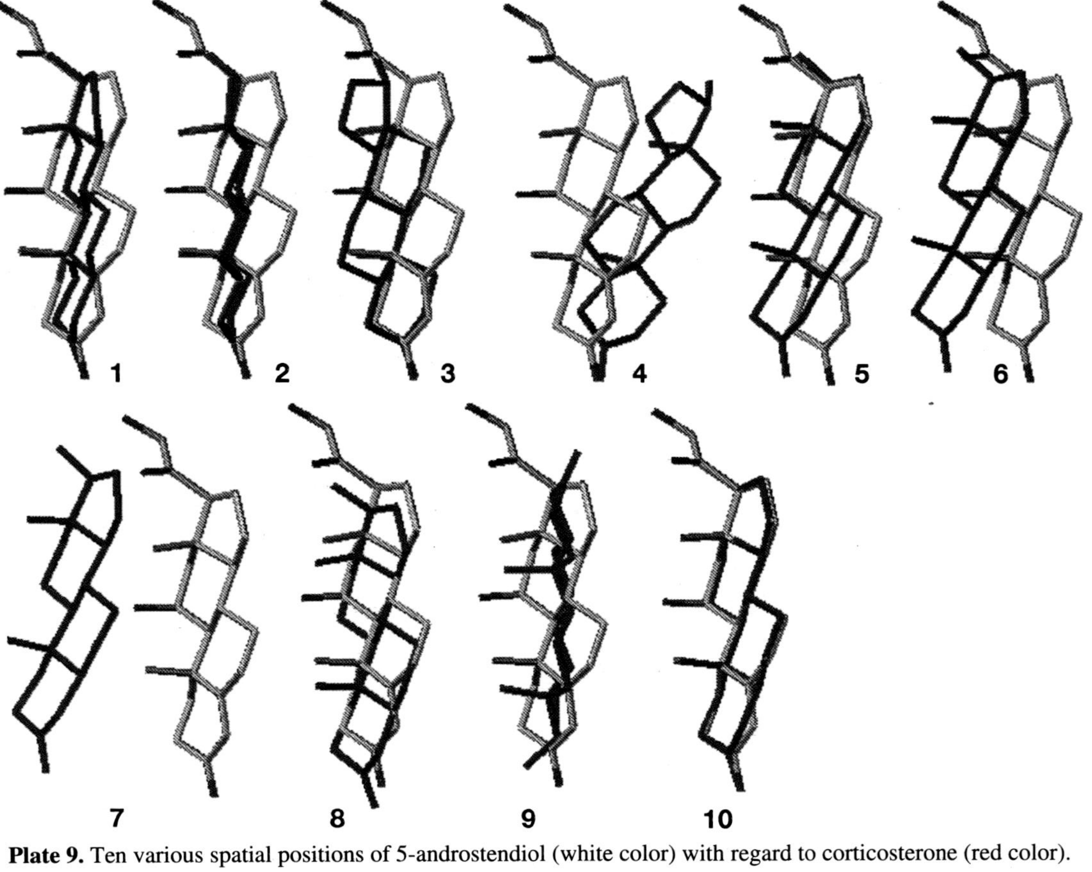

Plate 9. Ten various spatial positions of 5-androstendiol (white color) with regard to corticosterone (red color).

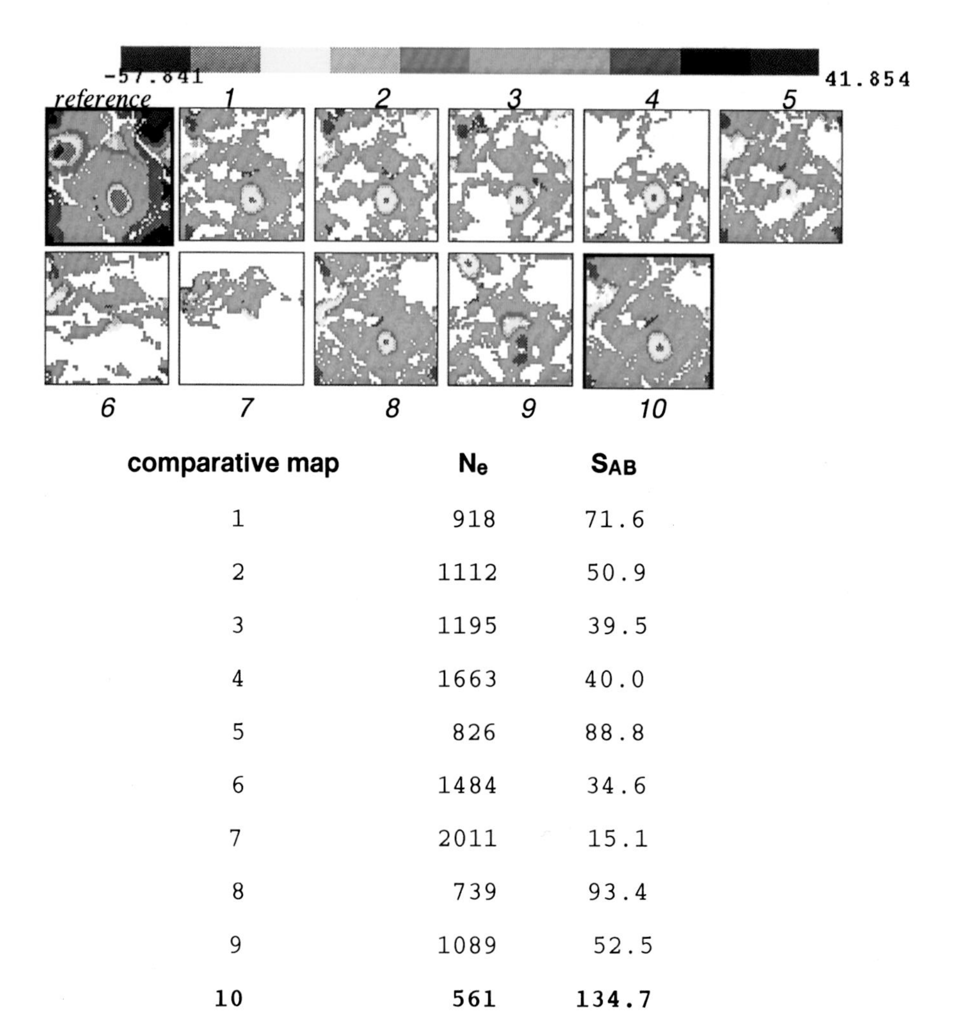

comparative map	N_e	S_{AB}
1	918	71.6
2	1112	50.9
3	1195	39.5
4	1663	40.0
5	826	88.8
6	1484	34.6
7	2011	15.1
8	739	93.4
9	1089	52.5
10	**561**	**134.7**

Plate 10. The maps obtained for ten various spatial positions of 5-androstendiol (comparative maps) using the van der Waals surface of corticosterone as the reference molecule. (N_e) = the number of empty neurons, (S_{AB}) = the similarity index.

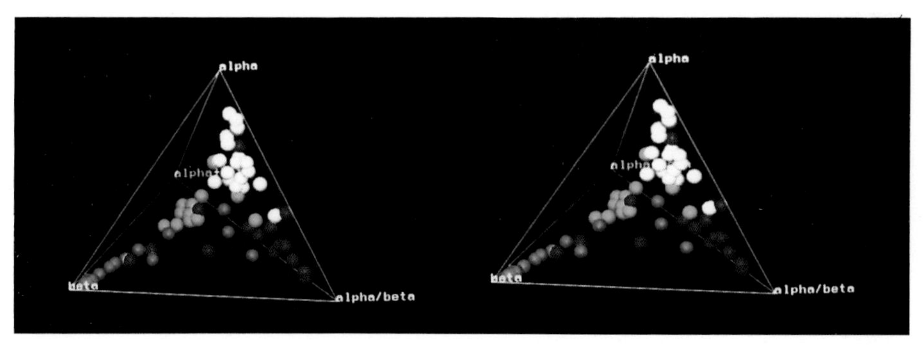

Plate 11. (a) Graphical portrayal of fuzzy-clustered protein data for the training set of 64 proteins. As can be seen from the figure, the data lies within a tetrahedron embedded in 3-dimensional euclidean space. Each of the vertices of the tetrahedron corresponds to one of the four structural classes of proteins – *all*-α *all*-β $\alpha+\beta$, and α/β. Individual proteins are represented by colored spheres: all-α (white), all-β (green), $\alpha+\beta$ (blue), and α/β (red). Clusters located near the center of the tetrahedron correspond to maximally fuzzy clusters. As a cluster moves towards one of the vertices, it becomes 'crisper' until at the vertex it is a crisp cluster.

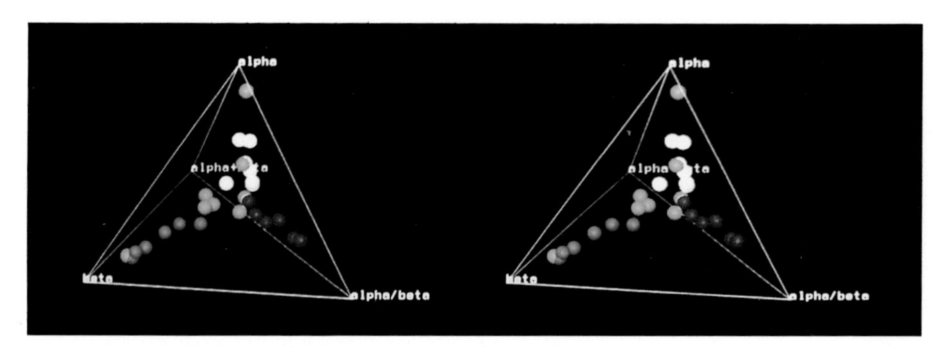

Plate 11. (b) Same as in (a), but for the test set of 27 proteins.

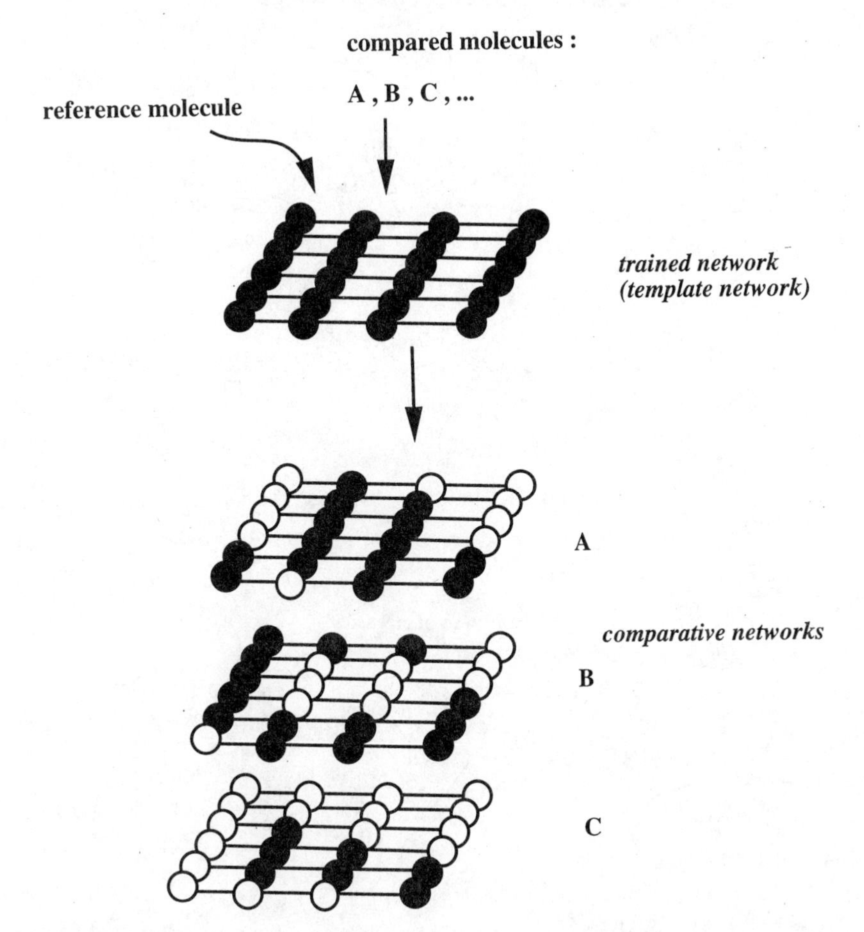

Figure 2 A model of the template approach. The reference molecule provides a template network, molecules that are to be compared are sent through this network leading to comparative networks.

superposition of the molecules is required for this approach. A model of the template approach is given in Figure 2.

FROM A 3D-SPACE TO A 2D-MAP

In order to give an idea of the correspondence between a 3D-space and its 2D-map, we show in Plate 2 an example of a projected Kohonen map of the

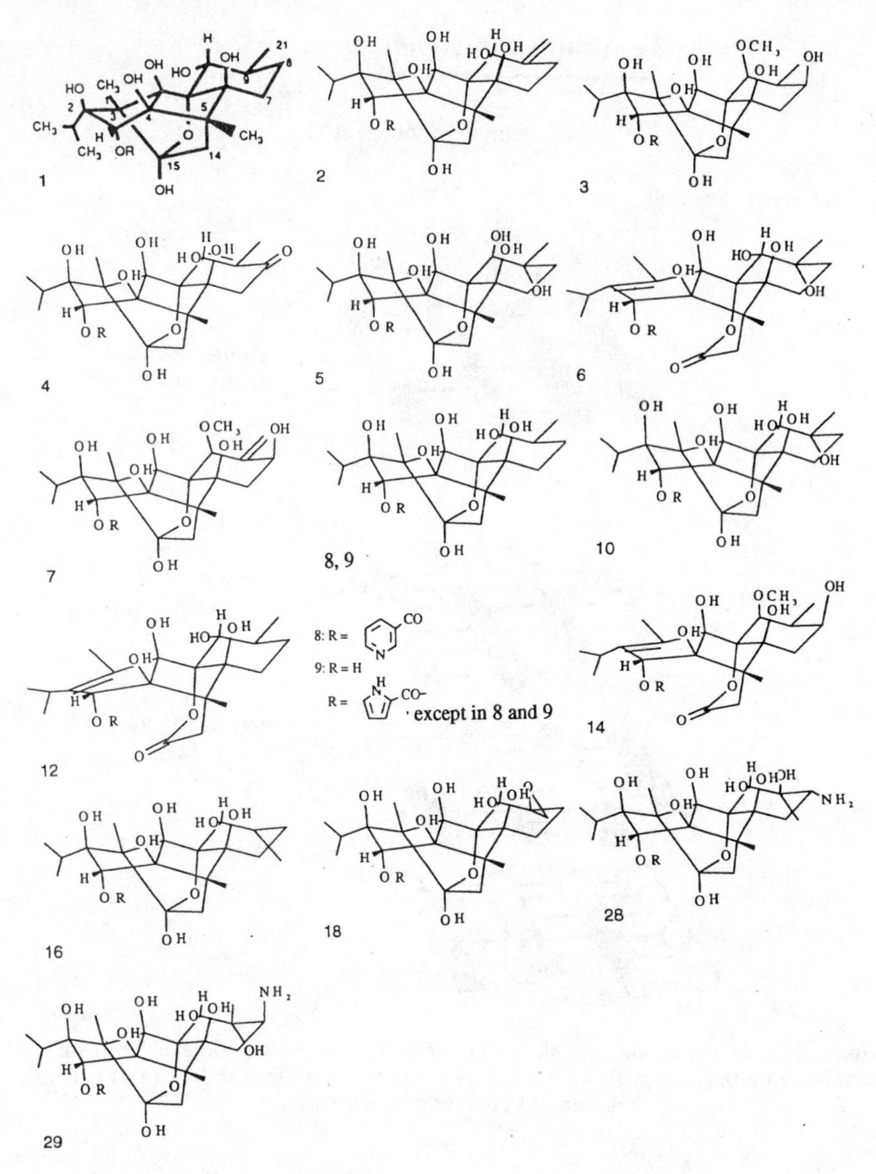

Figure 3 Ryanoid structures used to determine SARs.

van der Waals surface of ryanodine, a plant alkaloid. This representation permits one to realize the correspondence between the spatial arrangement of the functional groups in the molecule and areas in the 2D-map. The values of the electrostatic potential determine the colors of the map. Ryanodine is a rigid polycyclic system containing a cluster of several hydroxyl groups, a pyrrol-2-carboxylate group at position 3, and an isopropyl group at position 2 (see Figure 3). The MEP of ryanodine (Plate 2a) contains a large site with highly positive values, i.e., an area that traverses that of the polycyclic ring that includes the atoms 1, 2, 3, 15, the isopropyl group, and the area of the hydrogen atom of pyrrole nitrogen atom. Consistent with this, the Kohonen map consists of a large space with a blue–purple color for these sites (Plate 2b). The MEP of ryanodine also contains a large site with highly negative values, i.e., the area of the cluster of hydroxyl groups at positions 2, 4, 6, 10, 12, and 15, which is shown on the map as a red–yellow area.

The spatial distance of these groups is reflected by a difference in the positions of their spaces on the map. The white spaces (white lines and spots) of the map correspond to empty neurons. These are a result of a topological distortion of a torus (Li *et al.*, 1993).

CLUSTERING OF THE STRUCTURES BY AN INVESTIGATION OF THEIR MAPS

In the following example, 16 ryanoid derivatives were selected from literature (Welch *et al.*, 1994). Individual compounds and the experimental binding affinity data are shown in Figure 3 and Table I. Ryanoids are plant alkaloids, which bind to specific membrane proteins and alter the calcium permeability of intracellular membranes. Two groups, the isopropyl and the pyrrole group, a flexible part of the ryanodine, found by CoMFA (Welch *et al.*, 1994) to be critical to biological activity, are shown in Plate 3. CoMFA shows that the van der Waals interactions between the receptor and the isopropyl group and the pyrrole group are critical. The region formed by atoms 9, 10, and 21 can tolerate large changes in structure without destroying binding. Consequently, the hypothesis was proposed, that these regions are at the entrance of the binding pocket. The electrostatic forces of the hydroxyl group at position 15 (see Plate 3, O34) are highly correlated with ligand binding.

Conformational analysis yielded two local conformers for the orientation of the pyrrole nitrogen atom. Therefore, we considered both conformations for further investigations. Using the template approach, we trained a net for the most active compound (cpd. no. 2 see Table I) of this series. Thc resulting Kohonen maps are shown in Plate 4. It can be seen that with both sets of conformers, a clustering of maps into two groups is possible (see Plate 4 and Table I). The first nine maps (with K_D value of up to 240 nM) show similar MEP patterns and a nearly similar molecular shape (because of a few empty neurons) in relation to the reference compound dehydroryanodine. The seven further maps show more dissimilarity of the MEP patterns and/or the molecular shape to the reference molecule. The maps of molecules 28 and 29 show large spaces of white neurons (empty neurons). These molecules contain a polar amine group at positions 8α and 8β respectively, and thus,

Table I *Name used in the paper (Welch* et al. *(1994) and* K_D *for ryanodine analogs.*

Cpd. no.	Name used in the paper	K_D (nM) Rabbit
1	Ryanodine	7
2	Dehydroryanodine	7
16	9-Epiryanodine	13
18	9b,21-Epoxyryanodine	36
7	Ester D	51
10	9-Hydroxyryanodine	110
3	Ester A	110
5	Ester C1	210
4	Ester B	240
29	8b,9a-Dihydroxyryanodine	760
8	Ryanodyl nicotinate	1100
6	Ester C2	2600
12	Anhydroryanodine	2900
28	8a,9b-Dihydroxyryanodine	3800
14	Anhydroester A	5200
9	Ryanodol	12000

their electrostatic potential is strongly positive. In order to compare the electrostatic potential values of all compounds on a common color plate, we had to expand the color palette to a range between –113.0 and +210.6 kcal mol^{-1} for the electrostatic potential. However this distorted the color distinction of all the other maps. Therefore, the electrostatic potential values were limited to –98.11 to +43.69 kcal mol^{-1}. Thus, in molecules 28 or 29 quite a few electrostatic potential values were not taken into account and this resulted in a large blank space on their maps. In any case, the positive charge at atom 8 of the molecules 28 or 29 causes a very strong positive electrostatic potential of the whole molecular surface and consequently, a large difference in the MEP patterns to the first nine maps of affine compounds. Also, a strong correlation of positive charges at this position and decreased ligand affinity was found by CoMFA.

Altogether, the comparison of the structures of the molecules in the two clusters of Plates 4A and 4B shows four decisive modifications within this series of compounds that lead to a dramatic decrease of binding affinity:

a) A chemical modification of the hydroxyl group at position 15 to a carbonyl group of a lactone (molecules 6, 12, and 14, see Figure 3). The Kohonen maps of these molecules show quite similar MEP patterns and contain large spaces of strongly positive and negative potential values (see Plate 4).
b) The removal of the pyrrole group (molecule 9). The corresponding Kohonen map of molecule 9 shows a large blank area.

c) Exchange of the pyrrole to the pyridine ring at position 3 (molecule 8). Welch *et al.* (1994) maintained that too much bulk in the pyrrole region (molecule 8) is as deleterious as lack of bulk (molecule 9). However the corresponding map of molecule 8 indicates that the feature of the positive electrostatic potential area is different from those of the first nine maps. A comparison of the maps of the second conformation series (see Plate 4B), shows that the positive electrostatic potential area of the map of molecule 8 is less pronounced than in the maps of the active compounds.
d) A very strong positive MEP surface of the molecule decreases the binding affinity (molecules 29 and 28).

The backprojection of 2D-Kohonen maps onto a 3D-space is also possible. This is the space of the connection weights of a trained Kohonen net, which approximates the 3D-surface represented by the input data. This can be used in the template approach to identify the areas of the molecular surface where active and inactive compounds differ significantly.

Plate 5 shows the backprojection of the maps of molecules 2 and 9 (Table I). The backprojection of the blank areas (empty neurons) reveals the site where those two compounds differ: the carboxypyrrole group at position 3.

IN SEARCH OF THE BIOACTIVE CONFORMATION, THE BEST SUPERPOSITION, SAR

The identification of the bioactive conformer of a flexible molecule is an important task in drug design. This conformational analysis is usually done by a systematic search procedure. The use of a template approach in this context will be illustrated in the following example.

Two cardiac glycosides, bufalin and digitoxigenin (Figure 4) which bind to the digitalis receptor and inhibit the Na^+/K^+-ATPase are compared (Höltje and Anzali, 1992). A Kohonen net was trained using the Cartesian coordinates of points on the van der Waals surface of a known conformation of bufalin, a highly active and rigid compound. The molecule that is to be compared with this conformation of bufalin is represented by 36 (rotation around the C17–C20 bond in steps of 10°) of digitoxigenin, which also belongs to the highly active compounds. The maps obtained in this procedure are shown in Plate 6.

The low number of empty neurons in the comparative maps indicates the similarity of the surface of the two molecules. Figure 5 shows two local minima in the number of empty neurons from the comparative maps. Thus, the shape similarity, between the compared molecule and the reference molecule, can be determined by counting the empty neurons on the comparative

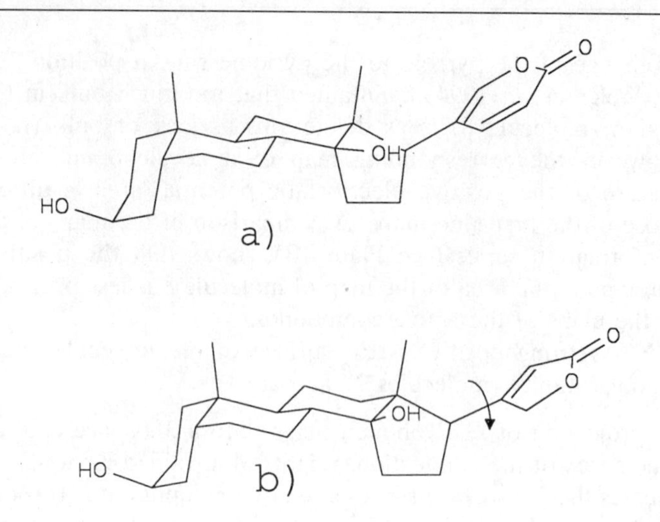

Figure 4 The chemical structures of bufalin (a) and digitoxigenin (b).

map (Figure 5). The two conformations of digitoxigenin found at a rotational angle of 80° and 280°, respectively, were discussed in previous studies (Fullerton *et al.*, 1986) as two distinct minimum energy conformations (14–21 and 14–22 conformations).

The following example shows a comparison between four compounds of the cardiac glycosides series (Plate 7). As an obvious shape dissimilarity exists between these molecules, the template approach can be presented clearly. The reference molecule is digitoxigenin, a highly active and semi-rigid cardenolide. The three compared compounds are in order of their affinities, 22-methoxy-digitoxigenin, 17-βH-digitoxigenin and estradiol. The experimental binding affinity data are taken from the literature (Schönfeld *et al.*, 1986).

For comparison of the Kohonen maps we used the following two measures:

I_{AB}: Scaled number of neurons, a_{ij}, from the reference map A and neurons, b_{ij}, from the compared map that have similar electrostatic potential values as expressed by the same color range ($c(a_{ij})$, $c(b_{ij})$).

D_{AB}: Scaled mean difference of neurons, a_{ij}, from the reference map A and neurons, b_{ij}, from the compared map that have different electrostatic potential values.

These are given by Eqs. (1) and (2):

$$I_{AB}[\%] = \frac{100}{n \cdot m} \cdot i_{AB} \text{ with } i_{AB} = \sum_{i}^{n} \sum_{j}^{n} \delta_{ij} \text{ and } \delta_{ij} = \left| \begin{array}{l} 1 \text{ for } c(a_{ij}) = c(b_{ij}) \\ 0 \text{ for } c(a_{ij}) \neq c(b_{ij}) \end{array} \right. \quad (1)$$

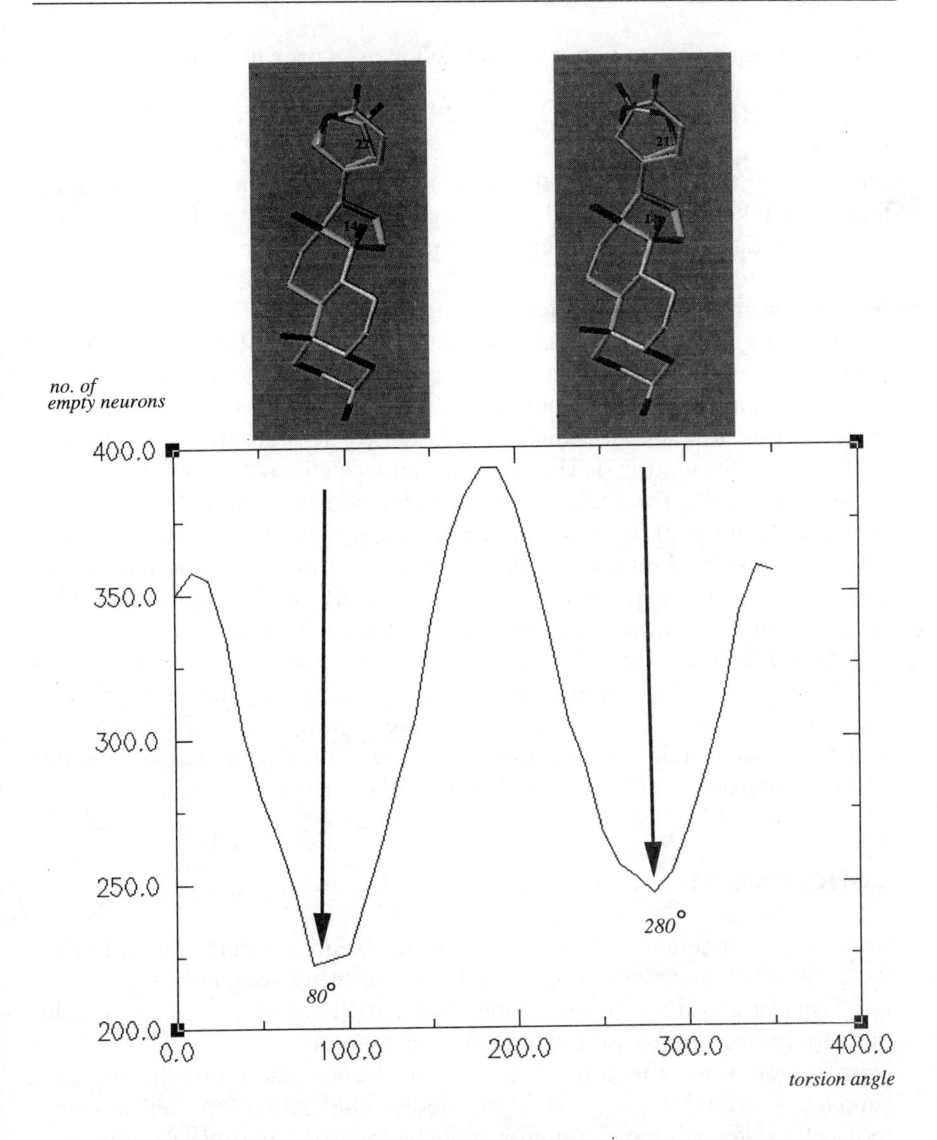

Figure 5 Two local minima of the number of the empty neurons from the comparative maps obtained from the van der Waals surfaces of 36 conformers of digitoxigenin.

$$D_{AB}[\%] = \frac{100}{n \cdot m - i_{AB}} \cdot \frac{1}{NC} \sum_{i}^{n} \sum_{j}^{n} |c(a_{ij}) - c(b_{ij})| \tag{2}$$

with NC = number of colors
$n \cdot m$ = total number of neurons

Thus a similarity index between two Kohonen maps can be calculated by:

$$S_{AB}[\%] = \frac{I_{AB}}{D_{AB}} \cdot 100$$

Through the similarity index the MEP and the molecular shape are simultaneously taken into account. Plate 8 shows the obtained Kohonen maps together with the number of empty neurons (Ne) and the similarity indices (S_{AB}). This result shows that the MEP and the molecular shape factor are both important for the binding on the digitalis receptor. The obtained results correspond quite well to the experimental binding affinities.

The next example shows the use of a template approach for finding the best alignment of two structures. Here, we compare two steroid molecules from a series of compounds that bind to the corticosteroid binding globulin (CBG) (Good *et al.*, 1993). The reference molecule is corticosterone, a highly active compound. The molecule to be compared is 5-androstendiol, a low active compound. It is taken in ten different spatial positions generated manually by a random transformation of each of the three coordinate axes (see Plate 9). The maps obtained are shown in Plate 10 together with their count of empty neurons. The best superposition (compare Plates 9 and 10), is deduced from the similarity index and by comparison of the number of empty neurons of the comparative maps. Thus, the best alignment between the compared and the reference molecule can be determined from the number of white neurons of the comparative map: the smaller the number of white neurons, the better are the molecular surfaces matched.

CONCLUSIONS

Using a Kohonen neural network, molecular surface properties of ryanodine derivatives are compared. The trained maps of this compound's series could be clustered according to their biological activities. In cases like ryanodine derivatives, the comparison of trained maps of each molecule allows direct classification of whether a molecule is active or not. Therefore, the approach supplies a straightforward tool to predict the activity of an unknown molecule. Using a template approach, the comparison of molecular shape by a reference molecule within the ryanodine derivatives, is taken into account.

Also, a similarity index used in this approach, could supply the best alignment possibility between two molecules with respect to their molecular electrostatic potential and shape. The results obtained are encouraging towards development of an automatic procedure.

Two conformers of digitoxigenin similar to bufalin as the reference molecule are identified.

The backprojection technique is the link between the nonintuitive form of information encoded in the Kohonen maps and the 3D world of molecular structures. The results open promising perspectives for the fast and accurate comparison of large datasets of molecules by Kohonen neural networks.

Acknowledgement

The authors acknowledge the Bundesminister für Bildung, Wissenschaft, Forschung und Technologie (BMBF), FRG, (ÄBAV project) grant # 01IB305D for financial support.

REFERENCES

Fullerton, D.S., Ahmed, K., From, A.H.L., McParland, R.H., Rohrer, D.C., and Griffin, J.F. (1986). Modelling the cardiac steroid receptor. In, *Topics in Molecular Pharmacology*, Vol. 3 (A.S.V. Burgen, G.C.K. Roberts, and M.S. Tute, Eds.). Elsevier, Amsterdam, pp. 257–284.

Gasteiger, J. and Li, X. (1994). Mapping the electrostatic potential of muscarinic and nicotinic agonists with artificial neural networks. *Angew. Chem. Int. Ed. Engl.* **33,** 643–646.

Gasteiger, J., Li, X., Rudolph, C., Sadowski, J., and Zupan, J. (1994b). Representation of molecular electrostatic potentials by topological feature maps. *J. Am. Chem. Soc.* **116,** 4608–4620.

Gasteiger, J., Li, X., and Uschold, A. (1994a). The beauty of molecular surfaces as revealed by self-organizing neural networks. *J. Mol. Graph.* **12,** 90–97.

Gasteiger, J. and Marsili, M. (1980). Iterative partial equalization of orbital electronegativity: A rapid access to atomic charges. *Tetrahedron* **36,** 3219–3228.

Good, A.C., So, S.S., and Richards, W.G. (1993). Structure–activity relationships from molecular similarity matrices. *J. Med. Chem.* **36,** 433–438.

Höltje, H.-D. and Anzali, S. (1992). Molecular modelling studies on the digitalis binding site of the Na^+/K^+-ATPase. In, *Trends in QSAR and Molecular Modelling 92* (C.G. Wermuth, Ed.). Escom Science Publishers B.V., Leiden, pp. 180–185.

Kohonen, T. (1982). Analysis of a simple self-organizing process. *Biol. Cybern.* **43,** 59–62.

Li, X., Gasteiger, J., and Zupan, J. (1993). On the topology distortion in self-organizing feature maps. *Biol. Cybern.* **70,** 189–198.

Purcell, W.P. and Singer, J.A. (1967). A brief review and table of semiempirical parameters used in the Hückel molecular orbital method. *J. Chem. Eng. Data* **12,** 235–246.

Sadowski, J., Rudolph, C., and Gasteiger, J. (1992). The generation of 3D-models of host-guest complexes. *Anal. Chim. Acta* **265,** 233–241.

Schönfeld, W., Schönfeld, R., Menke, K.-H., Weiland, J., and Repke, K.R.H. (1986). Origin of differences of inhibitory potency of cardiac glycosides in Na^+/K^+-transporting ATPase from human cardiac muscle, human brain cortex and guinea-pig cardiac muscle. *Biochem. Pharmacol.* **35,** 3221–3231.

Welch, W., Ahmed, S., Airey, J.A., Gerzon, K., Humerickhouse, R.A., Besch, H.K., Ruest, Jr.L., Deslongchamps, P., and Sutko, J.L. (1994). Structural deteminants of high-affinity binding of ryanoids to the vertebrate skeletal muscle ryanodine receptor: A comparative molecular field analysis. *Biochem.* **33,** 6074–6085.

10 A New Nonlinear Neural Mapping Technique for Visual Exploration of QSAR Data

D. DOMINE[1*], D. WIENKE[2], J. DEVILLERS[1], and L. BUYDENS[2]

[1]*CTIS, 21 rue de la Bannière, 69003 Lyon, France*

[2]*Catholic University of Nijmegen, Laboratory for Analytical Chemistry, Toernooiveld 1, 6525 ED Nijmegen, The Netherlands*

A new nonlinear neural mapping (N2M) technique based on the combined use of Kohonen self-organizing map (KSOM), minimum spanning tree (MST), and nonlinear mapping (NLM) is introduced. KSOM results are enhanced by the visualization of the actual distances between the loaded neurons from MST and NLM. N2M is then illustrated from different case studies dealing with the analysis of sensor data and the design of optimal test series for quantitative structure–activity relationship (QSAR) purposes.

KEY WORDS: *Nonlinear neural mapping; Kohonen self-organizing map; minimum spanning tree; nonlinear mapping; sensor data; selection of test series.*

INTRODUCTION

QSAR studies which involve the use of different molecular descriptors, require examination of multidimensional spaces which are not perceivable by humans. To solve this problem, linear and nonlinear multivariate techniques are generally employed in order to represent data in lower-dimensional spaces (i.e., 2 or 3D) so that structure–activity relationships (SAR) can be more easily derived. If the linear methods (Devillers and Karcher, 1991; van de Waterbeemd, 1994, 1995), have been the most popular until the end of the 1980s, nonlinear methods such as nonlinear mapping (Court *et al.*, 1988;

*Author to whom all correspondence should be addressed.

In, *Neural Networks in QSAR and Drug Design* (J. Devillers, Ed.)
Academic Press, London, 1996, pp. 223–253.
ISBN 0-12-213815-5

Livingstone *et al.*, 1988; Hudson *et al.*, 1989; Livingstone, 1989; Domine *et al.*, 1993a, 1994a,b, 1995; Chastrette *et al.*, 1994; Devillers, 1995) and neural networks (e.g., Aoyama *et al.*, 1990; Rose *et al.*, 1990, 1991; Aoyama and Ichikawa, 1991a,b,c, 1992; Andrea and Kalayeh, 1991; de Saint Laumer *et al.*, 1991; Liu *et al.*, 1992a,b; Livingstone and Salt, 1992; Cambon and Devillers, 1993; Chastrette *et al.*, 1993; Devillers, 1993; Devillers and Cambon, 1993; Domine *et al.*, 1993b; Feuilleaubois *et al.*, 1993; Livingstone and Manallack, 1993; Wiese and Schaper, 1993; Bienfait, 1994; Manallack and Livingstone, 1994; Devillers *et al.*, 1996a) are increasingly used for SAR purposes. They support the evidence that biological and physicochemical phenomena are not necessarily linked to molecular descriptors by means of linear relationships. Neural network techniques comprise a wide range of paradigms (Hopfield, 1982; Rumelhart *et al.*, 1986; Kohonen, 1989a; Pao, 1989; Eberhart and Dobbins, 1990; Carpenter and Grossberg, 1991) which can be classified as supervised and unsupervised methods. Among the supervised techniques, the backpropagation neural network (Rumelhart *et al.*, 1986) has received the most attention in the SAR and structure–property relationship (SPR) literature these last few years. Kohonen self-organizing map (KSOM) (Kohonen, 1989a,b, 1990), which belongs to the unsupervised class of methods, is considered to be the second after the backpropagation neural network in terms of number of applications (Eberhart and Dobbins, 1990) but is only emerging in the chemical and SAR literature (Arrigo *et al.*, 1991; Ferran and Ferrara, 1991, 1992a,b; Rose *et al.*, 1991; Tusar *et al.*, 1992; Ferran and Pflugfelder, 1993; Kateman and Smits, 1993; Melssen *et al.*, 1993; Simon *et al.*, 1993; Zupan and Gasteiger, 1993; Bienfait, 1994; Thoreau, 1994; Zupan *et al.*, 1994; Barlow, 1995; Anzali *et al.*, 1996). This is surprising since KSOM represents a powerful nonlinear tool for classification and graphical representation of multivariate data. KSOM allows the projection of n-dimensional patterns into lower m-dimensional arrays of neurons (i.e., $m = 1$ to 3). However, if the obtained maps preserve the topology (i.e., spatial arrangements of input vectors), distances between points are not preserved (Wienke and Hopke, 1994a,b; Wienke *et al.*, 1994, 1995; Martin-del-Brio *et al.*, 1995). To partially solve this problem, Wienke and coworkers (Wienke and Hopke, 1994a,b; Wienke *et al.*, 1994, 1995) recently proposed a new method called 3MAP combining KSOM with minimum spanning tree (MST) (Devillers and Doré, 1989) in order to improve the results obtained with KSOM by visualizing the shortest distances between the neurons on the self-organized topological maps. In the present study, we introduce the use of nonlinear mapping (NLM) for visualizing all the distances separating the loaded neurons of a self-organized topological map. This method, called nonlinear neural mapping (N2M), allows one to cluster n-dimensional data sets on m-dimensional arrays of neurons and to visualize the actual relationships between the neurons. The heuristic potency of N2M in structure–activity and structure–property relationship studies is explored from two different case studies.

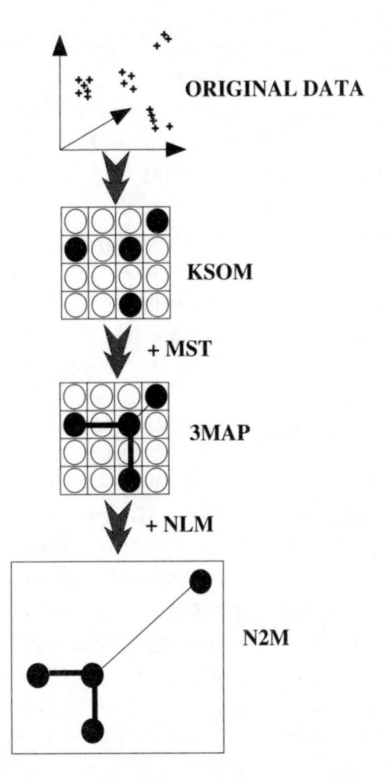

Figure 1 N2M algorithm flow diagram. The thick lines indicate the shortest distances.

BACKGROUND

The N2M algorithm (Figure 1) combines three distinct pattern recognition methods. First, the Kohonen map forms an initial step for data compression of p multivariate samples defined in an n-dimensional space into υ clusters (loaded neurons). The data reduction to a few clusters provides a simpler data structure. However, in Kohonen maps, the problem is that information about the correct distances between the neurons disappears during the projection onto the 1, 2, or 3D array of nodes. One possibility to overcome this distance problem is to add information that allows the simultaneous visualization of the distances between the loaded neurons in the map. The technique recently chosen by Wienke and coworkers (Wienke and Hopke, 1994a,b; Wienke *et al.*, 1994, 1995) and also in the present study is a calculation of a MST (Prim, 1957; Devillers and Doré, 1989) between the loaded neurons of a trained KSOM in order to visualize the shortest distances between them. These two steps constituted the basis of the 3MAP algorithm designed by Wienke and coworkers (Wienke and Hopke, 1994a,b; Wienke *et al.*, 1994, 1995). As there remains information not represented, about the correct distances between all the loaded neurons, in the third step of a N2M analysis, a NLM of the loaded neurons is performed in order to visualize all the distances separating them. Note that in N2M, we chose to use NLM for its ability to represent best on a 2-dimensional map the distances between a set of points (Sammon, 1969; Kowalski and Bender, 1972, 1973; Domine *et al.*, 1993a; Devillers, 1995). However, in some instances, it might be possible to apply a principal components analysis, if the percentages of variance carried by the first two principal components is sufficiently high to explain most of the information contained in the weight data matrix.

Kohonen self-organizing map (KSOM)

The Kohonen model of self-organization is based on the idea that the brain tends to compress and organize sensory data, spontaneously (Zeidenberg, 1990). In this sense, although 'it is possible to ask just how 'neural' the map is' (Kohonen, 1990), it is more biologically oriented than other artificial neural networks such as backpropagation neural networks. A Kohonen network consists of two layers of neurons (i.e., input and output) (Figure 2). The input layer contains n neurons corresponding to the n variables describing the samples. The output layer is a 1, 2, or 3-dimensional geometrical arrangement of u neurons. Most often, 2D arrays of neurons presenting a square, hexagonal, or rectangular shape are used. Note that Kohonen does not recommend one to use circles. Kohonen also suggested that the dimensions of the neuron arrays can be determined by initially performing an NLM analysis (Kohonen *et al.*, 1995). The n neurons of the input layer are all connected to each of the u neurons of the output layer so that these u neurons can be considered as u n-dimensional weight vectors.

The essence of the Kohonen algorithm is a repeated comparison of the p n-dimensional samples with the u weight vectors using any distance metric

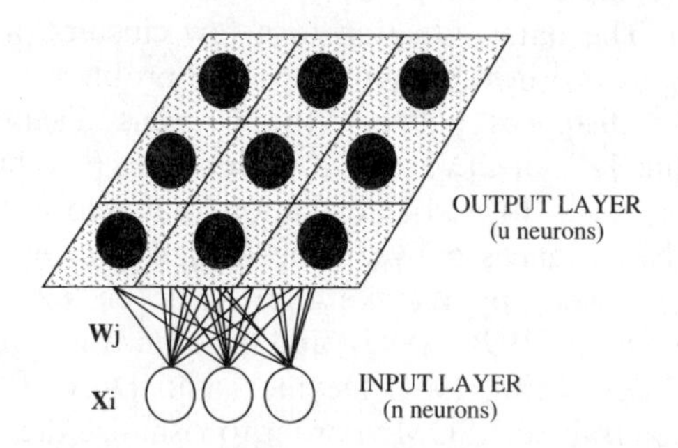

Figure 2 Architecture of a Kohonen network.

such as the Euclidean distance. In the classical algorithm, the weight vectors $\mathbf{W}_j$ of the neurons j on the Kohonen map are initialized with small random numbers. In the present study, we used a slightly different method called convex combination (Wasserman, 1989; Wienke and Hopke, 1994a,b) in which all weights are initialized with a constant value. Then, during each comparison, a winner ($\mathbf{W}_c$) among the u weight vectors can be found that has the highest degree of similarity (i.e., smallest distance) with the i-th considered sample $\mathbf{X}_i$.

Then, the n weights connected to the winning neuron $\mathbf{W}_c$ are updated in order to become closer to the considered sample using the Kohonen learning rule:

$$\mathbf{W}_c^{\text{new}} = \mathbf{W}_c^{\text{old}} + \eta(\mathbf{X}_i - \mathbf{W}_c^{\text{old}}) \qquad (1)$$

where $\mathbf{W}_c^{\text{new}}$ and $\mathbf{W}_c^{\text{old}}$ represent the weight vectors of the winning neuron after and before the update, $\mathbf{X}_i$ is the n-dimensional vector of sample i, and η is the learning rate with $0 < \eta < 1$. Together with the winning neuron, $\mathbf{W}_c$, the nodes that are topologically close in the neuron array up to a certain distance (i.e., within a limited topological neighborhood of radius R) will activate each other in order to learn from the same input sample. Squares, hexagons, circles, bubbles have been used as topological neighborhood (Kohonen, 1989a,b, 1990; Pao, 1989; Wasserman, 1989; Bienfait, 1994; Wienke and Hopke, 1994a,b; Wienke *et al.*, 1994, 1995; Kohonen *et al.*, 1995). After this adaptation, the subset of weight vectors within R became slightly more similar to the presented input $\mathbf{X}_i$, in terms of the distance metric employed. One sequential comparison of all the p samples with all u neural weight vectors and their modification is called one epoch. From one epoch to the next, the neighborhood, R, is decreased by a small step. Kohonen (Kohonen, 1989b; Kohonen *et al.*, 1995) suggests to decrement R by small steps within the first training epochs ne_1. In a 2D map, for example, when a square topological neighborhood is chosen, $R = 1$ means, that around a winning neuron, four neurons (i.e., left, above, right, down) are trained by the learning rule together with the winner. Some authors have proposed to bend the Kohonen map into a torus so that the neurons on one edge of the plane are made neighbors of the neurons located on the opposite side (Zupan and Gasteiger, 1993; Lozano *et al.*, 1995; Anzali *et al.*, 1996). Sometimes, it is recommended to change the weights as function of the closeness of a neuron to the winner neuron. This so called inhibitor function can be chosen as a Gaussian-like function around the winner, as pyramidal or, as a step function (Kohonen, 1989a; Wasserman, 1989; Eberhart and Dobbins, 1990; Kohonen *et al.*, 1995). Simultaneously with decreasing R, the learning rate η over the number (t) of training epochs is also decremented using the following formulae:

$$\eta_1 = k_1 \times (1 - t/ne_1) \qquad (2)$$

$$\eta_2 = k_2 \times (1 - t/ne_2) \qquad (3)$$

over two phases of training with ne_1 and ne_2 epochs. In the present work, k_1 and k_2 equaled 0.1 and 0.008, respectively. Indeed, these values have been found to work well for very different numbers of training epochs *ne*, different sizes of neural maps and different dimensions of input data matrices described in the literature (Kohonen, 1989b; Wienke and Hopke, 1994a,b), and also in the present study. After repeating the training process over many epochs, *ne*, with $ne = ne_1 + ne_2$ whereby $ne > 500 \times u$ (Wasserman, 1989) for example, a self-organizing behavior of the *p* samples in the low *m*-dimensional neural array can be observed. They form a visual topological structure. The final result of this training process is an aggregation of subsets of input vectors dedicated to a number υ of weight vectors in the neural array with $\upsilon \leq u$. The trained weight vector for an aggregated cluster of input vectors comes numerically very close to their mean vector. Wienke and coworkers (Wienke and Hopke, 1994a,b; Wienke *et al.*, 1994, 1995) called this kind of neuron a loaded neuron. In spite of its unsupervised character, it may be noted that Kohonen networks can be improved and optimized by a supervised method when the number of sought after classes is known. This procedure is called 'learning vector quantization' (Kohonen, 1990, 1992; Bienfait, 1994). It can also be combined with the Grossberg outstar algorithm to perform supervised training. This kind of network is named counter-propagation neural networks (Hecht-Nielsen, 1988; Wasserman, 1989; Peterson, 1992, 1995; Zupan and Gasteiger, 1993).

Minimum spanning tree (MST)

The theory related to the construction of minimum spanning trees has been described by different authors (Zahn, 1971; Lebart *et al.*, 1979; Barthélemy and Guénoche, 1988; Devillers and Doré, 1989). It can be summarized as follows. A set of *p* objects that are to be classified in relation to each other, may be considered as a set of points in a space. The representation is classical if the objects are described by a series of *n* parameters. In that case, one has *p* points in the $\mathfrak{R}^n$ space. More generally, if one has only the values of an index of dissimilarity, which does not necessarily satisfy the properties of a distance, one may represent the objects as points of a plane, for example. Each couple of objects is then connected by a continuous line to which the value of the dissimilarity index is assigned. Thus, set of objects and the index values are represented by a complete valued graph. However, if the number of objects is too important, this type of representation becomes inextricable.

One may then try to extract from this graph a partial graph having the same nodes but fewer edges. Offering an easier representation, it may nevertheless, allow a satisfactory summary of the index values. Among all the partial graphs that can be constructed, those that have a tree structure are particularly interesting, since they may be represented on a plane. Thus, a tree is a connected graph (i.e., a line may connect any vertex couple).

Moreover, it has no cycle (i.e., one may not have in that tree an edge coming from and converging on the same point without taking twice the same edge).

One may consequently define, in the same way, a tree with p vertices as being a connected graph without a cycle and with $p-1$ edges. The length of a tree is the summation of the lengths of its edges. When it is minimal, one has constructed a MST. Many different algorithms have been proposed for the construction of trees of a minimal length (Jarnik, 1936; Florek *et al.*, 1951; Kruskal, 1956; Prim, 1957; Roux, 1975; Cheriton and Tarjan, 1976). For the N2M technique, we applied Prim's algorithm. In N2M, this algorithm finds the shortest possible connections between all υ loaded neurons. To find the $(\upsilon-1)$ branches, the constraint, that no closed connections are allowed, was used.

Nonlinear mapping (NLM)

NLM (Sammon, 1969) is based on a concept similar to multidimensional scaling (MDS) (Kruskal, 1964, 1971; Wish and Carroll, 1982), which is a classical nonlinear method allowing one to represent a set of individuals in a metric space from a measure of their dissimilarity. Like the other display methods, NLM allows one to represent a set of points defined in an n-dimensional space by a human perceivable configuration of the data in a lower m-dimensional space (m = 2 or 3) which we will call either display space or nonlinear map. NLM tries to preserve distances between points in the display space as similar as possible to the actual distances in the original space. The procedure for performing this transformation can be summarized as follows.

(i) Interpoint distances in the original space d_{ij} are computed. The Euclidean distance is the most widely used. Nevertheless, it has been stressed that any distance measure would be suitable for nonlinear mapping as long as it is monotonic and the derivative of the mapping error (E) exists (Kowalski and Bender, 1972, 1973).

(ii) An initial configuration of the points in the display space is chosen. Most often, the coordinates of points in the display space are set in a random manner (Sammon, 1969). Calculations are performed from several initial configurations to avoid the trapping into local minima which would lead to erroneous conclusions (Hudson *et al.*, 1989). Sammon (1969) underlined that 'in practice the initial configuration, ..., is found by projecting the L-dimensional data orthogonally onto a d-space spanned by the d original coordinates with the largest variances'. Thus, several authors have proposed to use the coordinates of points on the first principal components (PCs) as initial configuration. However, it is always highly recommended to perform several trials either with random configurations or with the other PC coordinates (Domine *et al.*, 1993a).

(iii) A mapping error (E) is calculated from distances in the two spaces. The original mapping error (E) calculation for NLM was devised by Sammon (1969) on the basis of the Euclidean distance but many other error formulae have been devised (Domine *et al.*, 1993a).

(iv) The coordinates of the points in the display space are iteratively modified by means of a nonlinear procedure so as to minimize the mapping error. The various NLM algorithms available in the literature also differ in the way of minimizing the error (Sammon, 1969; Polak, 1971; Kowalski and Bender, 1973; Everitt, 1978; Klein and Dubes, 1989; Hamad and Betrouni, 1995). Thus, for example, Sammon preferred the 'steepest descent procedure' while Kowalski and Bender (1973) adopted the Polak-Ribière method. Actually, any of the usual minimization procedures (Polak, 1971) can be used (Kowalski and Bender, 1972).

(v) The algorithm terminates when no significant decrease in the mapping error is obtained over the course of several iterations.

In the present study, the Euclidean distance was selected and the steepest descent procedure was used as minimization method. The algorithm was therefore the following.

Suppose the interpoint distances at the t-th configuration described by the Euclidean distance as shown below:

$$d_{ij}(t) = \left[\sum_{k=1}^{m} (x_{ik}(t) - x_{jk}(t))^2 \right]^{1/2} \tag{4}$$

and the corresponding error $E(t)$ as defined by Sammon (1969):

$$E(t) = \frac{1}{\sum_{i<j}^{p} d_{ij}^*} \sum_{i<j}^{p} \frac{\left[d_{ij}^* - d_{ij}(t) \right]^2}{d_{ij}^*} \tag{5}$$

Then, the steepest descent procedure proceeds as shown below. The coordinates in the $(t+1)$th configuration are given by:

$$x_{lq}(t+1) = x_{lq}(t) - (\mathrm{MF} \cdot \Delta_{lq}(t)) \tag{6}$$

where:

$$\Delta_{lq}(t) = \frac{\partial E(t)}{\partial x_{lq}(t)} \Bigg/ \left| \frac{\partial^2 E(t)}{\partial x_{lq}(t)^2} \right| \tag{7}$$

and MF is a magic factor empirically determined as = 0.3 or 0.4 (Sammon, 1969).

This process is carried out iteratively until a threshold fixed by the user is attained (i.e., minimal error or minimal difference between the error at step $t-1$ and step t in the iteration process). Precautions must be taken to prevent any two points in the m-dimensional space from becoming identical to avoid problems in the calculation of the partial derivatives. Points in the n-dimensional space must also be different. Additional information on the practical aspects of the NLM method, a review of its uses in QSAR studies and original examples of applications can be found in previous papers (Domine *et al.*, 1993a, 1994a,b,c, 1995; Chastrette *et al.*, 1994; Devillers *et al.*, 1994; Budzinski *et al.*, 1995; Devillers, 1995; Devillers and Domine, 1995; Domine and Devillers, 1995).

CASE STUDY I: ANALYSIS OF SENSOR DATA

Up to now, there is no better odor identifier than the human nose (Devillers *et al.*, 1996b). However, for cost and safety reasons for example, there is a need for devices allowing one to automatically detect and identify odors. For this purpose, semiconductor gas-sensors possessing different sensitivities to various gases have become available. However, tools allowing one to interpret the results obtained with gas-sensors and link their responses to the odor quality of the gas studied are required. Under these conditions, the aim of this first case study is to try to classify odors from plural gas-sensors responses. The data were retrieved from Abe and coworkers (Abe *et al.*, 1987) who measured the responses of eight sensors towards 30 different odorant chemicals (Table I) with known odor qualities (Table II) and tried to objectively identify their odors by means of the combined use of hierarchical cluster analysis and principal components analysis (results not shown). Because the percentage of variance on the first two principal components was not sufficient to allow a complete representation of the data on a sole plane, the authors had to inspect higher order axes. Due to the ability of NLM to summarize at best on a sole plane the information contained in a data table, the capacity of KSOM to cluster multivariate data, and the ability of MST to link the clusters, N2M was applied to the same data matrix in order to obtain a sole, easily interpretable, nonlinear neural map and to present in a didactic manner its heuristic potency.

The first step of a N2M analysis consists in performing the KSOM of the data. Figure 3 shows the results of the KSOM of the data presented in Table I after autoscaling. Note that several trials were performed with different network architectures and that the results obtained were stable. If we replace the chemical numbers by the odor qualities (Figure 4), it is interesting to note that KSOM generally allows one to separate the chemicals with respect to their odor qualities even if the clusters are not always obvious. Thus, for example, it is noteworthy that cluster C only contains chemicals presenting a pungent (P) character. In the same way, cluster D groups ethereal chemicals. Compared with the results obtained by Abe *et al.* (1987), the clusters formed are different, but from a general point of view, a better discrimination

Table I *Sensor data.*

Chemical no.*	Sensor 1	Sensor 2	Sensor 3	Sensor 4	Sensor 5	Sensor 6	Sensor 7	Sensor 8
1	0.186	0.063	0.159	0.168	0.183	0.056	0.104	0.081
2	0.181	0.061	0.153	0.168	0.179	0.069	0.106	0.083
3	0.18	0.044	0.168	0.176	0.19	0.057	0.099	0.086
4	0.177	0.055	0.154	0.169	0.182	0.065	0.09	0.109
5	0.172	0.037	0.16	0.177	0.192	0.077	0.11	0.075
6	0.168	0.089	0.155	0.16	0.173	0.061	0.108	0.088
7	0.175	0.093	0.157	0.16	0.173	0.061	0.105	0.077
8	0.17	0.122	0.146	0.156	0.165	0.063	0.101	0.079
9	0.162	0.094	0.153	0.176	0.196	0.071	0.086	0.062
10	0.151	0.103	0.153	0.18	0.193	0.073	0.078	0.069
11	0.152	0.068	0.17	0.198	0.226	0.068	0.062	0.058
12	0.155	0.076	0.159	0.194	0.208	0.071	0.073	0.063
13	0.121	0.084	0.177	0.191	0.221	0.075	0.066	0.066
14	0.135	0.093	0.142	0.196	0.205	0.096	0.074	0.06
15	0.143	0.074	0.143	0.191	0.201	0.111	0.075	0.063
16	0.147	0.072	0.145	0.193	0.202	0.104	0.075	0.062
17	0.17	0.024	0.174	0.19	0.212	0.065	0.094	0.072
18	0.164	0.035	0.17	0.189	0.212	0.07	0.093	0.068
19	0.148	0.023	0.162	0.191	0.204	0.07	0.118	0.085
20	0.14	0.003	0.173	0.202	0.224	0.074	0.113	0.071
21	0.158	0.02	0.179	0.204	0.232	0.066	0.08	0.062
22	0.145	0.015	0.186	0.199	0.238	0.063	0.084	0.065
23	0.158	0.009	0.168	0.204	0.236	0.067	0.097	0.061
24	0.14	0.018	0.175	0.214	0.251	0.06	0.077	0.062
25	0.129	0.003	0.168	0.216	0.235	0.096	0.089	0.064
26	0.201	0.011	0.185	0.186	0.207	0.056	0.076	0.078
27	0.144	0.019	0.145	0.16	0.169	0.105	0.094	0.16
28	0.097	0.006	0.177	0.187	0.206	0.103	0.116	0.108
29	0.166	0.149	0.146	0.165	0.177	0.063	0.071	0.064
30	0.16	0.146	0.126	0.163	0.17	0.088	0.083	0.065

* See Table II for chemical names.

Table II *Substances used and odor qualities.*

Chemical no.	Chemical name	Odor quality*
1	Methyl acetate	ethereal
2	Diethyl ether	ethereal
3	Acetone	ethereal
4	Ethyl acetate	ethereal
5	Acetyl acetone	ethereal, minty
6	Methyl formate	ethereal
7	Ethyl formate	ethereal
8	Acetaldehyde	ethereal, pungent
9	Allyl acetate	ethereal, pungent
10	Isobutyl methyl ketone	ethereal
11	Pentyl formate	ethereal
12	Isopropyl acetate	ethereal
13	*tert*-Pentanol	minty
14	2-Methylcyclohexanone	minty
15	3-Methylcyclohexanone	minty
16	4-Methylcyclohexanone	minty
17	Methyl butyrate	ethereal
18	Isobutylaldehyde	pungent
19	Menthone	minty
20	Acetic acid	pungent
21	Methyl valerate	ethereal, pungent
22	Furfural	pungent
23	Acrylic acid	pungent
24	Phenetole	pungent
25	Propionic acid	pungent
26	Dioxane	ethereal
27	Chloroform	ethereal
28	Pyridine	pungent
29	2-Methyl-3-butanone	ethereal, pungent
30	Pyrrole	ethereal

* Main characteristic as described by Abe and coworkers (Abe *et al.*, 1987).

seems to be obtained, and this indicates that it could be possible to identify odors by adjusting the sensors or adding new ones which could discriminate more finely the chemicals with regard to their odor qualities. In Figure 3, if the topology is preserved, information about the correct distances between the loaded neurons is lost. To solve this problem partially, the MST between the loaded neurons has been represented on the Kohonen map to give a map called 3MAP (Figure 3). The classical graphical display of the MST (Devillers and Doré, 1989) obtained from the loaded neurons has not been represented. If 3MAP provides information on the shortest distances between the loaded neurons and facilitates the interpretation, it can be significantly improved by performing the NLM of the weight vectors. Figure 5 shows the nonlinear map of the weight vectors corresponding to the Kohonen

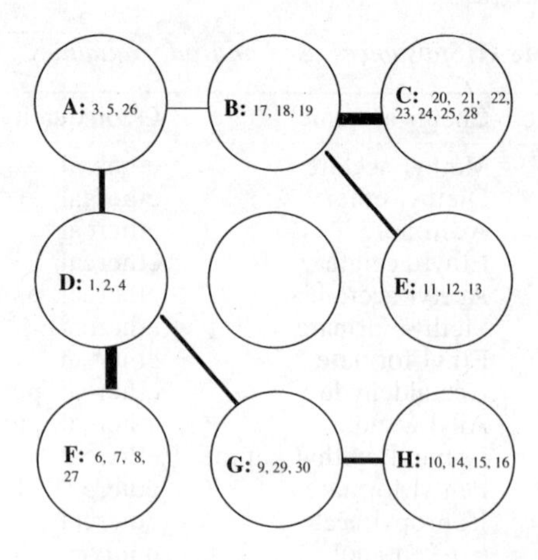

Figure 3 3MAP (i.e., Kohonen map + information on shortest distances provided by the MST) of sensor data. See Table II for correspondence between the numbers and the chemicals. The three different ticknesses of the lines are linked to the distances. The thicker the lines, the shorter the distances.

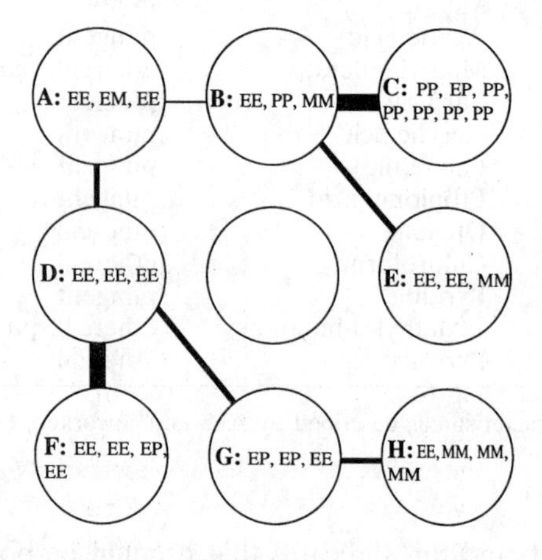

Figure 4 Representation of the odor quality memberships. EE: ethereal, MM: minty, PP: pungent. Chemicals presenting two of these qualities are indicated by two different letters (e.g., EP: ethereal-pungent).

map presented in Figure 3. Inspection of Figure 5 indicates that the NLM preserves the topology of the Kohonen map. Thus, for example, it can be noted that the alignment of neurons A, B, and C is preserved on the N2M display (Figure 5). One can also note that neuron E is located between neurons C and H. The same remark can be made for all other neurons on this figure. The representation of the MST directly on Figure 5 allows one

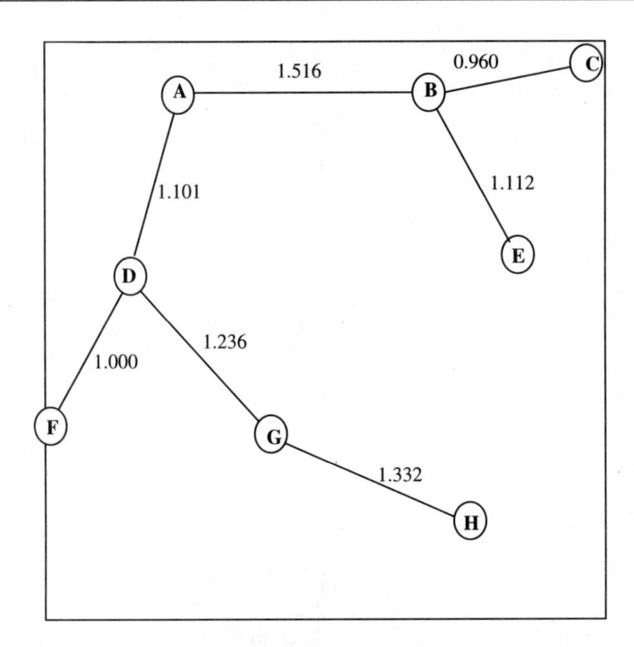

Figure 5 N2M display of the sensor data. The values reported are the actual distances in the MST. See Figure 3 for correspondence between the letters and the clusters.

to link the clusters exactly. This is an interesting feature of N2M, since on this map we can notice that by combining KSOM with MST and NLM, we benefit from the advantages of the three methods and we obtain a single, easily interpretable, map. In addition, as with a classical display method, graphical tools can then be used in order to facilitate the interpretation. Thus, for example, on Figure 6, we have represented for each cluster (i.e., loaded neuron) the values of the weights linked to each of the input descriptors. In Figure 6, the larger the square, the higher the weight value and the larger the circle the lower the weight value. Thus, it can be noted, for example, that the neurons located on the left-hand part of the map (i.e., neurons A, D, and F) are associated with high values of the weights linked to descriptor 1 (sensor 1) and low values for the weights attached to input descriptors 5 and 6 (sensors 5 and 6). Inspection of Figure 4 reveals that the chemicals belonging to these clusters are principally ethereal. In the same way, in order to interpret the nonlinear neural map in terms of the original variables, information concerning the mean values of the variables for the chemicals constituting each neuron can be plotted by means of squares (positive values) and circles (negative values) proportional in size to the mean values. Thus, for example, in Figure 7, it is possible to underline that cluster D is characterized by low values on sensor 5 (in black in Figure 7). In addition, it is possible to confirm by a comparison of Figures 6 and 7 that

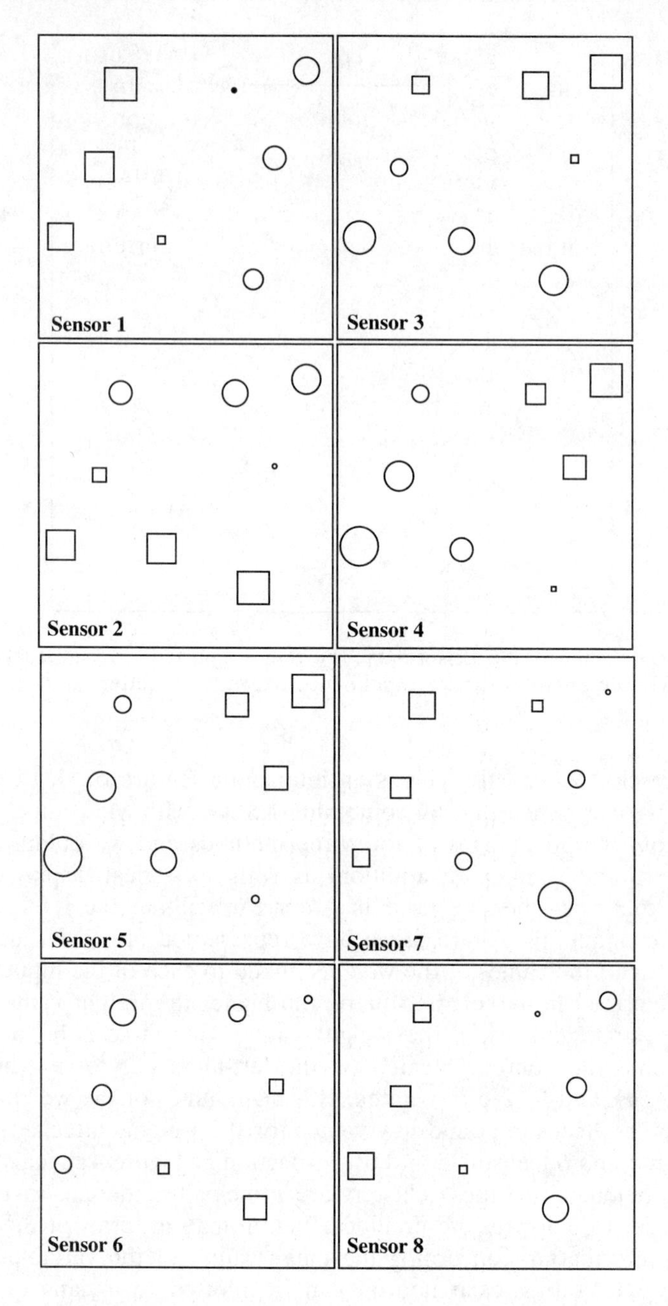

Figure 6 Plot of the weight values linked to the eight descriptors on each cluster of the N2M display (Figure 5). Squares (positive values) and circles (negative values) are proportional in size to the magnitude of the weights.

the neural weights are generally close to the mean values. However, this kind of interpretation must be done with some care. Indeed, the clusters group the most similar chemicals but some can have appreciable differences with respect to a given parameter. To quantify the similarity of the chemicals with their corresponding neurons, all the distances between the chemicals and the loaded neurons have been calculated (Table III). Thus, it is possible to confirm that the chemicals belonging to a cluster are always closer to this cluster than to the others. However, for some of them the distance can be appreciable (e.g., chemical 28 and cluster C).

The results suggest that plural semiconductor sensor systems can be elaborated in order to identify odors. The sensor responses can be simply interpreted from N2M and this method can guide improvements of the system by suggesting the addition or replacement of some sensors.

CASE STUDY II: OPTIMAL TEST SERIES DESIGN

The derivation of valuable structure–activity relationships requires the use of training and testing sets with compounds presenting a maximal 'meaningful variety'. In the same way, the research for new, biologically active, chemicals in medicinal chemistry and agrochemistry requires inspection of several thousands of candidates for which only limited information is available. Since it is practically impossible to fill this enormous data gap and test all the possible candidates, a decision must be taken as to where to start from. This implies the adoption of strategies for the selection of the most relevant compounds for preliminary biological testing. To meet this objective, the use of test series design is also of valuable help. A lot of work has been directed towards this aim and numerous methods have been proposed (e.g., Pleiss and Unger, 1990). Significant advances dealt with the use of linear multivariate methods (Hansch *et al.*, 1973) and their graphical plots allowing selection of test series by simple visual inspection of 2D maps summarizing the information content of a matrix of physicochemical properties (e.g., Dove *et al.*, 1980; Alunni *et al.*, 1983; van de Waterbeemd *et al.*, 1989; Tosato and Geladi, 1990). Even if these approaches have been successfully used, it must be pointed out that, from a practical point of view, none of them is completely satisfactory and there is a need for new tools providing complementary information or overcoming methodological problems. In this context, we recently proposed the use of an original graphical approach based on the nonlinear mapping (NLM) method (Domine *et al.*, 1994a,b; Devillers, 1995). We showed that it was possible to obtain easily interpretable nonlinear maps of aromatic and aliphatic substituent constants for the selection of test series and the derivation of SAR. In this case study, N2M has been applied to the data matrix of 103 aliphatic substituents (Table IV). The 103 aliphatic substituents (Table IV) were described by the hydrophobic constant for

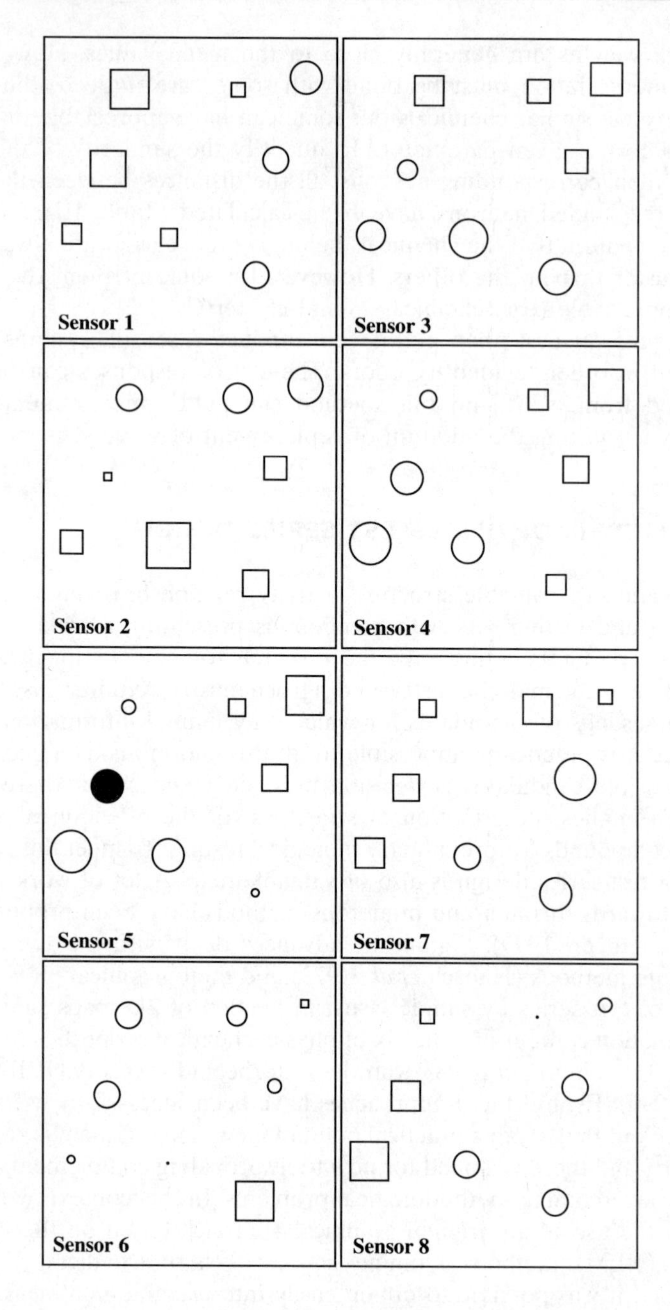

Figure 7 Plot of the mean values for the eight descriptors on each cluster of the N2M display (Figure 5). The mean values are calculated from the chemicals contained in the clusters. Squares (positive values) and circles (negative values) are proportional in size to the magnitude of the parameters.

Table III *Distances (calculated from scaled data) between the chemicals (1–30) and the clusters (A–H). Bold values indicate the memberships of chemicals to the clusters.*

Chemical no.	A	B	C	D	E	F	G	H
3	**0.7277**	2.067	3.004	1.336	2.816	2.183	2.423	3.431
5	**1.190**	2.003	2.805	1.395	2.442	2.155	2.166	3.097
26	**2.637**	2.861	3.402	3.425	3.624	4.251	3.982	4.382
17	1.355	**0.8626**	1.544	2.271	1.870	3.284	2.732	3.121
18	1.396	**0.5055**	1.247	2.070	1.312	3.037	2.306	2.588
19	2.054	**1.942**	2.341	2.492	2.498	3.251	3.003	3.664
20	2.962	1.994	**1.698**	3.627	2.499	4.511	3.860	4.029
21	2.994	1.741	**1.308**	3.693	2.095	4.611	3.664	3.410
22	3.319	2.026	**1.450**	4.021	2.464	4.933	4.053	3.866
23	2.752	1.638	**1.328**	3.496	2.102	4.434	3.617	3.559
24	4.091	2.732	**2.003**	4.689	2.816	5.534	4.503	4.044
25	4.300	2.941	**2.164**	4.713	2.654	5.455	4.343	3.844
28	4.965	4.317	**4.108**	5.086	4.365	5.517	5.054	5.306
1	1.417	2.827	3.764	**1.277**	3.328	1.705	2.412	3.560
2	1.532	2.840	3.750	**0.9910**	3.156	1.220	2.024	3.263
4	1.902	2.967	3.838	**1.476**	3.311	1.571	2.188	3.389
11	3.449	2.388	2.147	3.644	**1.951**	4.268	3.097	2.358
12	2.592	1.853	2.020	2.527	**1.160**	3.064	1.837	1.220
13	3.983	2.808	2.378	3.982	**2.209**	4.464	3.299	2.574
6	2.216	3.347	4.196	1.460	3.583	**1.158**	2.267	3.526
7	2.149	3.301	4.162	1.426	3.511	**1.222**	2.200	3.374
8	3.103	4.085	4.874	2.128	4.025	**1.321**	2.373	3.440
27	5.348	5.753	6.215	4.956	5.805	**4.740**	5.044	5.848
9	2.155	2.288	2.874	1.551	1.864	1.787	**0.9046**	1.314
29	3.791	4.186	4.748	2.938	3.732	2.458	**2.327**	2.522
30	4.550	4.950	5.460	3.547	4.306	2.815	**2.782**	2.978
10	2.620	2.475	2.905	1.978	1.815	2.069	0.9053	**0.8157**
14	4.089	3.392	3.313	3.660	2.321	3.782	2.542	**1.508**
15	4.206	3.640	3.635	3.798	2.703	3.904	2.793	**2.068**
16	3.815	3.222	3.238	3.455	2.272	3.640	2.455	**1.665**

aliphatic substituents Fr, the molar refractivity (MR), H-bonding acceptor (HBA) and donor (HBD) abilities and the inductive parameter F of Swain and Lupton, respectively. Fr, MR, and F values were centered (i.e., zero mean) and reduced (i.e., unit variance). HBA and HBD were not centered, and the '1' was replaced by a value yielding a unit variance. Different N2M analyses were performed. Indeed, several sizes of networks were assayed and run for 10 000 cycles (i.e., $ne_1 = 8000$ and $ne_2 = 2000$). During learning the topological radius was regularly decreased during the ne_1 epochs and trained for further 2000 cycles with different final radii (i.e., final $R = 0$, 1, and 2).

Figure 8 shows the nonlinear map of the 103 aliphatic substituents. A full interpretation of this map can be found in a previous paper (Domine *et al.*, 1994b). Briefly, we showed that the nonlinear map gave a full picture of the

Table IV *Aliphatic substituents.*

No.	Substituent	No.	Substituent
1	Br	2	Cl
3	F	4	I
5	NO_2	6	H
7	OH	8	SH
9	NH_2	10	CBr_3
11	CCl_3	12	CF_3
13	CN	14	SCN
15	CO_2^-	16	CO_2H
17	CH_2Br	18	CH_2Cl
19	CH_2I	20	$CONH_2$
21	CH=NOH	22	CH_3
23	$NHCONH_2$	24	OCH_3
25	CH_2OH	26	$SOCH_3$
27	OSO_2CH_3	28	SCH_3
29	$NHCH_3$	30	CF_2CF_3
31	C≡CH	32	CH_2CN
33	$CH{=}CHNO_2$-(*trans*)	34	$CH{=}CH_2$
35	$COCH_3$	36	$OCOCH_3$
37	CO_2CH_3	38	$NHCOCH_3$
39	$C{=}O(NHCH_3)$	40	CH_2CH_3
41	OCH_2CH_3	42	CH_2OCH_3
43	SOC_2H_5	44	SC_2H_5
45	$CH_2Si(CH_3)_3$	46	NHC_2H_5
47	$N(CH_3)_2$	48	CH=CHCN
49	Cyclopropyl	50	COC_2H_5
51	$CO_2C_2H_5$	52	$OCOC_2H_5$
53	$EtCO_2H$	54	$NHCO_2C_2H_5$
55	$CONHC_2H_5$	56	$NHCOC_2H_5$
57	$CH(CH_3)_2$	58	C_3H_7
59	$OCH(CH_3)_2$	60	OC_3H_7
61	$CH_2OC_2H_5$	62	SOC_3H_7
63	SC_3H_7	64	NHC_3H_7
65	$Si(CH_3)_3$	66	2-Thienyl
67	3-Thienyl	68	$CH{=}CHCOCH_3$
69	$CH{=}CHCO_2CH_3$	70	COC_3H_7
71	$OCOC_3H_7$	72	$CO_2C_3H_7$
73	$(CH_2)_3CO_2H$	74	$NHCOC_3H_7$
75	$CONHC_3H_7$	76	C_4H_9
77	$C(CH_3)_3$	78	OC_4H_9
79	$CH_2OC_3H_7$	80	NHC_4H_9
81	$N(C_2H_5)_2$	82	$CH{=}CHCOC_2H_5$
83	$CH{=}CHCO_2C_2H_5$	84	C_5H_{11}
85	$CH_2OC_4H_9$	86	C_6H_5
87	OC_6H_5	88	$SO_2C_6H_5$
89	NHC_6H_5	90	2-Benzthiazolyl
91	$CH{=}CHCOC_3H_7$	92	$CH{=}CHCO_2C_3H_7$
93	COC_6H_5	94	$CO_2C_6H_5$
95	$OCOC_6H_5$	96	$NHCOC_6H_5$
97	$CH_2C_6H_5$	98	$CH_2OC_6H_5$
99	$CH_2Si(C_2H_5)_3$	100	$CH{=}CHC_6H_5$-(*trans*)
101	$CH{=}CHCOC_6H_5$	102	Ferrocenyl
103	$N(C_6H_5)_2$		

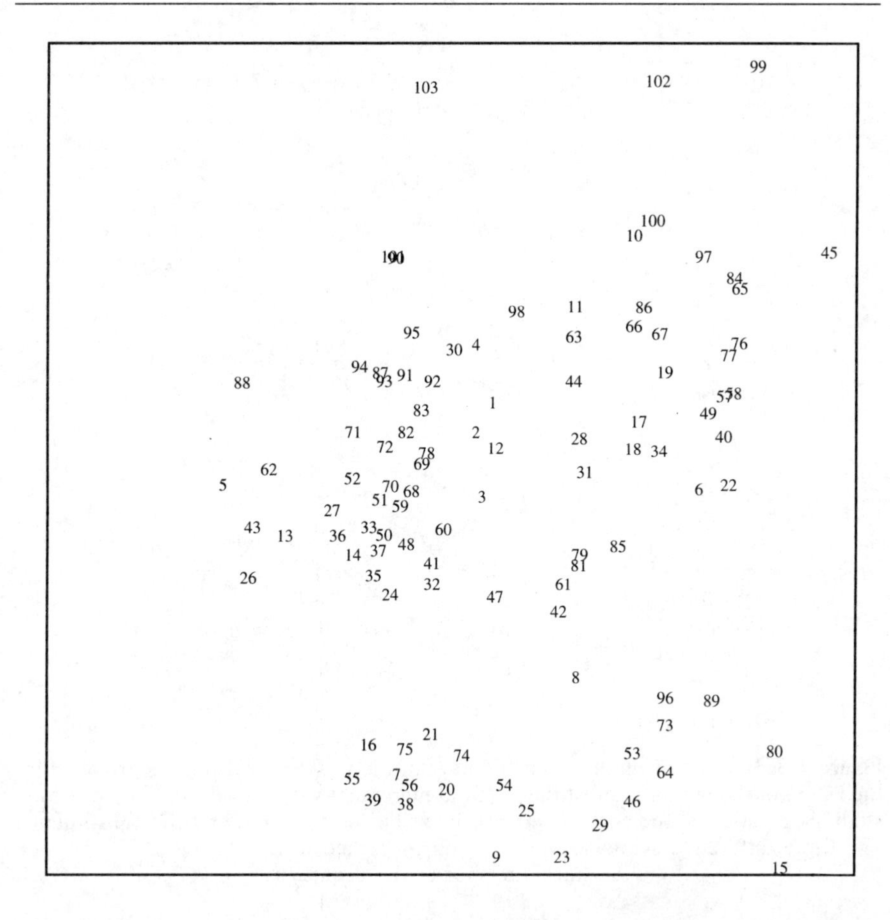

Figure 8 Nonlinear map of the 103 aliphatic substituents described by five substituent constants (Fr, HBA, HBD, MR, and F). See Table IV for correspondence between the numbers and the substituents.

data structure in the starting population and that it was chemically coherent, since the repartition of the substituents could be directly linked to their chemical structure. Figure 9 shows the 5 × 5 Kohonen map with a final topological radius of 0. The MST of the loaded neurons has also been represented to yield the 3MAP. From Figure 9, it is possible to select optimal test series by taking one or several substituents in each of the clusters formed. Comparatively with hierarchical cluster analysis (HCA) (Hansch and Leo, 1979), the Kohonen map provides information on the relationships between the clusters since it preserves the topology. In addition, representation of the MST to give the 3MAP provides supplementary information on the shortest distances between the loaded neurons. For comparison purposes, we have represented on the 3MAP, an example of substituent selection proposed in

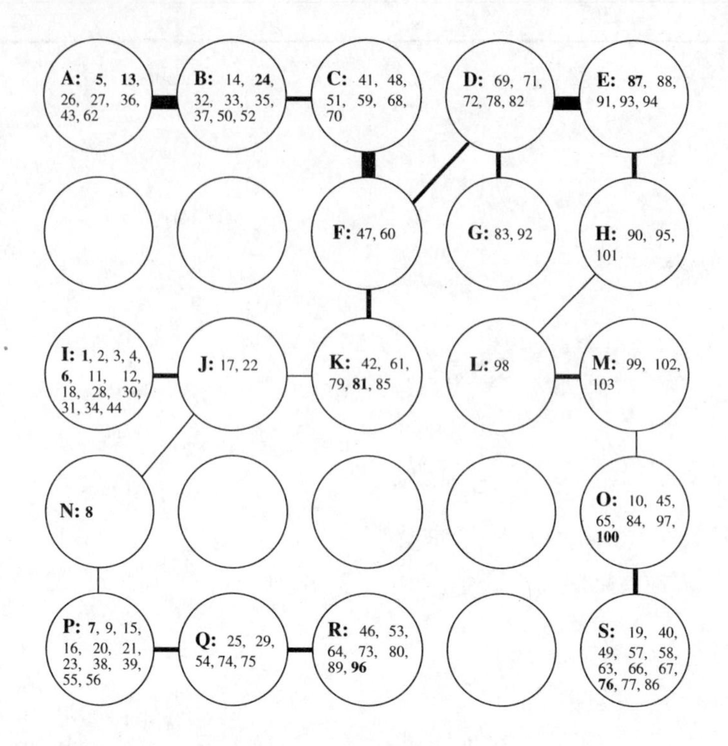

Figure 9 3MAP (i.e., Kohonen map + information on shortest distances provided by the MST) of the aliphatic substituents. Bold numbers are those selected in a previous study. See Table IV for correspondence between the numbers and the substituents. The three different thicknesses of the lines are linked to the distances. The thicker the lines, the shorter the distances.

a previous study (Domine *et al.*, 1994b). It is noteworthy that the results obtained are in accordance with those previously obtained since there is only one (or two) substituent(s) selected per neuron. There remain neurons with no selected substituents. These neurons can be used if a larger set of substituents is required. Figure 10, which shows the projection of the clusters on the nonlinear map (Figure 8), indicates that classical NLM and KSOM are complementary.

Indeed, KSOM allows us to define clusters which are generally coherent with the nonlinear map results. This may help to divide the nonlinear map into a given number of zones from which selection of optimal test series can be made, by picking up one or several substituents, depending on the population of the clusters and the repartition of the substituents on the map.

Figure 11 shows the nonlinear map of the weight vectors corresponding to the 3MAP presented in Figure 9. Inspection of Figure 11 indicates that the N2M display keeps the topology of the Kohonen map. Thus, for example,

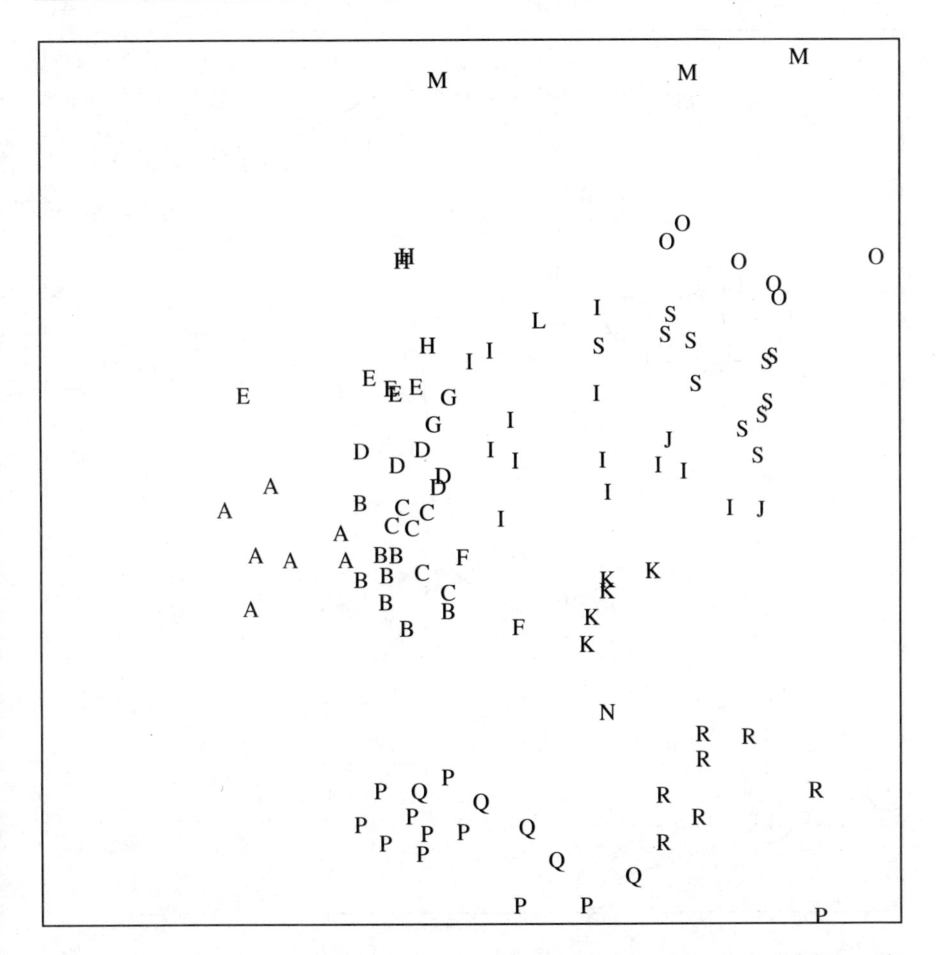

Figure 10 Representation of the KSOM clusters on the nonlinear map (Figure 8). See Figure 9 for correspondence between the letters and the clusters.

it can be noted that the alignment of neurons A to E is preserved on the NLM of the weight vectors. One can also note that neuron F is located between neurons C and K. The same remark can be made for all other neurons. The only slight divergence is observed for neuron I. Indeed, it can be noted that I is no longer between neurons A and N on the N2M display. In this case, the map, which preserved the actual distances between the neurons, lost the correct topology. However, it must be noted that this loss does not concern three adjacent neurons and that the alignment of the three adjacent neurons I, J, and K, for example, is kept.

In addition, the projection of the MST allows one to underline the shortest distances between the loaded neurons and therefore improves the visual

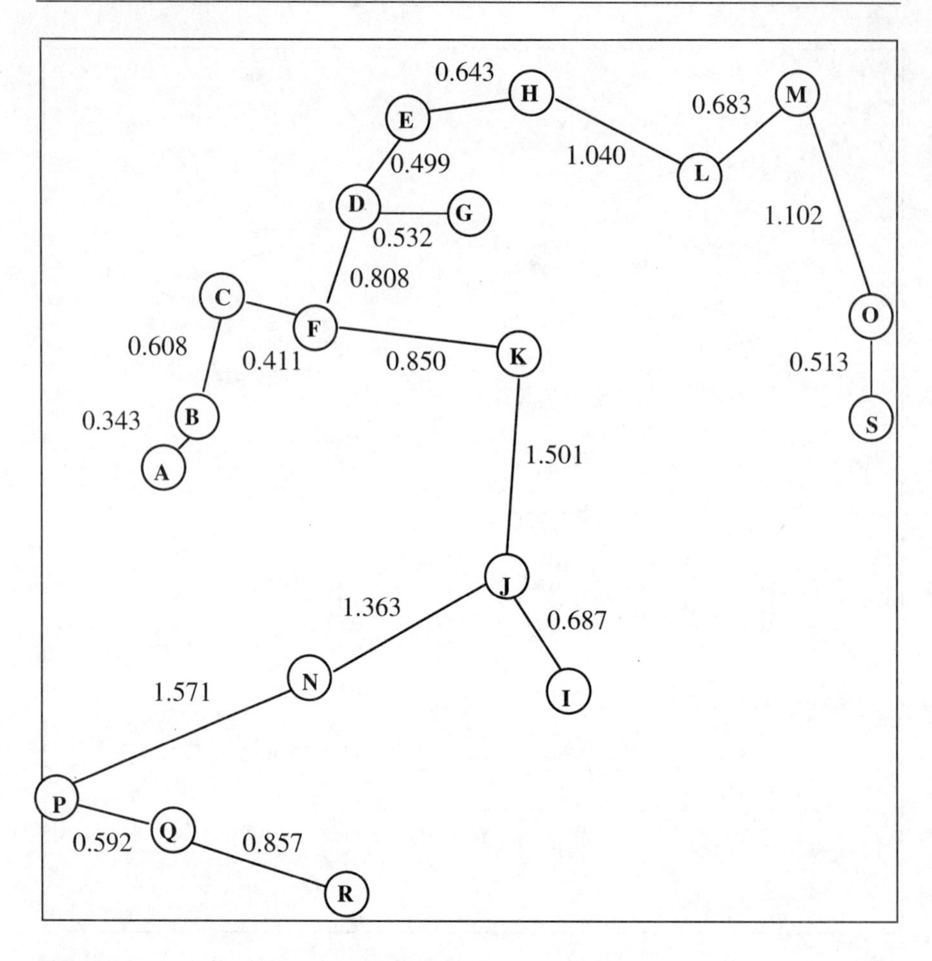

Figure 11 N2M display of aliphatic substituents. The values reported are the actual distances in the MST. See Figure 9 for correspondence between the letters and the clusters.

interpretation of the map. From a practical point of view, Figure 11 constitutes a useful basis for the simple visual selection of optimal test series. Indeed, by picking up one or several substituents in each of the clusters, it is possible to obtain a highly representative series of substituents in terms of physicochemical properties. In addition, if substituents present in a given cluster are difficult to synthesize or not useful for a QSAR study (due to their atypical character, as for cluster M), it is possible to select substituents in the nearest clusters from a simple visual inspection of the nonlinear neural map (Figure 11). Another advantage of the MST is that it can guide the precise selection of candidates. Thus, for example, if we consider that a

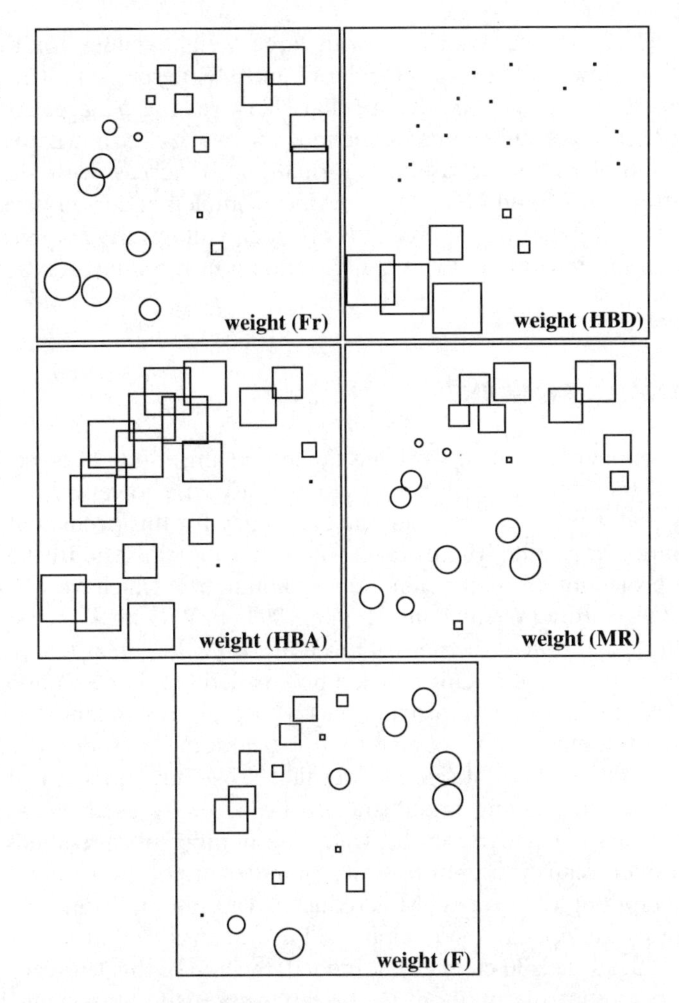

Figure 12 Plot of the weight values linked to the five descriptors on each cluster of the N2M display (Figure 11). Squares (positive values) and circles (negative values) are proportional in size to the magnitude of the weights.

substituent in cluster J (Figure 11) has to be replaced by a substituent in another cluster, the MST suggests cluster I. As underlined above, to facilitate the interpretation of the results and also the selection of optimal sets, graphical tools can be used. Thus, for example, it is possible to represent the weight vectors of neurons by means of collections of graphs where the larger the squares, the higher the values of the weights for each variable and the larger the circles, the lower the values of the weights (Figure 12). In Figure 12, we can underline that the neurons located in the bottom left-hand part of the map (i.e., P, Q, R) are associated with large weight values for the HBD descriptor and low weight values for the Fr and MR descriptors. The purpose of this case study was not to advance that N2M was the panacea for solving the complex task of substituent selection, but we have shown that it was a powerful tool providing useful information. It agrees with the results obtained from NLM and HCA and provides complementary information. It is also noteworthy that our approach is open, and allows the graphical representation of any information susceptible to help in the interpretation of the results.

CONCLUDING REMARKS

The pure Kohonen neural network has the interesting ability of preserving the topology of the high-dimensional variables space in a lower m-dimensional visual map. However, the cost that one has to pay for this projection is a loss of information regarding the distances between the clusters. In the present study, the projection of both a minimum spanning tree which was the basis of the 3MAP algorithm (Wienke and Hopke, 1994a,b; Wienke *et al.*, 1994, 1995) and the representation of the network weights using nonlinear mapping allow one to solve this problem. Our new method, called N2M, is an improvement of the 3MAP technique and provides easily interpretable maps. In addition, our results show that N2M, coupled to graphical tools, provides a useful and easy tool for the analysis of QSAR data tables. We have to admit that N2M requires a certain training in statistics to be correctly used. However, the method appears particularly useful when the number of individuals is high, since we obtain maps presenting a lower number of points (loaded neurons) and the computing time for NLM is reduced. Thus, in the former case study, N2M analysis can appear too complicated if we consider that the number of points is reduced (i.e., 30 chemicals). Indeed, the aim of this first example was basically to present the method. In the latter case study involving 103 substituents, N2M clearly shows its usefulness for simplifying the interpretation of large data matrices. Therefore, if we consider that in QSAR and drug design studies, one has to deal with larger and larger data tables, N2M will represent a useful tool in the kit of chemometric methods for treating these types of matrices.

REFERENCES

Abe, H., Yoshimura, T., Kanaya, S., Takahashi, Y., Miyashita, Y., and Sasaki, S.I. (1987). Automated odor-sensing system based on plural semiconductor gas sensors and computerized pattern recognition techniques. *Anal. Chim. Acta* **194,** 1–9.

Alunni, S., Clementi, S., Edlund, U., Johnels, D., Hellberg, S., Sjöström, M., and Wold,

S. (1983). Multivariate data analysis of substituent descriptors. *Acta Chem. Scand.* **B 37,** 47–53.

Andrea, T.A. and Kalayeh, H. (1991). Applications of neural networks in quantitative structure–activity relationships of dihydrofolate reductase inhibitors. *J. Med. Chem.* **34,** 2824–2836.

Anzali, S., Barnickel, G., Krug, M., Sadowski, J., Wagener, M., and Gasteiger, J. (1996). Evaluation of molecular surface properties using a Kohonen neural network. In, *Neural Networks in QSAR and Drug Design* (J. Devillers, Ed.). Academic Press, London, pp. 209–222.

Aoyama, T. and Ichikawa, H. (1991a). Basic operating characteristics of neural networks when applied to structure–activity studies. *Chem. Pharm. Bull.* **39,** 358–366.

Aoyama, T. and Ichikawa, H. (1991b). Obtaining the correlaton indices between drug activity and structural parameters using a neural network. *Chem. Pharm. Bull.* **39,** 372–378.

Aoyama, T. and Ichikawa, H. (1991c). Reconstruction of weight matrices in neural networks – A method of correlating outputs with inputs. *Chem. Pharm. Bull.* **39,** 1222–1228.

Aoyama, T. and Ichikawa, H. (1992). Neural network as nonlinear structure–activity relationship analyzers. Useful functions of the partial derivative method in multilayer neural networks. *J. Chem. Inf. Comput. Sci.* **32,** 492–500.

Aoyama, T., Suzuki, Y., and Ichikawa, H. (1990). Neural networks applied to structure–activity relationships analysis. *J. Med. Chem.* **33,** 2583–2590.

Arrigo, P., Giuliano, F., Scalia, F., Rapallo, A., and Damiani, G. (1991). Identification of a new motif on nucleic acid sequence data using Kohonen's self-organizing map. *Comput. Appl. Biosci.* **7,** 353–357.

Barlow, T.W. (1995). Self-organizing maps and molecular similarity. *J. Mol. Graphics* **13,** 24–27.

Barthélemy, J.P. and Guénoche, A. (1988). *Les Arbres et les Représentations des Proximités.* Masson, Paris, p. 239.

Bienfait, B. (1994). Applications of high-resolution self-organizing maps to retrosynthetic and QSAR analysis. *J. Chem. Inf. Comput. Sci.* **34,** 890–898.

Budzinski, H., Garrigues, P., Connan, J., Devillers, J., Domine, D., Radke, M., and Oudin, J.L. (1995). Alkylated phenanthrene distributions as maturity and origin indicators in crude oils and rock extracts. *Geochim. Cosmochim. Acta* **59,** 2043–2056.

Cambon, B. and Devillers, J. (1993). New trends in structure-biodegradability relationships. *Quant. Struct.-Act. Relat.* **12,** 49–56.

Carpenter, G.A. and Grossberg, S. (1991). *Pattern Recognition by Self-Organizing Networks*. MIT Press, Cambridge, Massachusetts, p. 691.

Chastrette, M., de Saint Laumer, J.Y., and Peyraud, J.F. (1993). Adapting the structure of a neural network to extract chemical information. Application to structure–odour relationships. *SAR QSAR Environ. Res.* **1,** 221–231.

Chastrette, M., Devillers, J., Domine, D., and de Saint Laumer, J.Y. (1994). New tools for the selection and critical analysis of large collections of data. In, *New Data Challenges in Our Information Age. Thirteenth International CODATA Conference, Beijing, China, October 1992* (P.S. Glaeser and M.T.L. Millward, Eds.). CODATA, Ann Arbor, pp. C29–C35.

Cheriton, D. and Tarjan, R.E. (1976). Finding minimum spanning trees. *SICOMP* **5,** 724–742.

Court, J.P., Murgatroyd, R.C., Livingstone, D.J., and Rahr, E. (1988). Physicochemical characteristics of non-electrolytes and their uptake by *Brugia pahangi* and *Dipetalonema viteae. Mol. Biochem. Parasitol.* **27,** 101–108.

de Saint Laumer, J.Y., Chastrette, M., and Devillers, J. (1991). Multilayer neural networks applied to structure–activity relationships. In, *Applied Multivariate Analysis in SAR and Environmental Studies* (J. Devillers and W. Karcher, Eds.). Kluwer Academic Publishers, Dordrecht, pp. 479–521.

Devillers, J. (1993). Neural modelling of the biodegradability of benzene derivatives. *SAR QSAR Environ. Res.* **1,** 161–167.

Devillers, J. (1995). Display of multivariate data using nonlinear mapping. In, *Chemometric Methods in Molecular Design* (H. van de Waterbeemd, Ed.). VCH, Weinheim, pp. 255–263.

Devillers, J. and Cambon, B. (1993). Modeling the biological activity of PAH by neural networks. *Polycyclic Aromatic Compounds* **3 (supp),** 257–265.

Devillers, J. and Domine, D. (1995). Deriving structure–chemoreception relationships from the combined use of linear and nonlinear multivariate analyses. In, *QSAR and Molecular Modelling: Concepts, Computational Tools and Biological Applications* (F. Sanz, J. Giraldo, and F. Manaut, Eds). J.R. Prous Science Publishers, Barcelona, pp. 57–60.

Devillers, J., Domine, D., and Bintein, S. (1994). Multivariate analysis of the first 10 MEIC chemicals. *SAR QSAR Environ. Res.* **2,** 261–270.

Devillers, J., Domine, D., and Boethling, R.S. (1996a). Use of neural networks and autocorrelation descriptors for predicting the biodegradation of organic chemicals. In, *Neural Networks in QSAR and Drug Design* (J. Devillers, Ed.). Academic Press, London, pp. 65–82.

Devillers, J. and Doré, J.C. (1989). Heuristic potency of the minimum spanning tree (MST) method in toxicology. *Ecotoxicol. Environ. Safety* **17,** 227–235.

Devillers, J., Guillon, C., and Domine, D. (1996b). A neural structure–odor threshold model for chemicals of environmental and industrial concern. In, *Neural Networks in QSAR and Drug Design* (J. Devillers, Ed.). Academic Press, London, pp. 97–117.

Devillers, J. and Karcher, W. (1991). *Applied Multivariate Analysis in SAR and Environmental Studies*. Kluwer Academic Publishers, Dordrecht, p. 530.

Domine, D. and Devillers, J. (1995). Nonlinear multivariate SAR of Lepidoptera pheromones. *SAR QSAR Environ. Res.* **4,** 51–58.

Domine, D., Devillers, J., and Chastrette, M. (1994a). A nonlinear map of substituent constants for selecting test series and deriving structure–activity relationships. 1. Aromatic series. *J. Med. Chem.* **37,** 973–980.

Domine, D., Devillers, J., and Chastrette, M. (1994b). A nonlinear map of substituent constants for selecting test series and deriving structure–activity relationships. 2. Aliphatic series. *J. Med. Chem.* **37,** 981–987.

Domine, D., Devillers, J., Chastrette, M., and Doré, J.C. (1995). Combined use of linear and nonlinear multivariate analyses in structure–activity relationship studies: Application to chemoreception. In, *Computer-Aided Molecular Design. Applications in Agrochemicals, Materials, and Pharmaceuticals* (C.H. Reynolds, M.K. Holloway, and H.K. Cox, Eds.). ACS Symposium Series No. 589, American Chemical Society, Washington, DC, pp. 267–280.

Domine, D., Devillers, J., Chastrette, M., and Karcher, W. (1993a). Nonlinear mapping for structure–activity and structure–property modelling. *J. Chemometrics* **7,** 227–242.

Domine, D., Devillers, J., Chastrette, M., and Karcher, W. (1993b). Estimating pesticide field half-lives from a backpropagation neural network. *SAR QSAR Environ. Res.* **1,** 211–219.

Domine, D., Devillers, J., Garrigues, P., Budzinski, H., Chastrette, M., and Karcher, W. (1994c). Chemometrical evaluation of the PAH contamination in the sediments of the Gulf of Lion (France). *Sci. Total Environ.* **155,** 9–24.

Dove, S., Streich, W.J., and Franke, R. (1980). On the rational selection of test series. 2. Two-dimensional mapping of intraclass correlation matrices. *J. Med. Chem.* **23,** 1456–1459.

Eberhart, R.C. and Dobbins, R.W. (1990). *Neural Network PC Tools. A Practical Guide.* Academic Press, San Diego, p. 414.

Everitt, B.S. (1978). *Graphical Techniques for Multivariate Data.* Heinemann Educational Books, London, p. 117.

Ferran, E.A. and Ferrara, P. (1991). Topological maps of protein sequences. *Biol. Cybern.* **65,** 451–458.

Ferran, E.A. and Ferrara, P. (1992a). Clustering proteins into families using artificial neural networks. *Comput. Appl. Biosci.* **8,** 39–44.

Ferran, E.A. and Ferrara, P. (1992b). A neural network dynamics that resembles protein evolution. *Physica A* **185,** 395–401.

Ferran, E.A. and Pflugfelder, B. (1993). A hybrid method to cluster protein sequences based on statistics and artificial neural networks. *Comput. Appl. Biosci.* **9,** 671–680.

Feuilleaubois, E., Fabart, V., and Doucet, J.P. (1993). Implementation of the three-dimensional-pattern search problem on Hopfield-like neural networks. *SAR QSAR Environ. Res.* **1,** 97–114.

Florek, K., Lukaszewicz, J., Perkal, H., Steinhaus, H., and Zubrzycki, S. (1951). Sur la liaison et la division des points d'un ensemble fini. *Colloq. Math.* **2,** 282–285.

Hamad, D. and Betrouni, M. (1995). Artificial neural networks for nonlinear projection and exploratory data analysis. In, *Artificial Neural Nets and Genetic Algorithms* (D.W. Pearson, N.C. Steele, and R.F. Albrecht, Eds.). Springer-Verlag, Wien, pp. 164–167.

Hansch, C. and Leo, A. (1979). *Substituent Constants for Correlation Analysis in Chemistry and Biology*. John Wiley & Sons, New York, p. 339.

Hansch, C., Unger, S.H., and Forsythe, A.B. (1973). Strategy in drug design. Cluster analysis as an aid in the selection of substituents. *J. Med. Chem.* **16,** 1217–1222.

Hecht-Nielsen, R. (1988). Applications of counterpropagation networks. *Neural Networks* **1**, 131–139.

Hopfield, J.J. (1982). Neural networks and physical systems with emergent collective computational abilities. *Proc. Natl Acad. Sci. U.S.A.* **74,** 2554–2558.

Hudson, B., Livingstone, D.J., and Rahr, E. (1989). Pattern recognition display methods for the analysis of computed molecular properties. *J. Comput.-Aided Mol. Design* **3,** 55–65.

Jarnik, V. (1936). Ojistém problém minimalnim. *Prace Mor. Prirodovèd Spol. Brne* **6,** 63–75.

Kateman, G. and Smits, J.R.M. (1993). Colored information from a black box? Validation and evaluation of neural networks. *Anal. Chim. Acta* **277,** 179–188.

Klein, R.W. and Dubes, R.C. (1989). Experiments in projection and clustering by simulated annealing. *Pattern Recognition* **22,** 213–220.

Kohonen, T. (1989a). *Self-Organization and Associated Memory*. Springer-Verlag, Heidelberg.

Kohonen, T. (1989b). Speech recognition based on topology-preserving neural maps. In, *Neural Computing Architectures. The Design of Brain-Like Machines* (I. Aleksander, Ed.). MIT Press, Cambridge MA, pp. 26–40.

Kohonen, T. (1990). The self-organizing map. *Proc. IEEE* **78,** 1464–1480.

Kohonen, T. (1992). Learning vector quantisation and the self organising map. In, *Theory and Applications of Neural Networks* (J.G. Taylor and C.L.T. Mannion, Eds.). Springer-Verlag, London, pp. 235–242.

Kohonen, T., Hynninen, J., Kangas, J., and Laaksonen, J. (1995). *SOM_PAK. The Self-Organizing Map Program Package,* Version 3.1 (April 7, 1995). SOM Programming Team of the Helsinki University of Technology, Laboratory of Computer and Information Science, Espoo, Finland.

Kowalski, B.R. and Bender, C.F. (1972). Pattern recognition. A powerful approach to interpreting chemical data. *J. Am. Chem. Soc.* **94,** 5632–5639.

Kowalski, B.R. and Bender, C.F. (1973). Pattern recognition. II. Linear and nonlinear methods for displaying chemical data. *J. Am. Chem. Soc.* **95,** 686–692.

Kruskal, J.B. (1956). On the shortest spanning subtree of a graph and the travelling salesman problem. *Proc. Amer. Math. Soc.* **7,** 48–50.

Kruskal, J.B. (1964). Multidimensional scaling by optimizing goodness of fit to a nonmetric hypothesis. *Psychometrika* **29,** 1–27.

Kruskal, J.B. (1971). Comments on 'a nonlinear mapping for data structure analysis'. *IEEE Trans. Comput.* **C20,** 1614.

Lebart, L., Morineau, A., and Fenelon, J.P. (1979). *Traitement des Données Statistiques. Méthodes et Programmes*. Dunod, Paris, p. 510.

Liu, Q., Hirono, S., and Moriguchi, I. (1992a). Application of functional-link net in QSAR. I. QSAR for activity data given by continuous variate. *Quant. Struct.–Act. Relat.* **11,** 135–141.

Liu, Q., Hirono, S., and Moriguchi, I. (1992b). Application of functional-link net in QSAR. II. QSAR for activity data given by ratings. *Quant. Struct.–Act. Relat.* **11,** 318–324.

Livingstone, D.J. (1989). Multivariate quantitative structure–activity relationship (QSAR) methods which may be applied to pesticide research. *Pestic. Sci.* **27,** 287–304.

Livingstone, D.J., Ford, M.G., and Buckley, D.S. (1988). A multivariate QSAR study of pyrethroid neurotoxicity based upon molecular parameters derived by computer chemistry. In, *Neurotox'88: Molecular Basis of Drug and Pesticide Action* (G.G. Lunt, Ed.). Elsevier, Amsterdam, pp. 483–495.

Livingstone, D.J. and Manallack, D.J. (1993). Statistics using neural networks: Chance effects. *J. Med. Chem.* **36,** 1295–1297.

Livingstone, D.J. and Salt, D.W. (1992). Regression analysis for QSAR using neural networks. *Bioorg. Med. Chem. Lett.* **2,** 213–218.

Lozano, J., Novic, M., Rius, F.X., and Zupan, J. (1995). Modelling metabolic energy by neural networks. *Chemom. Intell. Lab. Syst.* **28,** 61–72.

Manallack, D.T. and Livingstone, D.J. (1994). Neural networks and expert systems in molecular design. Neural networks – A tool for drug design. In, *Advanced Computer-Assisted Techniques in Drug Discovery* (H. van de Waterbeemd, Ed.). VCH, Weinheim, pp. 293–318.

Martin-del-Brio, B., Medrano-Marqués, N., and Blasco-Alberto, J. (1995). Feature map architectures for pattern recognition: Techniques for automatic region selection. In, *Artificial Neural Nets and Genetic Algorithms* (D.W. Pearson, N.C. Steele, and R.F. Albrecht, Eds.). Springer-Verlag, Wien, pp. 124–127.

Melssen, W.J., Smits, J.R.M., Rolf, G.H., and Kateman, G. (1993). 2-Dimensional mapping of IR spectra using a parallel implemented self-organizing feature map. *Chemom. Intell. Lab. Syst.* **18,** 195–204.

Pao, Y.H. (1989). *Adaptive Pattern Recognition and Neural Networks*. Addison-Wesley Publishing Company, Reading, p. 309.

Peterson, K.L. (1992). Counter-propagation neural networks in the modeling and prediction of Kovats indices for substituted phenols. *Anal. Chem.* **64,** 379–386.

Peterson, K.L. (1995). Quantitative structure–activity relationships in carboquinones and benzodiazepines using counter-propagation neural networks. *J. Chem. Inf. Comput. Sci.* **35,** 896–904.

Pleiss, M.A. and Unger, S.H. (1990). The design of test series and the significance of

QSAR relationships. In, *Comprehensive Medicinal Chemistry*, Vol. 4 (C.A. Ramsden, Ed.). Pergamon Press, Oxford, pp. 561–587.

Polak, E. (1971). *Computational Methods in Optimization, a Unified Approach*. Academic Press, New York.

Prim, R.C. (1957). Shortest connection networks and some generalizations. *Bell Syst. Technol. J.* **36,** 1389–1401.

Rose, V.S., Croall, I.F., and MacFie, H.J.H. (1991). An application of unsupervised neural network methodology (Kohonen Topology Preserving Mapping) to QSAR analysis. *Quant. Struct.–Act. Relat.* **10,** 6–15.

Rose, V.S., Hyde, R.M., and MacFie, H.J.H. (1990). U.K. usage of chemometrics and artificial intelligence in QSAR analysis. *J. Chemometrics* **4,** 355–360.

Roux, M. (1975). Note sur l'arbre de longueur minima. *Rev. Stat. Appl.* **23,** 29–35.

Rumelhart, D.E., Hinton, G.E., and Williams, R.J. (1986). Learning representations by back-propagating errors. *Nature* **323,** 533–536.

Sammon, J.W. (1969). A nonlinear mapping for data structure analysis. *IEEE Trans. Comput.* **C18,** 401–409.

Simon, V., Gasteiger, J., and Zupan, J. (1993). A combined application of two different neural network types for the prediction of chemical reactivity. *J. Am. Chem. Soc.* **115**, 9148–9159.

Thoreau, E. (1994). *Modélisation par Homologie du Domaine de Liaison à l'Hormone des Récepteurs Hormonaux Nucléaires à Partir de l'Hypothèse 'Serpin'. Contributions au Développement de la Méthode HCA*. Ph.D. Thesis, Paris VII, France.

Tosato, M.L. and Geladi, P. (1990). Design: A way to optimize testing programmes for QSAR screening of toxic substances. In, *Practical Applications of Quantitative Structure-Activity Relationships (QSAR) in Environmental Chemistry and Toxicology* (W. Karcher and J. Devillers, Eds.). Kluwer Academic Publishers, Dordrecht, pp. 317–341.

Tusar, M., Zupan, J., and Gasteiger, J. (1992). Neural networks and modelling in chemistry. *J. Chim. Phys.* **89,** 1517–1529.

van de Waterbeemd, H. (1994). *Advanced Computer-Assisted Techniques in Drug Discovery*. VCH, Weinheim.

van de Waterbeemd, H. (1995). *Chemometric Methods in Molecular Design*. VCH, Weinheim, p. 359.

van de Waterbeemd, H., El Tayar, N., Carrupt, P.A., and Testa, B. (1989). Pattern recognition study of QSAR substituent descriptors. *J. Comput.-Aided Mol. Design* **3,** 111–132.

Wasserman, P.D. (1989). *Neural Computing: Theory and Practice*. Van Nostrand Reinhold, New York, p. 230.

Wienke, D., Gao, N., and Hopke, P.K. (1994). Multiple site receptor modeling with a minimal spanning tree combined with a neural network. *Environ. Sci. Technol.* **28,** 1023–1030.

Wienke, D. and Hopke, P.K. (1994a). Projection of Prim's minimal spanning tree into a Kohonen neural network for identification of airborne particle sources by their multielement trace patterns. *Anal. Chim. Acta* **291,** 1–18.

Wienke, D. and Hopke, P.K. (1994b). Visual neural mapping technique for locating fine airborne particles sources. *Environ. Sci. Technol.* **28,** 1015–1022.

Wienke, D., Xie, Y., and Hopke, P.K. (1995). Classification of airborne particles by analytical SEM imaging and a modified Kohonen neural network (3MAP). *Anal. Chim. Acta* **310,** 1–14.

Wiese, M. and Schaper, K.J. (1993). Application of neural networks in the QSAR analysis of percent effect biological data: Comparison with adaptive least squares and nonlinear regression analysis. *SAR QSAR Environ. Res.* **1,** 137–152.

Wish, M. and Carroll, J.D. (1982). Multidimensional scaling and its applications. In, *Handbook of Statistics,* Vol. 2 (P.R. Krishnaiah and L.N. Kanal, Eds.). North-Holland Publishing Company, Amsterdam, pp. 317–345.

Zahn, C.T. (1971). Graph-theoretical methods for detecting and describing gestalt clusters. *IEEE Trans. Comput.* **C20,** 68–86.

Zeidenberg, M. (1990). *Neural Networks in Artificial Intelligence*. Ellis Horwood Limited, Chichester, p. 268.

Zupan, J. and Gasteiger, J. (1993). *Neural Networks for Chemists. An Introduction.* VCH, Weinheim, p. 305.

Zupan, J., Novic, M., Li, X., and Gasteiger, J. (1994). Classification of multicomponent analytical data of olive oils using different neural networks. *Anal. Chim. Acta* **292,** 219–234.

11 Combining Fuzzy Clustering and Neural Networks to Predict Protein Structural Classes

G.M. MAGGIORA*, C.T. ZHANG†, K.C. CHOU, and D.W. ELROD
Upjohn Laboratories, Kalamazoo, MI 49007–4940, USA

A procedure for predicting protein structural classes (viz., all-α, all-β, α + β, and α/β) from amino acid composition, based upon the combined use of fuzzy c-means (FCM) clustering and multi-layer perceptrons (MLP) is described. Unlike the case in classical, crisp clusters, objects can belong, with varying degrees between zero and one, to more than one fuzzy cluster. The prediction procedure is implemented in two phases, training and testing. In the training phase, FCM clustering is first applied to a training set of proteins, where each protein is described by a 20-component vector in 'amino-acid composition' space. After clustering, each protein is represented by a 4-component vector whose elements provide a measure of the degree that a given protein belongs to one of the four fuzzy clusters which represent the structural classes. The 4-component vectors are now used as input to an MLP, and the network is trained. In the testing phase, a new set of proteins is presented to the combined FCM-clustering/MLP system which generates a prediction in a single step. The results obtained by this hybrid 'neuro-fuzzy' procedure are comparable in quality to the best results obtained by other workers. Moreover, addition of the MLP improves the strength of the predictions over those obtained using FCM clustering alone.

KEY WORDS: *fuzzy clusters; multi-layer perceptron; neural network; protein structural classes.*

---

*Author to whom all correspondence should be addressed
†Permanent address: Department of Physics, Tianjin University, Tianjin, China

In, *Neural Networks in QSAR and Drug Design* (J. Devillers, Ed.)
Academic Press, London, 1996, pp. 255–279.
ISBN 0-12-213815-5

INTRODUCTION

As knowledge of the human genome continues to grow, a watershed of protein sequence data has become available. Now more than thirty thousand sequences (Anthonsen *et al.*, 1994) can be searched and analyzed. However, while sequence data are quite useful, knowledge of the three-dimensional structural features of proteins is considerably more useful, especially from the perspective of structure–function relationships. Even though there has been a significant increase in efforts to determine the three-dimensional structure of proteins using sophisticated techniques such as X-ray crystallography and multi-dimensional NMR, the gap between primary and tertiary structural data continues to widen at an increasing rate. Methods are available, however, which can provide lower resolution structural information that may still be of value in many instances. One such type of low resolution data is that relating to a protein's structural class or folding type (e.g., all-α, all-β, and α/β) as distinct from its functional class (e.g., globins, dehydrogenases, and cytochromes). In fact, under some circumstances prediction of a protein's structural class may be an important first step in helping to classify it, particularly in situations of low sequence similarity.

A number of methods have been developed to predict a protein's structural class (Klein, 1986; Klein and Delisi, 1986; Nakashima *et al.*, 1986; Chou, 1989; Chou and Zhang, 1992; Genfa *et al.*, 1992; Zhang and Chou, 1992; Dubchak *et al.*, 1993; Chou and Zhang, 1994; Mao *et al.*, 1994; Chandonia and Karplus, 1995; Chou, 1995; Zhang *et al.*, 1995). Most of the methods rely on data related to amino-acid composition, although other descriptors are also used (Klein, 1986; Klein and Delisi, 1986; Dubchak *et al.*, 1993; Chandonia and Karplus, 1995). Surprisingly, relatively good predictions have been made based solely upon amino-acid composition. Recently, an investigation by Mao *et al.* (1994) showed the existence of considerable overlap of the structural classes in 'amino-acid composition' space, clearly indicating a lack of well-separated clusters. This is further supported by the work of Chou and Zhang (1994) on both real proteins and a large number of 'simulated' proteins, which showed that overlapping clusters in amino-acid composition space still remain a problem even for a variety of distance metrics. In light of these observations, fuzzy clusters may provide a more appropriate description of the underlying data relationships. Since the elements of fuzzy clusters, proteins here, admit to degrees of membership in more than one cluster, elements that lie in the boundary regions of clusters can be treated naturally – there is no need to specify membership in one and only one cluster.

A study by Zhang *et al.* (1995) using fuzzy c-means (FCM) clustering (Bezdek, 1981) supports this view. In fact, from their results these workers were able to make predictions of protein structural classes that rival those of the best methods developed to date (*vide supra*). The current work is an attempt to improve upon the results obtained in that study through the use of computational neural networks (CNN), where the CNN acts as a device for further processing the results obtained initially by the FCM clustering procedure.

Hybrid approaches have been employed in a number of neural network studies. In many cases, Kohonen self-organizing maps (SOMs) (Kohonen, 1989) are used first, and the output obtained is then fed through some type of CNN, usually a multi-layer perceptron (MLP) (Pham and Bayro-Corrochano, 1994). In these studies, the SOM carries out an unsupervised clustering of the data, grouping similar objects into similar regions of the map space, thereby effecting a reduction-in-dimension and simplification of the original data. The processed data is then generally passed through an MLP, which uses supervised learning to classify the objects. While this process is similar to that reported here, there is one significant difference. It is well known that the behavior of SOMs, since they learn sequentially, is dependent upon the order in which the data is presented (Duda and Hart, 1973). This is not the case with FCM clustering, which treats all of the data simultaneously. Edge effects also pose a problem for SOMs, since the topology of the nodal structure may not be fully compatible with the inherent structure in the data (Zupan and Gasteiger, 1993). This is not a problem with FCM clustering. Thus, on this basis alone, FCM clustering appears to be superior to SOMs as an initial stage of a two-stage procedure. One additional advantage of FCM clustering over SOMs is that relational data, such as similarities, dissimilarities, or distances, can be used in addition to vector data as input (Hathaway *et al.*, 1989; Hathaway and Bezdek, 1994). While such data are not used here, it nevertheless clearly shows the greater flexibility of FCM clustering over SOMs.

METHODOLOGY

Fuzzy set basics

A brief overview is provided of fuzzy sets, which are the basis for the fuzzy clustering utilized here. There are a number of books, at varying levels of difficulty, that can be consulted for more details (Kaufmann, 1975; Klir and Folger, 1988; Zimmermann, 1991; Bandemer and Nather, 1992; Ross, 1995).

Fuzzy sets differ from classical or *crisp* sets in that the elements of a fuzzy set admit to *degrees of membership* in the set. This is in sharp contrast to crisp sets, whose elements either belong to or do not belong to the set – fuzzy sets are generalizations of *crisp* sets. The degree of membership of an element x in a fuzzy set A is usually given in terms of a *membership function* $\mathrm{m}_A(x)$, all of whose values lie in the range [0,1], i.e., $0 \leq \mathrm{m}_A(x) \leq 1$.

Table I *Tabular representation of the membership function for fuzzy set B.*

	x_1	x_2	x_3	x_4
$m_B(x)$	0.3	0.0	0.7	1.0

Mathematically, a fuzzy set is given as a set of *ordered pairs* $(x, m_A(x))$

$$A = \{(x, m_A(x)) \mid x \in X, m_A(x) \in [0, 1]\} \tag{1}$$

where $X = \{ x \mid x \in X \}$ is a crisp set called the *universe of discourse* or *frame of discernment*.

As a concrete example, consider the fuzzy set B from the universe of discourse $X = \{x_1, x_2, x_3, x_4\}$

$$B = \{(x_1, 0.3), (x_2, 0), (x_3, 0.7), (x_4, 1)\} \tag{2}$$

As shown in Table I, Eq. (2) can also be written in a tabular form that illustrates the functional nature of the definition of B.

Essentially all of the set operations such as union ($\cup$), intersection ($\cap$), and complementation ($\bullet^c$) have fuzzy analogs, but their definitions are somewhat unfamiliar. Table II summarizes the results obtained when these definitions are applied to the general fuzzy sets A and B. Examples illustrating the application of these three set operations to two actual fuzzy sets, C and D, defined on the universe of discourse $X = \{x_1, x_2, x_3, x_4\}$ are given in Table III.

While continuous fuzzy sets are important in many application areas such as control systems, they are not important in FCM clustering and thus, will not be discussed. For detailed coverage of continuous fuzzy sets see any of the texts cited at the beginning of this section.

Table II *Tabular summary of three important fuzzy set operations.*

Union ($\cup$)	$m_{A \cup B}(x) = \text{Max}[m_A(x), m_B(x)]$
Intersection ($\cap$)	$m_{A \cap B}(x) = \text{Min}[m_A(x), m_B(x)]$
Complementation ($\bullet^c$)	$m_{A^c}(x) = 1 - m_A(x)$

Table III *Examples of fuzzy set union ($\cup$), intersection ($\cap$), and complementation ($\bullet^c$).*

	x_1	x_2	x_3	x_4
$m_C(x)$	0.2	0.7	0.0	0.5
$m_D(x)$	1.0	0.1	0.3	0.8
$m_{C \cup D}(x)$	1.0	0.7	0.3	0.8
$m_{C \cap D}(x)$	0.2	0.1	0.0	0.5
$m_{C^c}(x)$	0.8	0.3	1.0	0.5

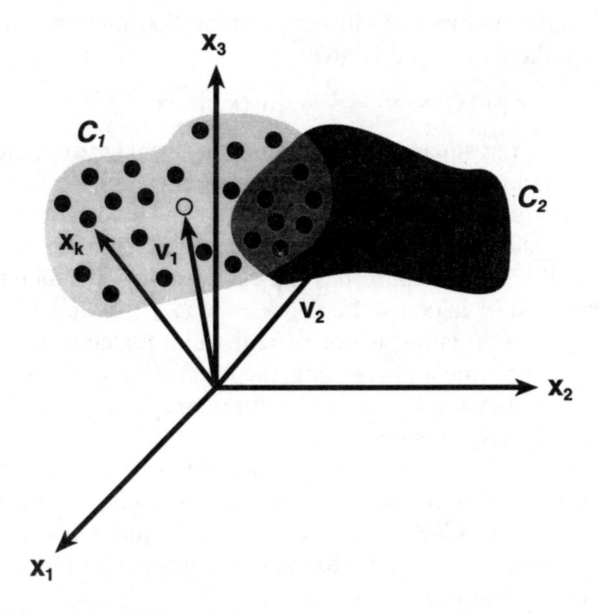

Figure 1 A portrayal of 'object' data in an abstract three-dimensional Euclidean feature space. In this space, an object is depicted as a point (•) described by a vector $x_k = [\mathbf{x}_{k1}, \mathbf{x}_{k2}, \mathbf{x}_{k3}]$, each of whose components represent a 'feature' of the object. The shaded areas represent two fuzzy clusters, C_1 and C_2. In contrast to classical clusters, there is a substantial region where the two clusters overlap. Every object in this region has a finite membership–function value for both fuzzy clusters. The centers of the two clusters depicted as open circles (○) are given by the vectors $\mathbf{v}_i = [v_{i1}, v_{i2}, v_{i3}]$, $i = 1, 2$. Note that the cluster centers need not fall on a data point.

Fuzzy c-means clustering

As discussed earlier, fuzzy ('soft') clustering as opposed to the usual crisp ('hard') clustering offers a number of advantages when there is significant 'overlap' of the data being clustered. Figure 1 illustrates this for a simple case, where the data are represented by feature vectors in a three-dimensional Euclidean space, $\mathbb{R}^3$. Each of the n datapoints is represented by vector $\mathbf{x}_k$, $k = 1, 2, \ldots, n$, with components x_{kl}, $l = 1, 2, 3$, while the centers of clusters C_1 and C_2 are represented by the vectors $\mathbf{v}_1$ and $\mathbf{v}_2$, respectively, with components v_{ij}, $i = 1, 2$; $j = 1, 2, 3$. As is clear from the figure, there is substantial overlap between the two fuzzy clusters C_1 and C_2. Note that the fuzzy clusters are fuzzy sets.

Fuzzy c-means clustering is similar in form to its crisp analog (Bezdek, 1981). To determine fuzzy clusters using this approach requires that a functional J be minimized with respect to the location of the cluster centers $\mathbf{v}_i$ and the magnitude of the cluster membership functions $\mathrm{m}_i(\mathbf{x}_k)$, where $i = 1,$

2, ... , c, c being the number of clusters, a value that must be chosen by the investigator. Mathematically, J is given by

$$J\,[\mathrm{m}(\mathbf{x}), \mathbf{v}, \kappa] = \sum_k \sum_i \,[\mathrm{m}_i(\mathbf{x}_k)]^\kappa \,\|\mathbf{x}_k - \mathbf{v}_i\|^2 \tag{3}$$

where $\|\mathbf{x}_k - \mathbf{v}_i\|^2$ is the square of the Euclidean distance from cluster center $\mathbf{v}_i$ to a given datapoint $\mathbf{x}_k$ and the parameter κ is called the *fuzziness index*. As $\kappa \rightarrow 1$ the clusters become more and more similar to crisp clusters, while as $\kappa \rightarrow \infty$ the clusters become fuzzier until an upper limit is reached where every object in the feature space has the same cluster membership-function value in every cluster: $\mathrm{m}_i(\mathbf{x}_k) = 1/c$ for $k = 1, 2, ... , n$ and $i = 1, 2, ... , c$. Obviously, the latter situation is not of real value for clustering since there is absolutely no discrimination among the clusters. Note also that objects whose cluster membership is maximally fuzzy are 'equi-distant' from all of the corresponding crisp clusters.

Based on our earlier work (Zhang *et al.*, 1995), four fuzzy clusters (i.e., $c = 4$) are also used here; each cluster corresponds to one of the four structural classes *all*-α, *all*-β, $\alpha+\beta$, and α/β (Levitt and Chothia, 1976). The value of the fuzziness index $\kappa = 1.4$ found to be optimal for protein structural-class prediction, is also used here.

Eq. (3) is usually solved under the constraints that

$$\sum_i \mathrm{m}_i(\mathbf{x}_k) = 1 \quad \text{for all } \mathbf{x}_k \tag{4a}$$

$$0 < \sum_k \mathrm{m}_i(\mathbf{x}_k) < n \quad \text{for all } C_i \tag{4b}$$

The first constraint requires that the sum of an object's membership-function values over all of the clusters equals unity. The second constraint essentially precludes the possibility that all of the objects will reside within a single cluster. Solving Eq. (3) for a fixed κ under the constraints given in Eq. (4) yields two equations, one for the membership functions and one for the cluster-center vectors, respectively:

$$\mathrm{m}_i(\mathbf{x}_k) = 1 \,/ \sum_j \,[\,\|\mathbf{x}_k - \mathbf{v}_i\| \,/\, \|\mathbf{x}_k - \mathbf{v}_j\|\,]^{2/\kappa-1} \tag{5}$$

$$\mathbf{v}_i = \sum_k \,[\mathrm{m}_i(\mathbf{x}_k)]^\kappa \bullet \mathbf{x}_k \,/ \sum_k \,[\mathrm{m}_i(\mathbf{x}_k)]^\kappa \tag{6}$$

Solution of the equations is obtained iteratively. An initial set of cluster centers is chosen. This can be accomplished, for example, by constructing an average or prototype amino-acid composition vector for each of the structural classes. These vectors are then substituted into Eq. (5) to obtain an initial set of cluster membership-function values that are then substituted into Eq. (6) to obtain an 'improved' set of cluster centers, and the process is repeated until self-consistency is reached with respect to either the cluster centers, the membership functions, or the functional J. The latter was used in the current work. Alternatively, the iterative process can be carried out beginning with an initial guess of the membership functions – the simplest approach would be to choose a crisp clustering as the initial guess. The iterative procedure has been shown to converge (Kim *et al.*, 1988; Wei and Mendel, 1994), but due to the nonlinearity of J there is no guarantee that the optimum reached is, in fact, the global optimum.

Multi-layer perceptrons

The use of computational neural networks has experienced a considerable resurgence since Werbos (1974) and Rumelhart *et al.* (1986) developed the backpropagation-of-error or 'backprop' algorithm for training multi-layer feedforward neural networks, now called multi-layer perceptrons (MLPs) or generalized perceptrons. These CNNs consist of an input layer, an output layer, and generally one or more 'hidden' layers lying between the input and output layers. As the name 'feedforward neural networks' implies, the flow of information in these CNNs moves in only one direction – from the input to the output layer. There is no feedback and hence, no underlying dynamics to these networks. Rather, MLPs learn by being trained on known data in a process called supervised learning.

Figure 2a depicts a simple MLP with two input nodes, a hidden layer of three nodes, and a single output node. Each layer of the network consists of a set of nodes as indicated by the circles, joined by weighted, unidirectional connections indicated by the arrows. The values of the two input variables, ξ_1 and ξ_2, are 'passed through' the input nodes without change. The remaining nodes, called processing elements (PE), carry out a summation (denoted by Σ) of the incoming signals. The value obtained in the summation is then transformed by a *nonlinear* transfer or activation function (denoted by f). The detailed form of the transfer function used in the current work, namely the hyperbolic tangent *tanh*, is given in Figure 2b. From the figure it is clear that *tanh* is quite sensitive to the value of the gain parameter γ; relatively small changes in γ can turn the function from one that closely resembles a step function to one that resembles a straight line. A value of $\gamma = 1.0$ was used in the present study. In addition to the weighted inputs summed by each PE, an additional threshold or bias term (denoted by θ_i) is added to the sum. The final output function $F(\xi_1, \xi_2)$ is then generated by appropriately scaling the value obtained from the output node. As the signals emanating from the output node lie in [–1, 1] (see Figure 2b), scaling is required to ensure that $F(\xi_1, \xi_2)$ corresponds to a meaningful value with respect to the function being estimated. Although not explicitly depicted in Figure 2b, threshold or bias terms shift the transfer function along the abscissa.

Multi-layer perceptrons are usually described by a shorthand notation that indicates the number of nodes in each layer. Thus, the three-layer MLP given in Figure 2, is designated by 2-3-1. In the current work all of the MLPs investigated are of the form 4-N-4, where N = 2, 3, 4, 5, or 10. Multi-layer

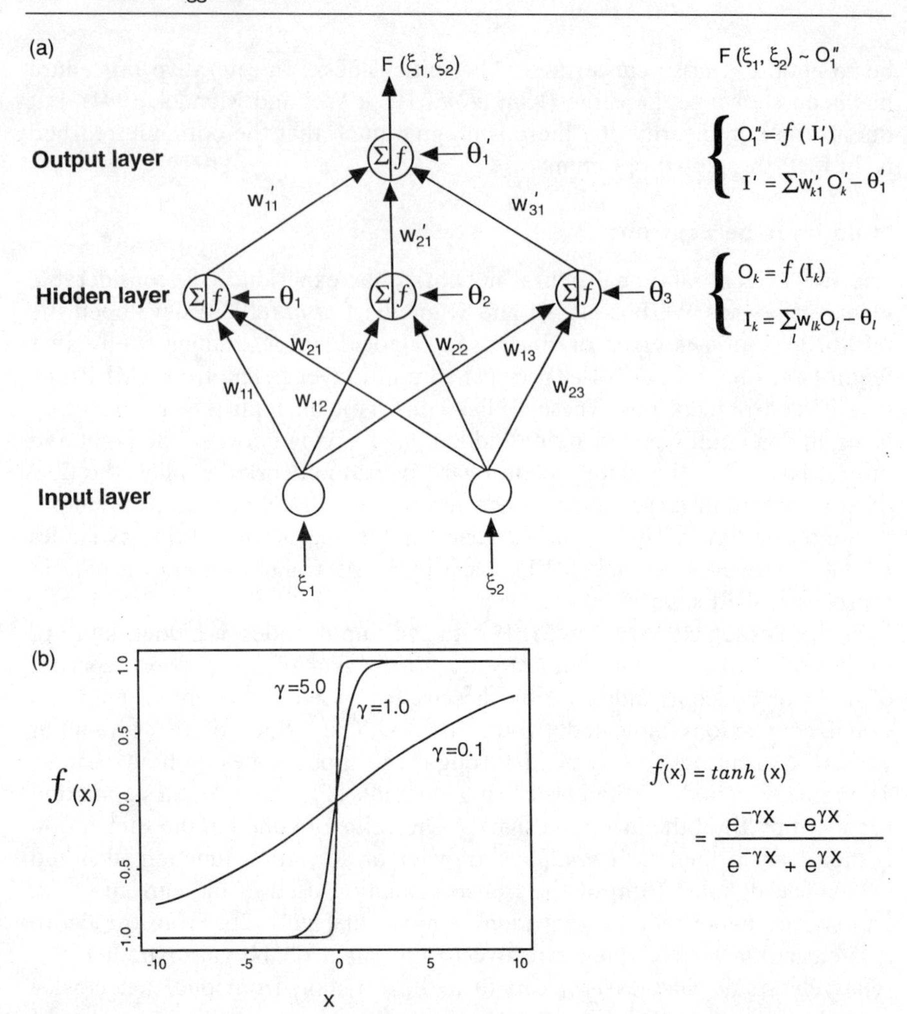

Figure 2 (a) Depiction of a 2-3-1 MLP. In the first layer, the two inputs (ξ_1, ξ_2) are 'passed through' the input nodes without change. Each of the nodes, k, in the hidden layer sums the appropriately weighted outputs obtained from the input layer, $I_k = \sum_l w_{lk} O_l - \theta_l$, where w_{lk} are the network weights and θ_l is the usual node bias. The sum is then subjected to a *nonlinear* transformation (see below) and is output, $O_k = f(I_k)$. The process is repeated for the output node, but the nonlinear output is scaled to ensure that the value of $F(\xi_1, \xi_2)$ lies in an appropriate range. (b) Graphical depiction of the hyperbolic tangent *tanh*, the nonlinear transformation used in this work. The gain parameter γ determines the steepness of the function. A value of $\gamma = 1.0$ was used here and represents a good compromise between the 'step' function, $\gamma = 5.0$, and 'linear' function, $\gamma = 0.1$. Note that $-1 \leq tanh \leq 1$.

perceptrons 'learn' by adjusting their weights in response to a set of training data so as to reduce the RMS error, E^{RMS}, of the set of network outputs, $\{F^{net}_{ij}\}$, compared to the set of desired values, $\{F^{obs}_{ij}\}$,

$$E^{RMS} = [(1/nm) \sum_i \sum_j (F^{obs}_{ij} - F^{net}_{ij})2]^{1/2}$$
$$i = 1, 2, ..., n; j = 1, 2, ..., m \qquad (7)$$

where n and m are, respectively, the number of datapoints and the number of output nodes. In the present work, each F^{obs}_{ij} is assigned a value of '1' if the component corresponds to the correct structural class; a value of '0' is assigned to each of the other components.

The actual weight adjustment procedure is carried out using the extended delta-bar-delta (EDBD) technique developed by Jacobs (1988), which is a variant of the usual backpropagation-of-error algorithm (Werbos, 1974; Rumelhart *et al.*, 1986). The data are repeatedly cycled through the network until the RMS error becomes constant. A value of 0.34 was obtained for the RMS error after 50 000 iterations of training (781 passes through the dataset or 'epochs') on a 4-5-4 MLP, a value typical of those observed in this work for all of the MLPs.

Because of the small size of the training datasets in many applications and the large number of weights found in a typical MLP, MLPs are usually *under-determined*, a situation that can lead to *overfitting* of the data. In cases where this obtains, MLPs cannot properly *generalize* from new data, and the usefulness of such networks may be limited (cf. Maggiora *et al.*, 1992). Procedures exist for addressing this problem, although they tend to be very computationally intensive. In the present work, generalization is tested by subjecting the trained MLP to data that was not used to train the network.

The protein dataset

The protein dataset employed in this work is the same as that used by Chou (1989) in his early attempts to predict protein structural classes from amino-acid composition data. Chou's dataset is made up of 64 proteins in four classes, *all*-α, *all*-β, α+β, and α/β. The first two classes consist of proteins whose secondary structure is primarily made up of α-helices or β-sheets, respectively. The latter two classes consist of proteins with mixtures containing significant amounts of both α-helices and β-sheets. In the case of α+β proteins there is no apparent sequential order to the α-helical and β-strand segments, while in α/β proteins there is usually alternation between the two types of segments, i.e., ... αβαβαβαβ... . Chou's paper (1989) should be consulted for additional details on the description of the structural classes (cf, Kneller *et al.*, 1990). The class of irregular structures has been omitted from the current work in order to keep this preliminary study as simple as possible. Issues related to the presence of multiple domains have also not been addressed here, nor have they been addressed in any of the earlier works in this area. While the issue is an important one, it goes beyond the scope of the current study. A testing dataset of 27 proteins from the same four classes, but independent of the training dataset, was chosen to assess the ability of the combined procedure to correctly classify the proteins into one of four structural classes (Chou and Zhang, 1992). Although there now exist several more extensive protein datasets (Hobohm *et al.*, 1992; Hobohm and Sander, 1994), the two datasets used in this preliminary work were chosen so that comparisons with our earlier work and that in several other laboratories could be made (see, e.g., Zhang *et al.*, 1995).

Each protein $\mathbf{x}_k$ is represented by an amino-acid composition vector defined as

$$\mathbf{x}_k = [\ x_k(\mathrm{A}), x_k(\mathrm{C}), x_k(\mathrm{D}), \ldots , x_k(\mathrm{W}), x_k(\mathrm{Y})\ ] \tag{8}$$

The uppercase letters in each of the components of $\mathbf{x}_k$ are given in alphabetic order and designate the specific amino acids based upon the standard one-letter amino-acid code (Perutz, 1992). For example, $x_k(\mathrm{A})$ is the fraction of alanine residues in $\mathbf{x}_k$.

Graphical portrayal of fuzzy-clustered protein data

Four fuzzy clusters are used to describe the *all*-α, *all*-β, α+β, and α/β structural classes investigated in this work. Due, however, to the constraint relationship (see also Eq. (4a))

$$\mathrm{m}_{\alpha}(\mathbf{x}) + \mathrm{m}_{\beta}(\mathbf{x}) + \mathrm{m}_{\alpha+\beta}(\mathbf{x}) + \mathrm{m}_{\alpha/\beta}(\mathbf{x}) = 1 \tag{9}$$

only three of the four membership functions determined for each protein are independent. Hence, it is possible to represent each protein as a point (X, Y, Z) in a 'reduced' membership-function space of three-dimensions in $\mathbb{R}^3$. The equations for carrying out this mapping are given in an earlier work (Zhang *et al.*, 1995) and are reproduced below:

$$\begin{aligned} X(\mathbf{x}) &= [(3)^{1/2}/4]\ \{2[\mathrm{m}_{\alpha}(\mathbf{x}) + \mathrm{m}_{\alpha+\beta}(\mathbf{x})] - 1\} \\ Y(\mathbf{x}) &= [(3)^{1/2}/4]\ \{2[\mathrm{m}_{\alpha}(\mathbf{x}) + \mathrm{m}_{\beta}(\mathbf{x})] - 1\} \\ Z(\mathbf{x}) &= [(3)^{1/2}/4]\ \{2[\mathrm{m}_{\alpha}(\mathbf{x}) + \mathrm{m}_{\alpha/\beta}(\mathbf{x})] - 1\}. \end{aligned} \tag{10}$$

This type of graphical representation of clustered data is quite powerful and, as will be described in the section 'results and discussion', provides a visually intuitive representation of the data that qualitatively explains why MLPs perform better than other simpler methods of classification with the current protein dataset.

The prediction algorithm

As portrayed in Figure 3, the overall prediction procedure is implemented in two phases, a *training phase* and a *testing phase*. In the training phase

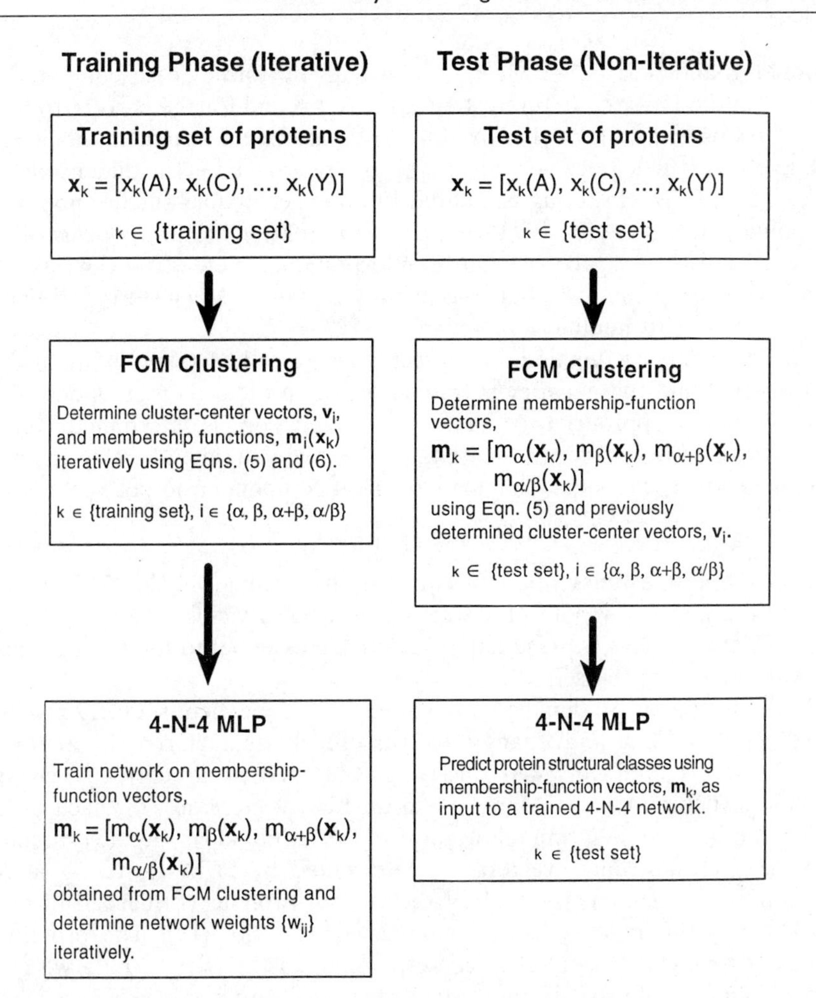

Figure 3 Scheme summarizing the essential features of the training and testing phases.

FCM clustering is carried out on a suitable training set of 64 proteins. In this phase, each protein $\mathbf{x}_k$ is represented as a 20-component 'amino-acid composition' vector (see Eq. (8)). Fuzzy clustering produces a *reduction in dimensionality* that maps this vector into a 4-component vector in 'membership-function' space given by

$$\mathbf{m}_k = [\mathrm{m}_\alpha(\mathbf{x}_k), \mathrm{m}_\beta(\mathbf{x}_k), \mathrm{m}_{\alpha+\beta}(\mathbf{x}_k), \mathrm{m}_{\alpha/\beta}(\mathbf{x}_k)] \quad (11)$$

The reduction in dimensionality brought about by FCM clustering considerably simplifies input to the MLP and removes or significantly reduces the degree to which the neural network is underdetermined. For example, the number of parameters in a 20-N-4 MLP is considerably more than the number found in a 4-N-4 MLP, $25N + 4$ in the former case compared to $9N + 4$ in the latter case. Interestingly, for $N \leq 5$ and for the current training set of 64 proteins, all of the $9N + 4$ MLPs are *overdetermined*, albeit to a relatively small degree. An advantage of the reduction in dimensionality effected by FCM clustering compared to other techniques such as nonlinear mapping (Domine *et al.*, 1993), is that in the former the *components* of the lower-dimensional space are more interpretable. New data can also be handled quite readily without having to redo the FCM clustering, which is not the case with nonlinear mapping.

The membership-function space vectors $\mathbf{m}_k$ are then taken as input to a 4-N-4 MLP, and the network is trained on the protein dataset as described earlier, until an appropriate set of network weights $\{w_{ij}\}$ is determined. During every cycle of the training process the MLP maps the membership-function vector $\mathbf{m}_k$ for each of the proteins into the 4-component output vector

$$\mathbf{F}_k^{\mathrm{net}} = [\mathrm{F}_{k,\alpha}^{\mathrm{net}}, \mathrm{F}_{k,\beta}^{\mathrm{net}}, \mathrm{F}_{k,\alpha+\beta}^{\mathrm{net}}, \mathrm{F}_{k,\alpha/\beta}^{\mathrm{net}}] \tag{12}$$

which is used to estimate the RMS error in the system (see Eq. (7)). In addition to the network weights, the training phase also yields a set of optimized fuzzy cluster centers $\{\mathbf{v}_i\}$ (see Eq. (6)) that are essential to the testing phase, as will be seen in the sequel.

In the testing phase depicted in Figure 3, an independent set of 27 proteins developed by Chou and Zhang (1992) is used to characterize the ability of the trained, combined FCM-clustering/MLP system to correctly classify protein structural classes. Membership-function vectors, $\mathbf{m}_k$, for each of the test proteins are determined using Eqs. (5) and (11). Eq. (5) requires the fuzzy cluster-center vectors, $\mathbf{v}_i$, determined by FCM clustering in the training phase. The $\mathbf{m}_k$ for each of the test-set proteins is then fed through the MLP yielding four output values shown in Eq. (12). The protein is assigned to the structural class corresponding to the output node with the largest value. Note that in the testing phase, the process is *non-iterative*, in contrast to the situation in the training phase. Once the fuzzy-cluster centers and network weights have been determined in the training phase, prediction of protein structural classes is accomplished in a single step.

RESULTS AND DISCUSSION

The results of FCM clustering on the training dataset of 64 proteins are given in Tables IV–VII. Table VIII contains the corresponding results for the 27 proteins in the testing set. All of the values are taken from our earlier study (Zhang *et al.*, 1995) and are reported here along with the MLP results to facilitate comparison.

Predictions in our earlier work (Zhang *et al.*, 1995) using FCM means clustering *and maximum-membership classification* are shown in boldface in

Table IV *Results obtained from FCM clustering and 4-5-4 MLP calculations on a training set of 19 all-α proteins.**

Protein	FCM Clustering†				4-5-4 MLP ‡			
	m_{α}	m_{β}	$m_{\alpha+\beta}$	$m_{\alpha/\beta}$	α	β	α+β	α/β
Ca-binding parvalbumin (carp)	**0.634**	0.054	0.129	0.183	**0.834**	0.019	0.170	−0.015
Cytochrome b_{562} (*E. coli*)	**0.627**	0.041	0.156	0.176	**0.725**	−0.014	0.167	0.042
Cytochrome *c* (tuna)	**0.423**	0.054	0.195	0.328	**1.050**	−0.077	0.050	−0.071
Cytochrome c_2 (*R. rubrum*)	**0.772**	0.021	0.074	0.137	**1.000**	0.125	0.007	−0.070
Cytochrome c_{550} (*P. denitrif.*)	**0.523**	0.030	0.177	0.271	**0.885**	−0.091	0.057	0.038
Cytochrome c_{555} (*C. thiosulfat.*)	**0.367**	0.133	0.265	0.236	**1.110**	0.013	0.078	−0.115
α-Deoxyhemoglobin (human)	**0.351**	0.039	0.261	0.349	**0.966**	−0.096	0.051	−0.007
β-Deoxyhemoglobin (human)	**0.351**	0.051	0.322	0.276	**1.070**	−0.001	0.205	−0.088
γ-Deoxyhemoglobin (fetal)	**0.443**	0.021	0.117	0.419	**0.642**	−0.114	0.021	0.305
Hemoglobin (*glycera*)	**0.487**	0.125	0.114	0.274	**1.110**	0.052	0.052	−0.117
Hemoglobin (lamprey)	**0.547**	0.043	0.091	0.319	**1.050**	−0.071	0.009	−0.070
Hemoglobin (midge larva)	**0.627**	0.076	0.094	0.203	**0.856**	0.028	0.218	−0.031
Hemerythrin B (*G. gouldii*)	**0.508**	0.050	0.098	0.346	**1.040**	−0.079	0.026	−0.064
Methemerythrin (*T. dyscritum*)	0.334	0.048	0.062	**0.557**	**0.791**	−0.109	0.041	0.148
Methemerythrin (*T. pyroides*)	**0.500**	0.100	0.192	0.207	**0.592**	−0.064	0.380	0.090
α-Methemoglobin (equine)	**0.523**	0.061	0.139	0.279	**1.070**	−0.043	0.052	−0.089
β-Methemoglobin (equine)	**0.384**	0.086	0.199	0.332	**1.090**	−0.049	0.073	−0.100
Myoglobin (*seal*)	**0.721**	0.021	0.058	0.200	**0.709**	−0.019	0.064	0.082
Myoglobin (*sperm whale*)	**0.759**	0.016	0.047	0.178	**0.853**	0.024	0.021	0.008

*Chou (1989).
†Numbers in boldface denote protein structural class predictions based upon maximum-membership classification.
‡Numbers in boldface denote protein structural class predictions based upon the output node with the maximum value.

Table V *Results obtained from FCM clustering and 4-5-4 MLP calculations on a training set of 15 all-β proteins.**

	FCM Clustering[†]				4-5-4 MLP ‡			
Protein	m_{α}	m_{β}	$m_{\alpha+\beta}$	$m_{\alpha/\beta}$	α	β	α+β	α/β
α-Chymotrypsin (bovine)	0.011	**0.898**	0.050	0.042	0.040	**0.927**	0.016	0.033
Concanavalin A (jack bean)	0.056	**0.556**	0.176	0.212	–0.069	**0.641**	0.080	0.285
Elastase (porcine)	0.021	**0.741**	0.164	0.074	–0.005	**0.815**	0.034	0.102
Erabutoxin B (sea snake)	0.083	**0.380**	0.348	0.189	–0.068	**0.555**	0.147	0.266
Ig F*ab* (V_H, C_H) (human)	0.014	**0.897**	0.044	0.045	0.058	**0.931**	0.011	0.021
Ig F*ab* (V_L, C_L) (human)	0.019	**0.876**	0.053	0.052	0.063	**0.923**	0.010	0.019
Ig B-J MCG (human)	0.012	**0.913**	0.039	0.036	0.024	**0.932**	0.021	0.045
Ig B-J REI (human)	0.067	**0.570**	0.235	0.129	–0.045	**0.719**	0.057	0.192
Penicillopepsin (*P. janthin.*)	0.015	**0.884**	0.061	0.039	0.005	**0.912**	0.028	0.066
Prealbumin (human)	0.100	0.169	0.167	**0.565**	–0.089	**0.551**	0.091	0.419
Protease A (*S. griseus*)	0.039	**0.712**	0.166	0.083	–0.008	**0.811**	0.035	0.106
Protease B (*S. griseus*)	0.052	**0.659**	0.180	0.109	–0.022	**0.770**	0.043	0.136
Rubredoxin (*C. pasteur.*)	0.223	0.133	0.309	**0.335**	–0.067	**0.480**	0.099	0.336
Superoxide dismutase (bovine)	0.145	0.178	0.269	**0.409**	–0.085	**0.573**	0.093	0.384
Trypsin (bovine)	0.020	**0.817**	0.103	0.060	0.027	**0.878**	0.021	0.055

*Chou (1989).
[†]Numbers in boldface denote protein structural class predictions based upon maximum-membership classification.
[‡]Numbers in boldface denote protein structural class predictions based upon the output node with the maximum value.

Table VI *Results obtained from FCM clustering and 4-5-4 MLP calculations on a training set of 14 α+β proteins.**

	FCM Clustering†				4-5-4 MLP ‡			
Protein	m_α	m_β	$m_{\alpha+\beta}$	$m_{\alpha/\beta}$	α	β	α+β	α/β
Actinidin (kiwi fruit)	0.038	0.164	**0.651**	0.148	–0.037	–0.003	**0.941**	0.077
Cytochrome b_5 (bovine)	0.249	0.086	0.178	**0.487**	**0.191**	0.069	0.074	0.187
Ferredoxin (*P. aerogenes*)	0.135	0.234	**0.421**	0.211	–0.027	0.006	**0.937**	0.058
High Pot. iron prot. (chromat.)	**0.472**	0.109	0.233	0.187	**0.499**	–0.068	0.431	0.133
Insulin (A, B chains) (porcine)	0.118	0.251	**0.401**	0.230	–0.031	0.008	**0.930**	0.066
Lysozyme (T_4 phage)	0.262	0.038	**0.356**	0.345	0.155	–0.110	**0.775**	0.175
Lysozyme (chicken)	0.077	0.170	**0.609**	0.144	–0.035	–0.000	**0.940**	0.072
Papain (papaya)	0.052	0.201	**0.590**	0.158	–0.033	0.001	**0.941**	0.068
Phospholipase A_2 (bovine)	0.135	0.174	**0.504**	0.187	–0.032	0.002	**0.937**	0.067
Ribonuclease S (bovine)	0.109	0.320	**0.377**	0.194	–0.036	0.102	**0.771**	0.094
Staph. nuclease (*S. aureus*)	**0.707**	0.018	0.074	0.201	**0.603**	–0.040	0.076	0.151
Subtilisin inhib. (streptomyces)	0.206	0.276	**0.249**	0.269	0.096	**0.455**	0.087	0.094
Thermolysin (*B. thermproteol.*)	0.050	0.335	**0.475**	0.140	–0.022	0.010	**0.936**	0.050
Trypsin inhib. (bovine)	0.159	0.107	**0.538**	0.196	–0.051	–0.016	**0.933**	0.116

*Chou (1989).
†Numbers in boldface denote protein structural class predictions based upon maximum-membership classification.
‡Numbers in boldface denote protein structural class predictions based upon the output node with the maximum value.

Table VII *Results obtained from FCM clustering and 4-5-4 MLP calculations on a training set of 16 α/β proteins.**

Protein	FCM Clustering†				4-5-4 MLP ‡			
	m_{α}	m_{β}	$m_{\alpha+\beta}$	$m_{\alpha/\beta}$	α	β	α+β	α/β
Adenylate kinase (porcine)	0.166	0.052	0.139	**0.643**	–0.123	0.067	0.028	**1.060**
Alcohol dehydr. (equine)	0.059	0.037	0.086	**0.818**	–0.125	–0.060	–0.002	**1.120**
Carbonic anhydrase B (human)	0.087	0.305	0.216	**0.392**	–0.081	**0.595**	0.095	0.349
Carbonic anhydrase C (human)	0.175	0.042	0.117	**0.665**	–0.124	0.007	0.017	**1.090**
Carboxypeptidase A (bovine)	0.034	**0.493**	0.301	0.172	–0.070	**0.639**	0.081	0.289
Carboxypeptidase B (bovine)	0.042	0.173	**0.604**	0.181	–0.038	–0.004	**0.942**	0.078
Dihydrofolate reduc. (*E. coli*)	0.100	0.039	0.336	**0.524**	–0.122	0.024	0.140	**1.020**
Flavodoxin (*Clostridium* MP)	0.168	0.087	0.270	**0.475**	–0.109	0.379	0.071	**0.698**
Glyceral. 3-P dehydr. (lobster)	0.085	0.034	0.080	**0.801**	–0.125	–0.063	–0.004	**1.120**
Glyceral. 3-P dehydr. (*B. ste.*)	0.238	0.051	0.144	**0.566**	–0.119	0.057	0.036	**0.993**
Lactate dehydr. (dogfish)	0.119	0.044	0.108	**0.730**	–0.124	–0.013	0.012	**1.100**
Phosphoglycer. kin. (equine)	0.362	0.008	0.037	**0.593**	0.108	–0.122	–0.003	**0.891**
Rhodanase (bovine)	0.103	0.058	0.203	**0.635**	–0.122	0.123	0.038	**1.010**
Subtilisin BPN' (*B. amylol.*)	0.062	**0.651**	0.151	0.136	–0.030	**0.748**	0.048	0.155
Thioredoxin (*E. coli*)	**0.605**	0.024	0.123	0.249	0.070	–0.107	0.254	**0.698**
Triose phos. isom. (chicken)	0.335	0.014	0.085	**0.566**	0.246	–0.120	0.007	**0.735**

*Chou (1989).
†Numbers in boldface denote protein structural class predictions based upon maximum-membership classification.
‡Numbers in boldface denote protein structural class predictions based upon the output node with the maximum value.

Tables IV–VIII. Maximum-membership classification is the simplest way to classify proteins into structural classes: each protein $\mathbf{x}_k$ is assigned to the class, say α, with the largest membership-function value, i.e., $m_\alpha(\mathbf{x}_k)$. The results of the predictions on both the training and test proteins are summarized in Table IX.

A value of $N = 5$ was chosen for the 4-N-4 MLP based upon the results presented in Tables X and XI, which are discussed further in the sequel. From the tables it is seen that maximum-membership classification yields reasonably predictive results for both the training and testing sets. A graphical representation of the distribution of training and test set proteins in membership-function space (see Eq. (8)) is given in Plates 11a and 11b, respectively. The plate shows that there is significant 'mixing' among the different structural classes. This suggests that it may be possible, using a more powerful classification methodology such as that afforded by CNNs, to improve the results obtained on the training and test sets of proteins.

Computational neural networks in general, and MLPs in particular, can provide improved classification of classes of objects with reasonably complicated boundaries. Table X summarizes the data for the 64 proteins in the training set obtained from five different 4-N-4 MLPs with $N = 2$, 3, 4, 5, and 10, respectively. The table clearly shows that improvements over maximum-membership classification are obtained to varying degrees for all of the MLPs. The best overall improvements are obtained for $N = 5$ and $N = 10$, with the latter being superior. However, without some form of validation, especially in the $N = 10$ case which is underdetermined, these results are meaningless.

Table XI summarizes the results obtained for the test set of 27 proteins. Here none of the neural network results exceed those obtained by maximum-membership classification. Also overfitting of the training data by the 4-10-4 MLP is now evident. The best choice, which is a compromise, appears to be the 4-5-4 MLP, and this is the network used in the comparative studies described in Tables IV–VIII. In evaluating this choice, the small size of the test dataset must be considered because it leads to changes in *% correct* of the predictions that are too sensitive to small changes in the number of correct predictions. For example, a single error in the α-protein test set reduces the *% correct* of the prediction from 100 to 86. This rather large change considerably exaggerates the actual predictivity of the hybrid method. On the other hand, an improvement of a single additional correct prediction changes the *% correct* of the predictions in the β-protein test set from 67 to 83. Clearly, a larger test set or an alternative validation procedure such as 'bootstrapping' (Effron and Tibshirani, 1993), which combines the training and test sets, should be used (*vide infra*). Such studies are on-going in our laboratory.

Keeping these factors in mind, it appears from the data in Tables IV–VIII that there is no significant overall improvement between maximum-membership and MLP classification. The 4-5-4 network does show some

Table VIII *Results obtained from FCM clustering and 4-5-4 MLP calculations on a test set of 27 proteins from all four structural classes.**

	FCM Clustering[†]				4-5-4 MLP ‡			
Structural class	m_{α}	m_{β}	$m_{\alpha+\beta}$	$m_{\alpha/\beta}$	α	β	α+β	α/β
all-α Proteins								
Cytochrome c_5	**0.345**	0.169	0.251	0.235	**1.110**	0.025	0.076	–0.117
Apoferritin	**0.393**	0.093	0.182	0.332	**1.090**	–0.041	0.074	–0.104
Cytochrome c_3 (*D. vulgar.*)	**0.474**	0.085	0.184	0.257	**1.080**	–0.015	0.077	–0.104
Cytochrome c_3 (*D. sulfur.*)	**0.425**	0.068	0.216	0.291	**1.080**	–0.049	0.051	–0.098
Cytochrome c_{551}	**0.592**	0.040	0.148	0.219	**0.232**	–0.088	0.297	–0.416
Cytochrome c'	**0.597**	0.080	0.143	0.181	**0.861**	0.031	0.233	–0.036
Erythrocruorin	**0.370**	0.088	0.205	0.336	**1.080**	–0.048	0.077	–0.101
all-β Proteins								
Actinoxanthin	0.124	**0.510**	0.221	0.145	–0.038	**0.733**	0.054	0.175
Neuraminidase	0.026	0.340	**0.486**	0.148	–0.026	0.007	**0.938**	0.056
Pepsin	0.028	**0.701**	0.176	0.094	–0.017	**0.782**	0.040	0.125
α-Lytic protease	0.055	**0.616**	0.217	0.112	–0.035	**0.744**	0.051	0.165
Acid protease	0.061	**0.686**	0.159	0.094	–0.006	**0.812**	0.034	0.103
Coat protein	0.116	0.315	**0.328**	0.241	–0.075	**0.597**	0.103	0.310

Table VIII *continued*

	FCM Clustering[†]				4-5-4 MLP[‡]			
Structural class	m_{α}	m_{β}	$m_{\alpha+\beta}$	$m_{\alpha/\beta}$	α	β	$\alpha+\beta$	α/β
$\alpha+\beta$ Proteins								
Ovomucoid (dom 3)	0.141	0.233	**0.422**	0.204	–0.025	0.008	**0.936**	0.055
Phospholipase A_2	0.127	0.158	**0.552**	0.162	–0.034	0.001	**0.938**	0.072
Ribonuclease	0.099	0.154	**0.532**	0.215	–0.040	–0.007	**0.940**	0.085
Ribonuclease ST	0.155	0.125	**0.376**	0.344	–0.062	–0.002	**0.890**	0.156
Ribonuclease T_1	0.029	**0.714**	0.187	0.070	–0.022	**0.803**	0.042	0.128
Hemaglutinin (HA2 chain)	0.166	0.037	**0.481**	0.316	–0.110	–0.086	**0.901**	0.614
Ribosomal L7/L12	**0.481**	0.106	0.159	0.255	**1.100**	0.010	0.088	–0.109
Superoxide dismutase, Mn	**0.870**	0.007	0.033	0.090	**1.000**	0.120	– 0.021	–0.063
α/β Proteins								
p-OH Benzoate hydroxylase	0.197	0.099	0.283	**0.421**	–0.103	0.446	0.078	**0.588**
Glycogen phosphorylase A	0.189	0.027	0.264	**0.520**	–0.124	0.018	0.019	**1.080**
Elongation factor T_u	0.157	0.032	0.183	**0.628**	–0.124	–0.012	0.013	**1.100**
Catalase	0.124	0.044	0.365	**0.467**	–0.117	–0.081	0.758	**0.808**
Aldolase	0.227	0.083	0.335	**0.355**	–0.095	0.214	0.263	**0.493**
Malate dehydrogenase	0.170	0.030	0.141	**0.658**	–0.124	–0.035	0.006	**1.110**

*Chou and Zhang (1992).
†Numbers in boldface denote protein structural class predictions based upon maximum-membership classification.
‡Numbers in boldface denote protein structural class predictions based upon the output node with the maximum value.

Table IX *Summary of the results from FCM clustering and maximum-membership classification on the training and test sets of proteins.*

	64 Training set proteins*				27 test set proteins†			
Pred.\Obs.	*all*-α	*all*-β	α+β	α/β	*all*-α	*all*-β	α+β	α/β
all-α	18	0	2	1	7	0	2	0
all-β	0	12	1	2	0	4	1	0
α+β	0	0	10	1	0	2	5	0
α/β	1	3	1	12	0	0	0	6
% Correct	95	80	71	75	100	67	63	100

*Chou (1989).
†Chou and Zhang (1992).

Table X *Summary of results obtained from 4-N-4 MLP-based classification of the training set of 64 proteins – reported in* % correct *predictions.*

	all-α	*all*-β	α+β	α/β	Total
Max. member.*	95	80	71	75	81
$N = 2$	95	93	71	63	81
$N = 3$	95	73	86	75	83
$N = 4$	95	**100**†	71	69	84
$N = 5$	**100**	**100**	71	75	88
$N = 10$	**100**	**100**	79	**81**	**91**

*Results for maximum-membership classification are included for comparison.
†Numbers in boldface denote the best predictions.

Table XI *Summary of results obtained from 4-N-4 MLP-based classification of the test set of 27 proteins – reported in* % correct *predictions.*

	all-α	*all*-β	α+β	α/β	Total
Max. member.*	**100** †	67	**63**	**100**	**82**
$N = 2$	86	67	50	83	70
$N = 3$	100	67	**63**	67	74
$N = 4$	100	**83**	**63**	83	81
$N = 5$	86	**83**	**63**	**100**	81
$N = 10$	86	67	50	83	70

*Results for maximum-membership classification are included for comparison.
†Numbers in boldface denote the best predictions.

improvement in *all*-β prediction, but this is compensated for by poorer prediction for the *all*-α structural class. A closer examination of Tables IV–VIII reveals quite clearly, however, that although MLP-based classification did not significantly improve the overall percentage of correct predictions, it did increase the *strength* of the predictions significantly. This can be seen, for example, by comparing the relative values for the membership functions of the α-protein *human α-deoxyhemoglobin* found in the training set with the output of the 4-5-4 MLP. Maximum-membership classification was based upon a membership-function value of 0.351 for the α class, a value that was barely greater than the membership-function value of 0.349 for the α/β class. This is definitely not the case for the MLP-determined values, where the α class is strongly predicted with a value of 0.966 that is significantly larger than the values produced at any of the other output nodes. Numerous examples in Tables IV–VIII exist that, although perhaps not as striking, nevertheless, illustrate this point quite well.

There are also examples of proteins with strong, *incorrect* FCM cluster-based predictions and strong, but *correct* MLP predictions. For example, the maximum-membership-function value for the β protein *human prealbumin* is 0.565, but for the α/β class. Its value for the β class is only 0.169. The MLP-based prediction of 0.551 correctly places the protein within the β class. Strong misprediction from FCM clustering does, however, usually lead to the expected result of a strong MLP misprediction. But, overall the strength of the classifications produced by the MLP is definitely an improvement over those produced by maximum-membership classification of FCM clusters alone.

CAVEATS AND CONCLUSIONS

The present work is preliminary; it is meant to illustrate that improvements can be achieved from a combination of FCM clustering and MLP classification. As seen from the data in Tables IV–VIII, the hybrid, multistage procedure described here did accomplish this goal, albeit in an unexpected way – the overall accuracy of prediction is about the same, but the strength and thus the reliability of the predictions is increased in almost all cases. Hence, it seems that further work in this direction is definitely warranted. A number of important issues must, however, be addressed in future work, including the size and composition, *vis-à-vis* its diversity, of the database, the selection of more or better 'predictive features', a clearer definition of the structural classes, and an assessment of cluster validity.

One of the most important issues is the size and nature of the database. The database used here (Chou, 1989) is rather small and not as diverse as the one developed recently by Sander and his group at the European Molecular Biology Laboratory (EMBL) (Hobohm *et al.*, 1992; Hobohm and

Sander, 1994). Nevertheless, the Chou database has been used in a number of earlier works, including some of our own (Chou and Zhang, 1992; Zhang and Chou, 1992; Zhang *et al.*, 1995) and thus was used here to facilitate comparisons. Future work in our laboratory will make use of the EMBL database, which should provide a sounder basis for analyzing protein structural-class predictions. In particular, problems arising from excessive homology among members of the training and testing sets will be significantly reduced. Moreover, due to the considerably larger number of samples available for training and testing, changes in the *percent* of correct predictions will not be so sensitive to small changes in the *number* of correct predictions as was the case for the test sample results reported here. Alternative types of validation procedures such as 'bootstrapping' (Effron and Tibshirani, 1993) or related techniques are also being investigated.

Feature selection is crucial to any pattern recognition/classification problem. The features used here, i.e., the fractional amino-acid composition, although quite simple, appear to capture enough of the essential features of the system to provide reasonable predictions. This has also been observed in a number of other studies (Chou, 1989; Chou and Zhang, 1992; Zhang and Chou, 1992; Mao *et al.*, 1994; Chou, 1995). Nevertheless, it is not unreasonable to assume that there is a certain amount of redundant information in the current feature set and that other types of features, such as the total number of amino acids or molecular weight or volume, might lead to significant improvements in predictions (see, e.g., Dubchak *et al.* (1993) and Chandonia and Karplus (1995)). An examination of possible improvements to the current feature set is ongoing.

The description of the structural classes used in the current work is identical to that given by Chou (1989) and is based on the percentage of α-helical and β-strand residues. Other descriptions are, of course, possible (see, e.g., Kneller *et al.* (1990) and Chou (1995)), and they are being evaluated in our laboratory. Moreover, even with computer algorithms of increasing sophistication, determining the exact number of residues of each secondary structural type from very accurate X-ray structural data is not straight forward in all cases, but this may not be a severe limitation given the rather 'soft' nature of the methods used for predicting structural classes.

Because only four structural classes were considered in the current investigation, four fuzzy clusters were used in this preliminary study, but there is no need to confine the number of clusters to four. In fact, other numbers of clusters may fit the data better. To determine whether a given number of clusters does a better job representing the inherent structure of the data requires some measure of cluster validity. A number of such measures exist (Dubes and Jain, 1980; Backer and Jain, 1981; Windham, 1982; Xie and Beni, 1991), although none appears to be without some faults. Generally, the range of values produced by a given measure tends to be quite narrow and thus, the discriminating power of the measure is limited, although one measure does appear to provide sufficient discriminating power to be of use (Xie and Beni, 1991). Using this measure, future work will examine the suitability of additional FCM clusters to represent the data and how the cluster number affects the ability of the hybrid approach to predict protein structural classes correctly. If more than four clusters are needed to describe the data optimally, the use of MLPs is of added benefit as maximum-membership classification cannot be used in a straight forward manner as before. For example, an M-N-4 MLP can be used, where $M > 4$ is the number of fuzzy clusters. In such cases, the additional clusters may hold new insights – some clusters may represent 'pure' structural classes while other clusters may represent structural classes of mixed 'parentage'.

REFERENCES

Anthonsen, H.W., Baptista, A., Drablos, F., Martel, P., and Petersen, S.B. (1994). The blind watchmaker and rational protein engineering. *J. Biotech.* **36,** 185–220.

Backer, E. and Jain, A.K. (1981). A clustering performance measure based on fuzzy set decomposition. *IEEE Trans. Pattern Anal. Machine Intell.* **3,** 66–75.

Bandemer, H. and Nather, W. (1992). *Fuzzy Data Analysis*. Kluwer Academic Publishers, Dordrecht.

Bezdek, J.C. (1981). *Pattern Recognition with Fuzzy Objective Function Algorithms*. Plenum, New York.

Chandonia, J.-M. and Karplus, M. (1995). Neural networks for secondary structure and structural class prediction. *Protein Sci.* **4,** 275–285.

Chou, P. (1989). Prediction of protein structural classes from amino acid compositions. In, *Prediction of Protein Structure and the Principles of Protein Conformation* (G.D. Fasman, Ed.). Plenum Press, New York, pp. 549–586.

Chou, K.C. (1995). A novel approach to predicting protein structural classes in a (20–1)D amino acid composition space. *Proteins: Struct. Funct. Genet.* **21,** 319–344.

Chou, K.C. and Zhang, C.T. (1992). A correlation-coefficient method to predicting protein-structural classes from amino acid compositions. *Eur. J. Biochem.* **207,** 429–433.

Chou, K.C. and Zhang, C.T. (1994). Predicting protein folding types by distance functions that make allowances for amino acid interactions. *J. Biol. Chem.* **269,** 22014–22020.

Domine, D., Devillers, J., Chastrette, M., and Karcher, W. (1993). Nonlinear mapping for structure–activity and structure–property modeling. *J. Chemometrics* **7,** 227–242.

Dubchak, I., Holbrook, S., and Kim, S. (1993). Prediction of folding class from amino acid composition. *Proteins: Struct. Funct. Genet.* **16,** 79–91.

Dubes, R. and Jain, A.K. (1980). *Clustering Methodology in Exploratory Data Analysis*. Academic Press, New York.

Duda, R. and Hart, P. (1973). *Pattern Classification and Scene Analysis*. John Wiley & Sons, New York.

Effron, B. and Tibshirani, R.J. (1993). *An Introduction to the Bootstrap*. Chapman and Hall, New York.

Genfa, Z., Xinhua, X., and Zhang, C.T. (1992). A weighting method for predicting protein structural class from amino acid composition. *Eur. J. Biochem.* **210,** 747–749.

Hathaway, R.J. and Bezdek, J.C. (1994). NERF c-means: Non-Euclidean relational fuzzy clustering. *Pattern Recognition* **27,** 429–437.

Hathaway, R.J., Davenport, J.W., and Bezdek, J.C. (1989). Relational duals of the c-means clustering algorithms. *Pattern Recognition* **22,** 205–212.

Hobohm, U. and Sander, C. (1994). Enlarged representative set of protein structures. *Protein Sci.* **3,** 522–524.

Hobohm, U., Scharf, M., Schneider, R., and Sander, C. (1992). Selection of representative protein datasets. *Protein Sci.* **1,** 409–417.

Jacobs, R.A. (1988). Incremental rates of convergence through learning rate adaptation. *Neural Networks* **1,** 295–307.

Kaufmann, A. (1975). *Introduction to the Theory of Fuzzy Subsets*. Academic Press, New York.

Kim, T., Bezdek, J.C., and Hathaway, R.J. (1988). Optimality tests for fixed points of the fuzzy c-means algorithm. *Pattern Recognition* **21,** 651–663.

Klein, P. (1986). Prediction of protein structural class by discriminant analysis. *Biochim. Biophys. Acta* **874,** 205–215.

Klein, P. and Delisi, C. (1986). Prediction of protein structural class from the amino acid sequence. *Biopolymers* **25,** 1659–1672.

Klir, G.J. and Folger, T.A. (1988). *Fuzzy Sets, Uncertainty, and Information*. Prentice Hall, Englewood Cliffs, New Jersey.

Kohonen, T. (1989). *Self-Organization and Associative Memory*, 3rd Edition. Springer-Verlag, Berlin.

Kneller, D.G., Cohen, F.E., and Langridge, R. (1990). Improvements in protein secondary structure prediction by an enhanced neural network. *J. Mol. Biol.* **214,** 171–182.

Levitt, M. and Chothia, C. (1976). Structural patterns of globular proteins. *Nature* **261,** 552–558.

Maggiora, G.M., Elrod, D.W., and Trenary, R.G. (1992). Computational neural networks as model-free mapping devices. *J. Chem. Inf. Comput. Sci.* **32,** 732–741.

Mao, B., Zhang, C.T., and Chou, K.C. (1994). Protein folding classes: A geometric interpretation of the amino acid composition of globular proteins. *Prot. Engineer.* **7,** 319–330.

Nakashima, H., Nishikawa, K., and Ooi, T. (1986). The folding type of a protein is relevant to the amino acid composition. *J. Biochem. (Tokyo)* **99,** 152–162.

Perutz, M. (1992). *Protein Structure: New Approaches to Disease and Therapy.* W.H. Freeman and Co., New York, p. 278.

Pham, D.T. and Bayro-Corrochano, E.J. (1994). Self-organizing neural-network based pattern clustering method with fuzzy outputs. *Pattern Recognition* **27,** 1103–1110.

Ross, T.J. (1995). *Fuzzy Logic with Engineering Applications.* McGraw-Hill, New York.

Rumelhart, D.E., Hinton, G.E., and Williams, R.J. (1986). Learning representations by backpropagation of errors. *Nature* **323,** 533–536.

Wei, W. and Mendel, J.M. (1994). Optimality tests for the fuzzy c-means algorithm. *Pattern Recognition* **27,** 1567–1573.

Werbos, P.J. (1974). *Beyond Regression: New Tools for Prediction and Analysis in the Behavioral Sciences* (Doctoral Dissertation, Applied Mathematics). Harvard University, Cambridge, Massachusetts.

Windham, M.P. (1982). Clustering validity for the fuzzy c-means clustering algorithm. *IEEE Trans. Pattern Anal. Machine Intell.* **4,** 357–363.

Xie, L.X. and Beni, G. (1991). A validity measure for fuzzy clustering. *IEEE Trans. Pattern Anal. Machine Intell.* **13,** 841–847.

Zhang, C.T. and Chou, K.C. (1992). An optimization approach to predicting protein structural class from amino acid composition. *Protein Sci.* **1,** 401–408.

Zhang, C.T., Chou, K.C., and Maggiora, G.M. (1995). Predicting protein structural classes from amino acid composition: Application of fuzzy clustering. *Prot. Engineer.* **8,** 425–435.

Zimmermann, H.J. (1991). *Fuzzy Set Theory and Its Applications.* Kluwer Academic Publishers, Dordrecht.

Zupan, J. and Gasteiger, J. (1993). *Neural Networks for Chemists.* VCH, Weinhein.

Index